THE BIRTH OF NASA

NASA SP-4105

THE BIRTH OF NASA
The Diary of T. Keith Glennan

with an Introduction by

Roger D. Launius

Edited by

J. D. Hunley

The NASA History Series

National Aeronautics and Space Administration
NASA History Office
Washington, DC   1993

Library of Congress Cataloging-in-Publication Data

Glennan, Thomas Keith, 1905-
       The birth of NASA : the diary of T. Keith Glennan / with an
introduction by Roger D. Launius ; edited by J.D. Hunley.
             p.   cm.--(NASA SP ; 4105)   (The NASA history series)
       Includes bibliographical references (p.   ) and index.
             1. United States. National Aeronautics and Space
Administration--History--Sources. 2. Space sciences--Research--
United States--History. 3. Glennan, Thomas Keith, 1905-   --
Diaries. 1. Hunley, J.D., 1941-   . II. Title. III. Series.
IV. Series: The NASA history series.
TL521.312.G58 1993
353.0087'77--dc20   93-31238
                                                                    CIP

For sale by the U.S. Government Printing Office
Superintendent of Documents, Mail Stop: SSOP, Washington, DC 20402-9328

CONTENTS

Illustrations

Preface

    When I first began keeping this journal or diary, I never thought that it might, one day, be published. When I was appointed as the first Administrator of the newly authorized National Aeronautics and Space Administration on 19 August 1958, I started to keep a hand-written diary of sorts but soon found that my time was all too limited for that task. When I went back to Cleveland for the year-end holidays in 1959, I found that my four children had become much interested in knowing more about my job. They were also developing an interest in national and international affairs that intrigued me. I resolved to record my activities using my daily appointment cards to remind me of the important meetings that had become a daily way of life. I had a small, battery-operated recording machine called a dictette, and I usually dictated a summary of the day's happenings before I turned off the light each night. I sent the tapes back to my office at Case Institute of Technology in Cleveland where my secretary, Barbara Helberg, transcribed and stored them. I never saw them until I returned to Case in early 1961. Nor did I or anyone else edit them until NASA's chief historian, Dr. Roger D. Launius, and Dr. J. D. Hunley of the NASA History Office undertook the task. I did retain all of the daily appointment record cards, however.

    In 1963 my wife and I decided to take a long holiday in Europe, and I took the dictette and appointment cards with me, intending to record the events of the days between 19 August 1958 and 1 January 1960. I soon found that my memory was a bit hazy; I therefore chose to provide the kids with synopses of relationships with individuals or groups rather than the hour-by-hour recitation mode I had used to record the events after 1 January 1960. Throughout, I had embellished the unfolding story with bits of personal feelings or philosophy when stimulated by significant meetings or events. I do regret that I did not record the full diary when I started in the new post. When I completed the diary proper in 1963, I decided to voice my concern over the "crash" nature of the Apollo program, although I recognize that my conservative nature certainly clouded my vision at the time. When the Apollo 11 astronauts landed on the moon on 20 July 1969, I was glued to a television screen at the Bohemian Grove north of San Francisco and was as thrilled and emotionally moved as anyone could be. The management of that program by Jim Webb, Hugh Dryden, Tom Paine, Bob Seamans and Bob Gilruth was in the best tradition of the great undertakings that have periodically marked our nation's history.

    I am grateful for the enduring support of my wife, Ruth, who after 62 years of married life continues to encourage me to be myself, and I hope that my children and grandchildren will absorb something of the majesty of man's first successful attempt to walk upon the surface of our earth's only natural satellite, the moon.

                                      T. Keith Glennan
                                      Mitchelville, Maryland
                                      17 July 1993

Introduction

Early in the morning of 4 October 1957, T. Keith Glennan went to work, just as he had for more than a decade, at the president's office of the Case Institute of Technology (CIT) in Cleveland, Ohio. As president of CIT, he had been instrumental in transforming it from a small commuter school turning out journeymen engineers and other technicians into one of the top 20 technical institutions of higher learning in the United States.[1] He was pleased with the results of his presidency, and quite happy to continue those efforts, but on that Friday Glennan's career path changed sharply. Because of the events of that day he soon became caught up in the vortex of superpower rivalries and projects to enhance international prestige. Less than a year later he would be in Washington, D.C., serving as head of the National Aeronautics and Space Administration (NASA), a newly-constituted, federal research and development agency charged with helping to define and execute a far-reaching space exploration effort.[2]

Glennan's move from CIT to NASA came about because of the Soviet Union's successful launch of Sputnik I on 4 October 1957, as part of the scientific activities associated with the International Geophysical Year (IGY). When news of the Soviet success became known, U.S. congratulations for the technical accomplishment followed, but what President Dwight D. Eisenhower and others feared was that the Soviet Union could now legitimately claim leadership in a major technological field. The international image of the Soviet Union, thus, was greatly enhanced overnight. More importantly, Sputnik I meant that longstanding rivalries between the United States and the Soviet Union had entered a new plane, one where Americans were at a disadvantage. Americans were shocked and incredulous about this achievement by a communist country, and the result was increased spending for aerospace endeavors, technical and scientific educational programs, and the chartering of new federal agencies to manage air and space research and development.[3]

One of the most important results of this event was the establishment of NASA on 1 October 1958. Glennan headed NASA from its inception until the

---

[1] On the success of Glennan's presidency at Case Institute of Technology see C.H. Cramer, *Case Institute of Technology: A Centennial History, 1880-1980* (Cleveland: Case Western Reserve University, 1980), pp. 171-208.

[2] On NASA's creation and early years see, Robert L. Rosholt, *An Administrative History of NASA, 1958-1963* (Washington, D.C.: National Aeronautics and Space Administration SP-4101, 1966); Enid Curtis Bok Schoettle, "The Establishment of NASA," in Sanford A. Lakoff, ed., *Knowledge and Power* (New York: The Free Press, 1966), pp. 162-270; Rip Bulkeley, *The Sputniks Crisis and Early United States Space Policy: A Critique of the Historiography of Space* (Bloomington: Indiana University Press, 1991).

[3] The standard work on how international rivalries between the United States and the Soviet Union were played out in the two nations' space program is Walter A. McDougall, . . . *The Heavens and the Earth: A Political History of the Space Age* (New York: Basic Books, 1985).

change of presidential administration in 1961. During this period he oversaw the definition of U.S. policies for operations in space, contributed to the development of goals and programs to further those policies, and consolidated the resources needed to carry them out. In the process he set the stage for both NASA's future accomplishments and its failures, established most of the methodologies and many of the strategies employed in America's exploration of space, and created the infrastructure that still supports NASA's space efforts.

Glennan came to the leadership of NASA naturally enough, perhaps, but his earlier career also shaped many of the priorities, limitations, and accomplishments of his tenure as NASA Administrator. He was born in Enderlin, North Dakota, on 8 September 1905, the son of a train dispatcher. His early life was not so very different from those of the other boys raised in the upper Midwest. He spent most of his youth in Eau Claire, Wisconsin, where he attended public schools. There, Glennan found he had a knack for mathematics; worked at a variety of jobs including one at Continental Clothiers, a local store; and in 1922 graduated from high school. After a short stint at the Wisconsin State Teacher's College, in September 1924 Glennan entered Sheffield Scientific School of Yale University as a sophomore. During his junior and senior years, Glennan received the Lord Strathcona Scholarship recognizing achievements of the children of railroaders. While in New Haven he made the acquaintance of Dr. Thomas Sewell Adams, a noted economist on the Yale University faculty, and began working for him as a driver, which also helped him offset the costs of his Yale education. He graduated cum laude with a B.S. in electrical engineering in 1927. This was his last earned degree, although he received several honorary Ph.D.s over the course of his career.[4]

Keith Glennan's experience at Yale University was important to him in several ways. Without question, his Midwestern background and the necessity of working his way through school provided important perspectives that shaped his later career. At the same time, a Yale education ensured that he would have more ready entrance to many opportunities—public service as well as business—that might not be afforded to individuals educated at less prestigious institutions. Similarly, the contacts that he made there, with both students and faculty, provided the beginning of a web of relationships that lasted a lifetime. Indeed, Thomas Sewell Adams became something of a mentor and friend, as well as Glennan's father-in-law on 20 June 1931 when he married Ruth Haslup Adams (Vassar College, 1931). Opportunities undoubtedly availed themselves because of Glennan's Yale background; what he made of them was up to him, but his working-class roots fostered driving ambition.[5]

---

[4] "Interview Highlights on T. Keith Glennan," 21 June 1947, Biographical Files, Collection 7PI, University Archives, Case Western Reserve University, Cleveland, and T. Keith Glennan to Roger D. Launius, Chief, NASA History Office, 8 May 1993.

[5] G. Edward White has described the formation of an eastern establishment in the late nineteenth century as a male order in which the progression from brahmin stock, to prep school, to Ivy League college, to men's clubs played a central role in defining an elite core of American leaders. While Glennan entered this world only at the Ivy League college level, his capabilities helped him to maintain and flourish as part of this establishment. See G. Edward White, *The Eastern Establishment and the Western Experience: The West of Frederic Remington, Theodore Roosevelt, and Owen Wister* (New Haven: Yale University Press, 1968), pp. 11-30.

Immediately after graduation in June 1927, Glennan hitched his fortunes to a rapidly expanding technology, the "talkie" motion picture industry. In one way or another he was connected to the film industry for the next 14 years. Most of that time he worked for Electrical Research Products, Inc. (ERPI), a subsidiary of Western Electric Co., which installed sound motion picture recording and reproducing equipment in theaters all over the world. His first installation was in an elegant Philadelphia theater, but after a year he was transferred to Great Britain where he served as the regional installation manager for Western Electric Company, Ltd. (WECO). In this position, Glennan had a staff of 85 engineers and technicians from the United States and another 200 British employees. As a 23-year-old, he was in his first management position, and he found he had both a taste and a talent for the administration of complex technical organizations. Even as the worldwide depression was deepening in early 1930, Glennan was assistant manager for continental WECO, setting up and running installation and service operations with about 1,000 employees in ten countries of the British Commonwealth and on the continent of Europe. He finally returned to the United States in March 1930 to head a succession of ERPI offices in New York City. Among them, significantly for his later career, Glennan served as assistant manager of the ERPI educational picture project, which produced educational films on science. He also managed the Audio Cinema (originally the Edison studio in the Bronx), when ERPI took it over on default during the depths of the depression. This was his introduction to commercial motion picture production, which he was involved in throughout the rest of the decade.

Glennan got more directly involved in feature-length motion pictures in March 1934 when he went to Hollywood as ERPI's vice president and general manager of General Service Studios, Inc. (GSSI). The next year he left Electrical Research Products to take a job as operations manager of Paramount Pictures in Hollywood. Four years later he became the studio manager, a position of considerable importance where he was responsible for budgeting productions, lighting, sound, set construction, wardrobe, art, and film processing. During his five and a half years at Paramount Glennan provided the logistics necessary to allow the studio's creative teams to stage their productions, working with such Hollywood notables as Cecil B. DeMille. Glennan was also credited with important innovations in the film industry during his time at Paramount, including the first full-fledged engineering department in the business and the first recognized industrial relations department. However, in one of the typical moves in the motion picture business, he was fired in 1940. After a short stint with the Vega Airplane Corporation in Burbank during the summer of 1941, Glennan became the studio manager of Samuel Goldwyn Studios.[6]

A major change to Glennan's career came, as it did for most other Americans, with the U.S. entry into World War II in December 1941. The summer thereafter he left Goldwyn to enter defense work, taking a position with the Navy Underwater Sound Laboratory, operated by Columbia University at New London,

---

6 "Interview Highlights on T. Keith Glennan," 21 June 1947, Biographical Files, Collection 7PI; letter cited above in note 4.

Connecticut. This laboratory reported to the Office of Scientific Research and Development (OSRD), an agency organized by Franklin D. Roosevelt in 1941 to coordinate scientific research and development on behalf of the war effort.[7] Under the overall direction of Vannevar Bush, the MIT scientific research organizer par excellence, the OSRD exercised broad influences over national defense research efforts until the end of the war.[8]

In working for the Underwater Sound Laboratory, Glennan became part of the science and technology team that went to war, the "Scientists against Time" of later fame. It also brought him into contact for the first time with the whole milieu of scientific research as a public service, and it fundamentally affected his outlook during the rest of his career.[9] He embodied the progressive trend in U.S. politics that emphasized professionalism and scientific or technological expertise over politics in the solving of national problems.[10]

When he became director of the laboratory in December 1942, Glennan found himself appointed to a variety of OSRD and other government committees, working closely with such eminent scientific leaders as Bush and James Bryant Conant, a prominent chemist who was president of Harvard University. He enjoyed the public service aspects of his work, and at the same time made important contacts with people and agencies that he periodically tapped for assistance in working on the varied challenges he faced in later positions. Glennan performed well in his assigned responsibilities and won the respect of those with whom he worked. It was

---

[7] The predecessor of this organization had been established as the National Defense Research Committee on 27 June 1940, following a conversation between Vannevar Bush and one of Roosevelt's key advisors, Harry Hopkins, over concerns the scientists had about the nation's lack of preparedness for war. In that meeting, Bush offered to take charge of an effort to organize scientific activities for victory. He was motivated to this end, he recalled, by "the threat of a possible atomic bomb [that] was in all our minds, and time might well determine whether it became ours or a means for our enslavement" (Vannevar Bush, *Pieces of the Action* [New York: William Morrow and Co., 1970], p. 34).

[8] The history of this organization is documented in James Phinney Baxter III, *Scientists Against Time* (Boston: Atlantic-Little, Brown, 1946). On earlier scientific activities in the Federal Government see, A. Hunter Dupree, *Science in the Federal Government: A History of Policies and Activities to 1940* (Cambridge, Massachusetts: Harvard University Press, 1957).

[9] Glennan's commitment to public service was consistent throughout his career. When he was serving as a commissioner for the Atomic Energy Commission in he early 1950s, one of his friends wrote to ask if there was a way he could help his son avoid the draft (Ed Frawley to T. Keith Glennan, 18 January 1951, Glennan Personal Papers, 19DD4, Case Western Reserve University Archives, Cleveland, OH). Glennan responded that he was not inclined to use his public position to influence such matters, but added, "frankly, I think the best thing to do is to let him take his chance with the Selective Service or whatever else is decreed as the normal course for all young men. It is my opinion that all of these youngsters are going to have to give some of their time to government service whether it be in the armed forces or in some other field." Such service should not be considered onerous, but was a responsibility all should be pleased to bear (T. Keith Glennan to Ed Frawley, 23 January 1951, Glennan Personal Papers, 19DD4).

[10] See Susan Curtis, *A Consuming Faith: The Social Gospel and Modern American Culture* (Baltimore, MD: Johns Hopkins University Press, 1991), pp. 270-85; Wayne K. Hobson, "Professionals, Progressives and Bureaucratization: A Reassessment," *The Historian*, 39 (August 1977): 639-58; Brian Balogh, "Reorganizing the Organizational Synthesis: Federal-Professional Relations in Modern America," *Studies in American Political Development*, 5 (Spring 1991): 119-72.

some of these same people, of course, who later recommended him for positions of responsibility at Case, the Atomic Energy Commission, and NASA. The mutually beneficial relationship cannot be underestimated in either preparing Glennan or helping him secure his later positions of trust.[11]

On 1 June 1945, just as World War II was coming to an end and demobilization was beginning, Glennan resigned from his directorship of the laboratory to return to business. He went to work as an assistant production manager of the Ansco Division, General Aniline and Film Corporation of Binghamton, New York. Less than a year later he was named manager of administrative services for Ansco. This, and a host of civic activities—chair of the advisory committee for the schools of applied arts and sciences in the Binghamton area, trustee of the YMCA, etc.—kept him busy for another year. But he was not content, and looked for new challenges.[12]

He found one in the vacant presidency of the Case Institute of Technology in April of 1947. A few days before Glennan was to attend a reunion in New York City of those who had worked at his laboratory during the war, he received a call from Chuck Williams, his wartime co-worker and a graduate of Case. Williams told him that Case was in need of a new president and suggested that Glennan had the right skills and temperament to fit the position. They agreed to meet at the Yale Club in New York just before the reunion to discuss the issue. Thereafter, Williams arranged a meeting between Glennan and several trustees of Case in Cleveland. At a luncheon at the Cleveland Athletic Club Lee Clegg, an important Cleveland businessman and a trustee of Case, asked Glennan "how I would go about raising money. I told him I had no idea. I had never raised a nickel in my life, but on second thought I said I guess it would be clear that unless I fully believed in what I was trying to sell, I couldn't raise any money." If he believed in the product, however, Glennan thought he could be an effective fund raiser. He made a positive impression, and the Case Board of Trustees offered Glennan the presidency in the summer of 1947. He accepted and moved to Cleveland in time for the beginning of the fall semester.[13]

When he arrived at Case Glennan found a somewhat down-at-the-heels engineering school with some 1,585 full-time undergraduates, 205 full-time graduate students, most of whom were commuters, studying under the G.I. Bill, and 702 part-time (evening) students, mostly undergraduates. Many of the buildings were

---

[11] Vannevar Bush, Director, Office of Scientific Research and Development, to T. Keith Glennan, 27 October 1942, 4 December 1943; T. Keith Glennan to Fern Sullivan, Office of Scientific Research and Development, 22 November 1944, all in President's Files, Collection 19DB4, Case Western Reserve University Archives.

[12] "Interview Highlights on T. Keith Glennan," 21 June 1947, Biographical Files, Collection 7PI; telephonic interview of T. Keith Glennan by J. D. Hunley, 8 July 1993.

[13] Interview of T. Keith Glennan by Ruth W. Helmuth, 19 April 1976, President's Collection, 19DB4, Case Western Reserve University Archives, Cleveland.

old and dilapidated and the coffers for refurbishment and rejuvenation were empty. The institution itself had a certain measure of respect as a tough engineering school, however, and Glennan was prepared to build on that foundation. He had a vision, expressed to a friend from Hollywood less than six months after his arrival, of remaking CIT into one of the top five or six engineering schools in the country. While conceding leadership in the discipline to MIT and Caltech, he opined:

> My task is not an easy one but I think that with some leadership the school will go forward at a rapid rate. We have a fine alumni body and good support from our trustees, and the faculty seems to be accepting the change with a very high attitude. We are presently engaged in surveying our entire operation with a view to assure ourselves of the adequacy of our attack on the problems of engineering and scientific education for the next few years.[14]

Based on what he found in this survey Glennan identified needs, prioritized initiatives, and set out on an expansion of Case's plant, program, and people.

Glennan, always active, aggressively raised funds to support this expansion. Among his other fund-raising techniques was one he borrowed from Robert A. Millikan, president of Caltech. Millikan had set up an associates program in which one hundred contributors gave $1,000 a year for ten years. Glennan recalled in 1976:

> I dreamt up a Case Associates program where the base payment would be $5000 a year. I request[ed] that they take a vow of intent to continue for five years, but to reexamine it each year to see that they continued to believe that we were making progress. I must say that the thought that I could get . . . that kind of money was not accepted by faculty, alumni or anybody else around here. . . . While the survey was going on however, I made the calls and in some instances went into the board rooms. I had a lovely invitation printed which looked like an invitation to become a member of the board of a bank and an unusually high percentage of the people that I called upon started out in the Case associates program.[15]

This program became a huge success and Glennan followed it with other fund-raising efforts to finance CIT's rapid post-war expansion.

These efforts made it possible for Glennan to begin an aggressive development program on campus. During his tenure Case built twelve new buildings and remodelled virtually all the earlier buildings. He also increased CIT's programs and enhanced its reputation as a scientific and technical institution through reorganizations of the curricula and departments on campus, the creation of an Engineering Division without as much of the normal separation of the discipline into subfields

---

[14] *Ibid.*; T. Keith Glennan to Charles Boren, 24 January 1948, T. Keith Glennan Personal Papers, 19DD4; T. Keith Glennan to J. D. Hunley, 28 June 1993.

[15] Interview of T. Keith Glennan by Ruth W. Helmuth, 19 April 1976, President's Collection, 19DB4.

as was usual elsewhere, the establishment of interdisciplinary research centers, and the expansion of the core undergraduate program emphasizing mathematics, physics, and chemistry. At the same time the enrollments increased, rising to 1,726 undergraduates and 879 graduate students in 1965-1966. Between 1947-1948 and 1965-1966 at Case, the institution grew rapidly in size, endowment, and prestige. Its budget rose from $1.95 to $14.9 million, its endowment more than doubled, and annual gifts and grants increased fifteen-fold. By almost any measure that could be applied to a college president, Glennan had been enormously successful in directing CIT's affairs.[16]

At the same time that Glennan was heading CIT in the late 1940s, he continued to take a strong interest in public service activities. In addition to a number of significant privately-organized civic activities in the Cleveland area, he was intensely interested in what was happening on the national scene. A moderate in politics who supported both Democratic and Republican candidates depending on the issues, he usually took a conservative stance toward larger questions of national importance. Two became immediately apparent during his early years at Case. First, he was an ardent "cold warrior," distrusting Stalin's Soviet regime and debating with others the propriety of the U.S. president treating the Russian leader the same way that he dealt with other heads of state. He also spoke up for the forceful prosecution of the war in Korea. "The lesson we are learning" in Korea, he confided to Rufus Day, a Cleveland community leader, in 1951, "is a costly one but I am hopeful that the results will be such as to make this nation so strong that attack by others is unthinkable. I am afraid that I am not at all optimistic about our ability to avoid war even though it may be delayed."[17] He advocated building a formidable defense capability and constructing a powerful nuclear deterrent force as the best hopes for ensuring the nation's peace.

Second, Glennan espoused a role for government in American society that was limited and less proactive than what was rapidly becoming the norm on the post-war scene. He challenged the rising amount of governmental regulation and direction in the affairs of individual Americans, and expressed a desire for a return to a less intrusive federal state. He told an audience in Akron, Ohio, in March 1950 that a new cold war was raging "between democratic practices and philosophies on the one hand and the practices and philosophy of a powerful central government on the other." He commented that "the area left to the exercise of private responsibility had steadily dwindled as government has been given control over more and more of our economic and personal affairs."[18]

---

[16] T. Keith Glennan to William E. Barbour, Jr., President, Tracerlab, Inc., 3 March 1952, Glennan Personal Papers, 19DD4; Cramer, *Case Institute of Technology*, pp. 171-200; "He Did It All," *CWRU Magazine*, February 1990, pp. 9-15; T. Keith Glennan to J. D. Hunley, 27 June 1993.

[17] T. Keith Glennan to Rufus Day, Jr., McAfee, Grossman, *et al.*, 16 January 1951, Glennan Personal Papers, 19DD4.

[18] "Nurturing Paternalism," Huntington (WV) *Advertiser*, 23 March 1950.

    Glennan also believed that it was the responsibility of the private sector to take the lead in a variety of areas, rather than abdicating responsibility and allowing government to fill the vacuum. One of those areas was the development of science and technology, perhaps a particularly appropriate area of concern for Glennan because of his background and position at Case. He told the president of a scientific research-oriented corporation in 1952, for instance, that "unless industry moves much farther and faster along the road of social responsibility (much fine progress has been made to date), I am inclined to believe that the omnipresent politician and big government man will step in with costly promises."[19] In concert with this position, Glennan deplored what he called the expansion of the "welfare state," and argued for a strong showing among private institutions so that government leaders would not feel compelled to get involved in too many aspects of people's lives.[20] Accordingly, he expressed well the apparent attitudes of many white middle-class, slightly conservative Americans who went to the polls in 1952 and elected Republican candidate Dwight D. Eisenhower as president. Both of these guiding beliefs in Glennan's life helped to shape his actions and contributed fundamentally to the direction he charted for NASA as its first administrator.

    While Glennan had strong beliefs about limiting the federal government's role in daily life, he tempered those ideological leanings with a certain pragmatism, informed by principle, that recognized some regulation and some activity was required by the federal government to ensure the safety of the citizens and the advance of the nation's technological and scientific base. One of those areas where he had little difficulty accepting the federal government's preeminence was in the management of nuclear power. This same area also afforded Glennan his first important opportunity to offer public service at the national level in the postwar period.

    That occasion came while he was on vacation in the summer of 1950, when Glennan called back to CIT and learned that the White House had been trying to reach him. Glennan found that outgoing Atomic Energy Commission (AEC) member Lewis L. Strauss had given his name to Donald Dawson, a Truman aide who was looking at candidates to serve on the AEC. Strauss had been Herbert Hoover's private secretary in the World War I era, became a successful Wall Street businessman in the 1920s, and turned his attention to philanthropic enterprises associated with scientific research in the latter 1930s. A member of the Naval Reserve since 1925, he had served as a Rear Admiral in World War II as a special assistant to Navy Secretary James V. Forrestal. During the war he had also learned of Glennan's work at the Naval Underwater Sound Laboratory, and when the time came to seek a replacement for his own seat on the AEC, Strauss remembered the

---

[19] Glennan to Barbour, 3 March 1952, Glennan Personal Papers, 19DD4.

[20] J.M. Telleen to T. Keith Glennan, 15 November 1956; T. Keith Glennan to J.M. Telleen, 19 November 1956, both in Glennan Personal Papers, 19DD4.

45-year-old Case president. While some of the Democratic congressmen involved in the search preferred the appointment of AEC General Counsel Joseph A. Volpe, Jr., Strauss used his connections in the White House and the Pentagon to gain Glennan's appointment. Glennan brought to the job, as the official history of the AEC commented, "a solid business background, some experience in Government, and a great interest in the role of science and technology in modern industry."[21]

After speaking with Dawson, Glennan visited Washington to learn more about the AEC position. He recalled that he met Dawson in the White House, where he was briefed on the activities of the Commission. Glennan told Dawson that he would wait to hear from him about whether or not President Truman wished him to serve. As he was preparing to leave, however, Dawson said, "wait a minute, Mr. Glennan, I think the president wants to see you." Dawson took Glennan into the Oval Office where he met the president and they spoke briefly about the AEC. Glennan allowed that he was not particularly well-qualified for the post, but Truman responded with an "I'm not sure you're the best judge of that."[22]

Following this discussion, Truman sent Glennan's name to the Senate for confirmation as one of five commissioners for the AEC. Glennan testified before the congressional committee handling the nomination on 16 August 1950, was confirmed handily, and began work for the AEC in Washington on 1 October. He served on the AEC for the next twenty-five months. Probably the most significant issue the AEC became involved in during Glennan's tenure was the development of the hydrogen bomb. Glennan resigned from the AEC on 30 October 1952, and the first hydrogen bomb test took place at Eniwetok Atoll in the Pacific Ocean on 1 November 1952.[23]

During the time that Glennan was at AEC, he was officially on a leave of absence from the presidency of Case Institute of Technology. Acting in his stead during those two years was Elmer Hutchinson, a CIT alumnus and an old Glennan associate from World War II. Glennan, never one to sit and vegetate, continued to assist Hutchinson in managing the affairs of Case. He often spent his weekends in Cleveland, especially early in his AEC tenure since his family remained behind to allow his children to stay in their schools. Whenever he did so, Glennan typically spent some time with Hutchinson discussing CIT business, usually answering questions about fund-raising strategies and plans underway.

---

[21] Richard G. Hewlett and Francis Duncan, *Atomic Shield, 1947-1952*, Volume II of *A History of the United States Atomic Energy Commission* (University Park: Pennsylvania State University Press, 1969), p. 468.

[22] "He Did It all," *CWRU Magazine*, February 1990, p. 12.

[23] Joint Committee on Atomic Energy, *Hearing on Confirmation of Thomas Keith Glennan To Be a Member of the Atomic Energy Commission, August 16, 1950* (Washington, D.C.: Government Printing Office, 1950); Albert L. Baker, President, Kellex Corp., to T. Keith Glennan, 2 November 1950, Glennan Personal Papers, 19DD4; "He Did It All," *CWRU Magazine*, February 1990, p. 13; Hewlett and Duncan, *Atomic Shield*, pp. 542-45.

The result was that Case did not suffer materially during its president's absence, often the case whenever an "acting president" heads an institution of higher learning. In fact, it may actually have benefitted the institution, for Glennan's experience broadened both his perspectives and his contacts with other influential people. It also brought renown to him and by extension to the school. "When I came back," he recalled of his experiences at AEC and later at NASA, "I could get hold of almost anybody in the industries which had been associated [with the AEC and NASA] and that meant most of the industries in this country: space and atomic energy. I kept up those relationships very well and I think they did us a lot of good at Case."[24]

During the years between the conclusion of his AEC service in 1952 and his acceptance of the NASA administrator's post in 1958 Glennan continued his expansion program at Case Institute of Technology and dabbled in public policy issues. In the latter arena, he participated in several government committees and task forces relative to higher education and to nuclear power. In 1955 he began serving on a committee under the direction of a Joint Committee (of Congress) on Atomic Energy, and helped to prepare a study on "The Impact of the Peaceful Uses of Atomic Energy." While on this committee Glennan developed additional associations with members of Congress, among them Senate wheelhorse and science and technology benefactor, Clinton P. Anderson (Democrat-New Mexico).[25] He also provided advice to Lewis Strauss, by that time back as chair of the Atomic Energy Commission, on policy direction and energy for peaceful purposes.[26] He was appointed to the board of the National Science Foundation in 1956. He joined with several other college and university presidents in an 18-day visit to the Soviet Union to study the educational programs of the USSR under the sponsorship of the Scaife Foundation in 1958.[27] Because of this sort of activity, when the Eisenhower administration sought a person to head up the newly created NASA, Glennan was quite well known in business, higher education, science and technology, and public service circles.

When Sputnik burst onto the national scene, there was a rapid and sustained whir of public opinion condemning the Eisenhower administration for neglecting the American space program. The Sputnik crisis reinforced for many

24 Interview of T. Keith Glennan by Ruth W. Helmuth, 19 April 1976, President's Files, Collection 19DB4; Cramer, *Case Institute of Technology*, pp. 200-202.

25 Walter Hamilton, Secretary to the Panel, of Congress' Joint Committee on Atomic Energy, to T. Keith Glennan, 30 June 1955; T. Keith Glennan to Walter Hamilton, 1 June 1955; T. Keith Glennan to Senator Clinton P. Anderson, 9 February 1956, all in Glennan Personal Papers, 19DD4.

26 T. Keith Glennan to Lewis L. Strauss, 21 January 1957, 25 February 1957; Lewis L. Strauss to T. Keith Glennan, 15 February 1957, with enclosure; all in Glennan Personal Papers, 19DD4.

27 *Report On Higher Education in the Soviet Union* (Pittsburgh: University of Pittsburgh Press, 1958); Alan T. Waterman, Director of National Science Foundation, to T. Keith Glennan, 11 July 1958, President's Collection, DB14; Glennan to Launius, 8 May 1993.

people the popular conception that Eisenhower was a smiling incompetent; it was another instance of a "do-nothing," golf-playing president mismanaging events.[28] G. Mennen Williams, the Democratic governor of Michigan, even wrote a poem about it:

> Oh little Sputnik, flying high
> With made-in-Moscow beep,
> You tell the world it's a Commie sky
> and Uncle Sam's asleep.
>
> You say on fairway and on rough
> The Kremlin knows it all,
> We hope our golfer knows enough
> To get us on the ball.[29]

It was a shock, creating the illusion of a technological gap and providing the impetus for a variety of remedial actions.

The more serious reaction to Sputnik came from Senator Lyndon B. Johnson (Democrat-Texas), who opened hearings by the Senate Armed Services Committee's preparedness subcommittee on 25 November 1957 to review the whole spectrum of American defense and space programs in the wake of the Sputnik crisis. One of Johnson's concerns, of course, was that a nation capable of orbiting satellites was also capable of developing technology to support the arms race. This subcommittee found serious underfunding and incomprehensible organization for the conduct of U.S. space activities, and that worried it all the more. It blamed the president and the Republican party. Johnson spoke for many Americans when he remarked in two speeches in Texas that the "Soviets have beaten us at our own game—daring, scientific advances in the atomic age." Since those cold war rivals had already established a foothold in space, Johnson proposed to "take a long careful look" at what had gone wrong in the U.S. space and missile program and to chart a course that would lead to U.S. parity in space.[30]

---

28 This proved incorrect, however, and Fred I. Greenstein demonstrated the fact in *The Hidden-Hand Presidency: Eisenhower as Leader* (New York: Basic Books, 1982). He argued that Eisenhower worked behind the scenes while giving the appearance of inaction, and in most instances his indirect approach to leadership was highly effective. This has been extended to Eisenhower's space program in R. Cargill Hall, "Eisenhower, Open Skies, and Freedom of Space," IAA-92-0184, paper delivered on 2 September 1992 to the International Astronautical Federation, Washington, D.C.

29 G. Mennen Williams, quoted in William E. Burrows, *Deep Black: Space Espionage and National Security* (New York: Random House, 1987), pp. 94-94. See also Derek W. Elliott, "Finding an Appropriate Commitment: Space Policy Under Eisenhower and Kennedy," Ph.D. Diss., George Washington University, 1992.

30 Speeches of Lyndon B. Johnson, Tyler, TX, 18 October 1957, and Austin, TX, 19 October 1957, both in Statements file, Box 22, Lyndon B. Johnson Presidential Library, Austin, TX. On the preparedness subcommittee, see McDougall, *Heavens and the Earth*, pp. 142, 151-155, 162, 166, 214, 387; Robert Dallek, *Lone Star Rising: Lyndon Johnson and His Times, 1908-1960* (Oxford: Oxford University Press, 1991), pp. 529-531.

Emerging from this investigation was a policy to make space exploration a concerted effort both for technological development and for the national prestige it would engender in the context of the cold war. Johnson's subcommittee assessed the nature, scope, and organization of the nation's long-term efforts in space, and was a force behind the Senate's vote on 6 February 1958 to create a Special Committee on Space and Aeronautics whose charter was to frame legislation for a permanent space agency. The House of Representatives soon followed suit. With Congress leading the way, it was obvious that some government organization to direct American space efforts would emerge before the end of the year.[31]

While this was taking place in Congress, Eisenhower was not inactive. Asking his new science advisor, James R. Killian, Jr., to convene the President's Science Advisory Committee (PSAC), established in the wake of Sputnik, to come up with a plan, in March 1958 the PSAC proposed that all non-military space efforts be assigned to a strengthened and renamed National Advisory Committee for Aeronautics (NACA). Although established in 1915 to foster aviation research in the United States, the NACA had already moved into space-related areas of research and engineering. Its civilian character; its recognized excellence in technical activities; and its quiet, research-focused image all made it an attractive choice. It could fill the requirements of the job without exacerbating cold war tensions with the Soviet Union.[32]

President Eisenhower accepted the PSAC's recommendations and a new National Aeronautics and Space Administration (NASA) emerged from them. The National Aeronautics and Space Act of 1958 set forth a broad mission for the agency to "plan, direct, and conduct aeronautical and space activities"; to involve the nation's scientific community in these activities; and to disseminate widely information about these activities. An administrator, appointed by the president, would head the new agency. Lyndon Johnson inserted into the proposed legislation language that provided for the creation of a Space Council of no more than nine members charged with working out "the aeronautic and astronautic policies, programs and projects of the United States." Required seats on the Council included the head of NASA, the secretaries of State and Defense, and the chief of the AEC. Eisenhower signed the act into law on 29 July 1958. The new organization started functioning on 1 October.[33]

---

[31] Alison Griffith, *The National Aeronautics and Space Act: A Study of the Development of Public Policy* (Washington, D.C.: Public Affairs Press, 1962), pp. 19-24.

[32] Rosholt, *Administrative History of NASA*, pp. 8-12. On the history of the NACA see, Alex Roland, *Model Research: The National Advisory Committee for Aeronautics, 1915-1958* 2 vols. (Washington, D.C.: National Aeronautics and Space Administration, NASA SP-4103, 1985).

[33] "National Aeronautics and Space Act," *Hearings before the Senate Committee on Space and Astronautics, 85th Cong., 2d Sess.* (Washington, D.C.: Government Printing Office, 1958), pp. 91, 258; Griffith, *National Aeronautics and Space Act*, p. 93; Rosholt, *Administrative History of NASA*, pp. 10-17; U.S. Congress, House, Select Committee on Astronautics and Space Exploration, *Astronautics and Space Exploration, Hearings on H.R. 11881, 85th Cong., 2d. Sess.* (Washington, D.C.: Government Printing Office, 1958), pp. 3-5, 11-15, 967-69; Schoettle, "Establishment of NASA," pp. 229-39; "Analysis of S.3609 with Proposed Modifications," 8 May 1958; Senate Special Committee Memo, 11 April 1958, both in Richard B. Russell Papers, ser. 9, A, box 3, Johnson Library; Lyndon B. Johnson, *The Vantage Point: Perspectives of the Presidency, 1963-1969* (New York: Holt, Rinehart & Winston, 1971), p. 277; James R. Killian, Jr., *Sputnik, Scientists, and Eisenhower: A Memoir of the First Special Assistant to the President for Science and Technology* (Cambridge, Massachusetts: MIT Press, 1977), pp. 137-38.

Even before Eisenhower signed this act creating NASA, individuals within his administration began a search for the agency's first administrator. Most people on the Washington scene thought that the NACA's Director, Dr. Hugh L. Dryden, would be named to the post. Dryden, a career civil servant and an aerodynamicist by discipline, was also considered one of the government's top science and technology managers. He had frequently incurred the wrath of Congress in the post-Sputnik era, however, for his cautious stance on competing with the Soviets in any space race. Representatives and senators saw NASA as a weapon to be wielded in order to "leapfrog" the cold war rivals, if only its head had the ability and the will to employ it effectively. Several sources reported that Dryden's candidacy was "vetoed" early in the search process by the House Select Committee on Astronautics and Space Exploration because he did not seem to possess either that ability or will.[34]

Since Dryden was not an acceptable candidate to head NASA, Killian looked elsewhere and quickly hit on Glennan as a possible candidate. Glennan recalled that Killian asked him to come to Washington early in August 1958 and took him to meet Eisenhower, to whom he had been introduced for the first time in 1955. The president told him that he needed a person to run NASA who would not be spooked by the cold war crisis atmosphere present in the nation but would build a firm foundation upon which to carry out a far-reaching, reasonable program aimed at the exploration of space using "cutting-edge" technology. Eisenhower asked Glennan to accept the job, and Glennan agreed provided that Dryden was named as his deputy. After weathering a senatorial hearing on the two nominations on 14 August 1958, they were confirmed by the whole Senate the next day, and were sworn in on 19 August. Glennan worked out another leave of absence from Case and reported for duty in Washington on 9 September 1958.[35]

Glennan fit perfectly into the Eisenhower administration. He was a Republican with a fiscally conservative bent, an aggressive businessman with a keen sense of public duty and an opposition to government intrusion into the lives of Americans, and an administrator and an educator with a rich appreciation of the role of science and technology in an international setting. His values and perspectives found themselves replicated in NASA as he began to direct its affairs in the fall of 1958. First, Glennan worked for the development of a well-rounded space

[34] On the "leapfrog" concept see, U.S. Congress, House Select Committee on Astronautics and Space Exploration, *Establishment of the National Space Program, H. Rprt. 1770 on H.R. 12575, 85th Cong., 2d Sess.* (Washington, D.C.: Government Printing Office, 1958), p. 4. U.S. Congress, House Select Committee on Astronautics and Space Exploration, *Authorizing Construction for the National Aeronautics and Space Administration, Hearings on H.R. 13619, 85th Cong., 2d. Sess.* (Washington, D.C.: Government Printing Office, 1958), pp. 9, 12; *New York Times,* 6 August 1958; *Washington Post,* 9 August 1958.

[35] Rosholt, *Administrative History of NASA,* pp. 40-42; "He Did It All," *CRWU Magazine,* February 1990, p. 13; McDougall, *Heavens and the Earth,* p. 196; Senate Special Committee on Space and Astronautics, *Hearing on Nomination of T. Keith Glennan To Be Administrator, National Aeronautics and Space Administration and the Nomination of Dr. Hugh L. Dryden to Be Deputy Administrator, National Aeronautics and Space Administration, August 14, 1958* (Washington, D.C.: Government Printing Office, 1958); *Washington Post,* 9 August 1958, 11 August 1958; Killian, *Sputnik, Scientists, and Eisenhower,* pp. 138-40.

program that did not focus on "spectacular" missions designed to "one-up" the Soviets. While he was an ardent cold warrior and understood very well the importance of the space program as an instrument of international prestige, Glennan emphasized long-range goals that would yield genuine scientific and technological results. Second, he believed that the new space agency should remain relatively small, and that much of its work would of necessity be done under contract to private industry and educational institutions. This was in line with his concerns about the growing size and power of the federal government. Third, when it grew, as he knew it would, Glennan tried to direct it in an orderly manner. Along those lines, he tenaciously worked for the incorporation of the non-military space efforts being carried out in several other federal agencies—especially in the Department of Defense—into NASA so that the space program could be brought together into a meaningful whole.[36]

About 170 employees of the new space organization gathered in the courtyard of the Dolly Madison House near the White House on 1 October 1958 to listen to Glennan as he charted the course for the space agency. The newly-appointed NASA administrator announced the bold prospects being considered for space exploration. Glennan was presiding over a NASA that had absorbed the NACA intact; its 8,000 employees and an annual budget of $100 million made up the core of the new NASA. When Glennan arrived NASA consisted of a small headquarters staff in Washington that directed operations, plus three major research laboratories—the Langley Aeronautical Laboratory established in 1917, the Ames Aeronautical Laboratory activated near San Francisco in 1939, and the Lewis Flight Propulsion Laboratory built at Cleveland, Ohio, in 1940—and two small test facilities, one for high-speed flight research at Muroc Dry Lake in the high desert of California and one for sounding rockets at Wallops Island, Virginia. The scientists and engineers who came into NASA from the NACA brought a strong sense of technical competence, a commitment to collegial in-house research conducive to engineering innovation, and a definite apolitical perspective.[37]

---

[36] These themes are well developed in T. Keith Glennan's diary that follows. See also, "Glennan Announces First Details of the New Space Agency Organization," 5 October 1958, NASA Historical Reference Collection, NASA History Office, National Aeronautics and Space Administration, Washington, D.C.; Killian, *Sputnik, Scientists, and Eisenhower*, pp. 141-44; James R. Killian, Jr., Oral History, 23 July 1974, NASA Historical Reference Collection. Eisenhower's concerns about this aspect of modern America are revealed in "Farewell Radio and Television Address to the American People," 17 January 1961, *Papers of the President, Dwight D. Eisenhower 1960-61* (Washington, D.C.: Government Printing Office, 1961), pp. 1035-40.

[37] On these institutions see, Roland, *Model Research*, I:283-303; James R. Hansen, *Engineer in Charge: A History of the Langley Aeronautical Laboratory, 1917-1958* (Washington, D.C.: NASA SP-4305, 1987); Elizabeth A. Muenger, *Searching the Horizon: A History of Ames Research Center, 1940-1976* (Washington, D.C.: NASA SP-4304, 1985); Virginia P. Dawson, *Engines and Innovation: Lewis Laboratory and American Propulsion Technology* (Washington, D.C.: NASA SP-4306, 1991); Richard P. Hallion, *On the Frontier: Flight Research at Dryden, 1946-1981* (Washington, D.C.: NASA SP-4305, 1984); Joseph Adams Shortal, *A New Dimension: Wallops Island Flight Test Range, The First Fifteen Years* (Washington, D.C.: NASA Reference Publication 1028, 1978). On the technical culture of the NACA see, Howard E. McCurdy, *The Two NASAs: High Technology and Organizational Change in the U.S. Space Program* (Baltimore, Maryland: Johns Hopkins University Press, 1993); Nancy Jane Petrovic, "Design for Decline: Executive Management and the Eclipse of NASA," Ph.D. Diss., University of Maryland, 1982.

Within a short time after NASA's formal organization, Glennan incorporated several organizations involved in space exploration projects from other federal agencies into NASA to ensure that Eisenhower's desires for a viable scientific program of space exploration could be reasonably conducted over the long-term. One of the important ingredients consisted of the 150 personnel and resources associated with Project Vanguard at the Naval Research Laboratory, located along the Potomac River just outside of Washington. Officially becoming a part of NASA on 16 November 1958, this project remained under the operational control of the Navy until 1960 when it was transferred en masse from Navy facilities to a newly established NASA installation, the Goddard Space Flight Center, in suburban Maryland. Those who had been associated with the Naval Research Laboratory brought a similar level of scientific competence and emphasis on in-house research and technical mastery to that of the NACA elements.[38]

It would be superfluous to add further details about the early history of NASA because Glennan tells that story himself in the diary that follows. He relates, for example, how in December 1958 NASA acquired control of the Jet Propulsion Laboratory, a contractor facility operated by the California Institute of Technology (Caltech) in Pasadena, California. A contractor for the Army, this oddly-named institution had been specializing in the development of weaponry since World War II.[39] He also explains the struggle with the Army leading to NASA's acquisition of the development operations division within the Army Ballistic Missile Agency (ABMA)—a part of the Redstone Arsenal, located at Huntsville, Alabama—and its Saturn rocket program, presided over by one of the nation's foremost space advocates, German postwar immigrant Wernher von Braun. This rocket team brought to NASA a strong sense of technical competence, a keen commitment to the goal as defined by von Braun, and an especially hardy group identity.[40] Glennan also explains some of the background behind the founding of the Manned Spacecraft Center in Houston in 1961 (after he left NASA) and of the later Kennedy Space Center at Cape Canaveral soon thereafter.[41]

---

[38] The transfer of several DOD programs to NASA was ordered in William J. Hopkins, White House Executive Clerk to T. Keith Glennan, 2 October 1958, w/enclosures, NASA Historical Reference Collection. On Goddard's creation see, Alfred Rosenthal, *Venture into Space: Early Years of Goddard Space Flight Center* (Washington, D.C.: NASA SP-4301, 1968).

[39] On JPL's history see also Clayton R. Koppes, *JPL and the American Space Program: A History of the Jet Propulsion Laboratory* (New Haven: Yale University Press, 1982).

[40] See McCurdy, *Inside NASA*, esp. pp. 16, 36; U.S. Senate Committee on Aeronautical and Space Sciences, NASA Authorization Subcommittee, 86th Cong., 2d Sess., *Transfer of Von Braun Team to NASA* (Washington, D.C.: Government Printing Office, 1960); Rosholt, *Administrative History of NASA*, pp. 46-47, 117-20.

[41] On the creation of the Houston center see Henry C. Dethloff, *"Suddenly Tomorrow Came...": A History of the Johnson Space Center* (Washington, D.C.: NASA SP-4307, 1993). On the Kennedy Space Center see, Charles D. Benson and William Barnaby Faherty, *Moonport: A History of Apollo Launch Facilities and Operations* (Washington, D.C.: NASA SP-4204, 1978).

In the course of the diary, Glennan narrates among other things the history of the Mercury program and its astronaut corps.[42] The diary reveals how clearly Glennan understood the tenor of the cold war atmosphere of the latter 1950s and the seemingly life-and-death struggle between the two superpowers. He believed that Project Mercury was more a means to an end than something to be done because it would yield important scientific results. "I came to realize," he recalled in 1990, "that we wouldn't have a program at all if we didn't have one that was exciting to people. That was the reason for manned space flight. But I was interested in what the law required us to do for the benefit of all mankind. And I think that what has been done without man is much more for the benefit of mankind than all we did in getting ahead of the Russians [with human spaceflight]."[43] As a result he fashioned a program that incorporated a healthy human spaceflight element with a solid space science and applications base. In this, he had the strong support of Dryden.

The two of them opted for a deliberate program with clear objectives and a long timetable. The Eisenhower administration's goal, as Pulitzer Prize-winning historian Walter A. McDougall concluded, was to refrain from beginning a race against the Soviets that "might kick off an orgy of state-directed technological showmanship that would be hard to stop, might spill over into other policy arenas, and would relinquish to the Soviets the initiative in defining the fields of battle for the hearts and minds of the world."[44]

In 1960 Senator John F. Kennedy (Democrat-Massachusetts) ran for president with Lyndon B. Johnson as his running mate. Using the slogan, "Let's get this country moving again," Kennedy charged the Republican Eisenhower administration with doing nothing about the myriad social, economic, and international problems that festered in the 1950s. He was especially hard on Eisenhower's record in international relations, taking a hard-line position on a supposed "missile gap" (which turned out not to be the case) wherein the United States lagged far behind the Soviet Union in ICBM technology. The Republican candidate, Richard M. Nixon, who had been Eisenhower's vice president, tried to defend his mentor's record but when the results were in, Kennedy was elected by a narrow margin of 118,550 out of more than 68 million popular votes cast. The change in administration ensured that Glennan would be released from his work at NASA and be allowed to return to CIT. He resigned effective 20 January 1961, the day of Kennedy's inaugural, and moved back to Cleveland after spending 28 months in Washington.[45]

[42] "Glennan Looks to Moon, But With Purpose in Mind," *Times Herald*, 4 February 1960. See also "Space Death Wouldn't Halt U.S. Effort, Glennan Says," *Baltimore Sun*, 11 April 1960; "Glennan Has Goal in Space," *New York World Telegram*, 5 February 1960; "Capital Circus," *New York Times*, 30 December 1959.

[43] "He Did It All," *CWRU Magazine*, February 1990, p. 14.

[44] T. Keith Glennan to James R. Killian, Jr., 27 May 1959, NASA Historical Reference Collection; McDougall, *Heavens and the Earth*, p. 202.

[45] T. Keith Glennan to President Dwight D. Eisenhower, 28 December 1960; President Dwight D. Eisenhower to T. Keith Glennan, 29 December 1960, both in NASA Historical Reference Collection; "NASA Post is Resigned by Glennan," *Baltimore Sun*, 30 December 1960.

During the time that Glennan headed NASA he worked to assure the viability of the space agency as an organization that could carry out an exceptionally complex and arduous set of scientific and technological tasks. In an irony of massive proportions, Glennan was in large part responsible for positioning NASA to be able to accomplish the type of large-scale, federally-operated technological enter-prise that he believed was not in the nation's best interests. He also positioned NASA so that it could serve as the vehicle for com-peting with the Soviet Union in a so-called space race, which he also believed was not in the nation's best interest. In some respects, Glennan had succeeded too well in establishing NASA as viable agency, for it was his organization that accomplished a goal he had eschewed—an accelerated Project Apollo to race the Soviets to the Moon. In some respects he was a tragic figure; he played Frederick Wilhelm I (who built the Prussian Army into

T. Keith and Ruth Glennan in Sydney, Australia, in 1961 for a meeting of the International Air Transport Association, where Dr. Glennan discussed the future of supersonic transports (SSTs). NASA had been active with the FAA and the DOD in developing the concept of an SST in its early stages. In the background is the Quantas V-jet on which the Glennans arrived.

the finest fighting force in Europe in the early eighteenth century but was reluctant to use it) to successor James E. Webb's Frederick the Great (who used the Prussian Army to conquer and defend Silesia in the middle decades of the eighteenth century, thereby raising Prussia to great power status).[46]

---

46 This story is told in Gordan A. Craig, *The Politics of the Prussian Army, 1640-1945* (New York: Oxford University Press, 1955), pp. 11-14; Sidney B. Fay, *The Rise of Brandenburg-Prussia to 1786*, revised by Klaus Epstein (New York: Holt, Rinehart and Winston, Inc., 1964), pp. 101-111; and E. J. Feuchtwanger, *Prussia, Myth and Reality: The Role of Prussia in German History* (Chicago: Henry Regnery Company, 1970), pp. 52-73.

From his seat at Case Institute of Technology Glennan watched the activities at NASA during the Kennedy administration with great interest. For a time Kennedy seemed quite happy to allow NASA to execute Project Mercury at a deliberate pace, but that changed in April 1961 because of two important events. On 12 April Soviet Cosmonaut Yuri Gagarin became the first human in space with a one-orbit mission aboard the spacecraft Vostok 1. The effort to place a human in space before the Soviet Union did so had now failed. Glennan wrote a letter to Robert Gilruth, the head of NASA's Space Task Group, offering consolation. "Even though we all expected the Russians to get there . . . first," he commented, "I am sure that each of us had a fervent hope that Mercury would make the grade in time." Even so, he urged Gilruth "not, under any circumstances, [to] deviate from the path you have chosen. There is now even greater reason for applying your engineering judgment to all phases of the project. . . . Keep your chin up!"[47] Within a few days, the aborted invasion of Cuba at the Bay of Pigs heaped international censure on the U.S.

While Gilruth and company succeeded in launching Alan Shepard on a 15-minute suborbital flight on 6 May 1961, that did not salve the open wound to U.S. pride. The perception of American technical inferiority to the Soviet Union worried the Kennedy administration because of what it would mean in the larger cold war environment. To reestablish the nation's credibility as a technological leader before the world, Kennedy unveiled an accelerated Project Apollo among other proposals in a 25 May 1961 speech on "Urgent National Needs." After clearly laying out the cold war origins of the project, he stated, "I believe that this Nation should commit itself to achieving the goal, before this decade is out, of landing a man on the moon and returning him safely to the earth. No single space project in this period will be more impressive to mankind, or more important for the long-range exploration of space; and none will be so difficult or expensive to accomplish."[48]

Glennan's reaction was immediate. He began corresponding with several people who had been members of the Eisenhower administration, including Ike himself, and expressed misgivings about the commitment to race the Soviets.[49] For instance he told Eisenhower, then in retirement at Gettysburg, Pennsylvania, that "this is a very bad move—that we are entering into a competition which will be

---

[47] T. Keith Glennan to Robert Gilruth, 16 April 1961, Glennan Personal Papers, 19DD4.

[48] U.S. Congress, Senate Committee on Aeronautical and Space Sciences, *Documents on International Aspects of the Exploration and Uses of Outer Space, 1954-1962,* 88th Cong., 1st Sess., 1963, Senate Document 18, pp. 202-204; *Public Papers of the Presidents of the United States: John F. Kennedy . . ., 1961* (Washington, D.C.: Government Printing Office, 1962), pp. 396-406, quotation from p. 404. The standard work on this decision, while more than twenty years old, remains John M. Logsdon, *The Decision to Go to the Moon: Project Apollo and the National Interest* (Cambridge, Massachusetts: MIT Press, 1970).

[49] Richard E. Horner, Northrop Corp., to T. Keith Glennan, 1 June 1961; T. Keith Glennan to J.B. Lawrence Chair, International Fact Finding Inst., 16 May 1961; T. Keith Glennan to Richard M. Nixon, 14 June 1961; T. Keith Glennan to Dwight D. Eisenhower, 14 June 1961, 13 November 1961; T. Keith Glennan to George Kistiakowsky, 4 December 1961; T. Keith Glennan to James R. Killian, Jr., 4 December 1961; James R. Killian to T. Keith Glennan, 19 December 1961; T. Keith Glennan to Neil McElroy, 22 September 1961, all in Glennan Personal Papers, 19DD4.

exceedingly costly and which will take up an increasingly large share of that small portion of the nation's budget which might be called controllable."[50] Glennan harped on this concern throughout the rest of the year, never quite able to accept the view of the Kennedy administration that large expenditures for science and technology in the form of a race to the moon against the Soviets could have much positive benefit for the nation.

Four former NASA administrators at Pad 39 of Kennedy Space Center on 10 April 1981 as they awaited the lift-off of the first space shuttle. In order clockwise from the upper left of the photo are Thomas O. Paine (1969-1970), Robert A. Frosch (1977-1981), T. Keith Glennan (1958-1961), and James E. Webb (1961-1968). The shuttle flight had to be scrubbed that day, but two days later Columbia did lift off for the first space transportation mission, which ended on 14 April with the first landing of an airplane-like craft from orbit for reuse. The photo was taken by James J. Harford, then the executive secretary of the American Institute of Aeronautics and Astronautics.

Glennan also maintained contact with his successor at NASA, James Webb, and expressed to him his dismay at Kennedy's mandate. He told Webb in July 1961:

> I have no doubt at all as to the desirability and inevitability of manned flight to the moon. And I would accept—not willingly—a national decision to beat the Russians to the moon if such a decision resulted in a truly "crash" program with no effort spared or held back. No one knows the intentions of the Soviet Union but all of us understand the ability they have to dedicate

---

[50] T. Keith Glennan to Dwight D. Eisenhower, 31 May 1961, Glennan Personal Papers, 19DD4.

men and facilities and treasure to that particular effort they believe desirable or necessary. To enter a "race" against an adversary under such conditions and to state that no additional taxes are necessary—indeed to suggest tax reductions—does not seem to me to be facing facts nor to be completely frank about the on-going program. . . .

There can be only one real reason for such a "race". That reason must be "prestige". The present program without such a "race" but with full intention of accomplishing whatever needs to be accomplished in lunar and planetary exploration, unmanned and/or manned, is a vigorous and costly one. It will produce most of the significance technology and essentially all of the scientific knowledge that will be produced under the impetus of the "race" and at the lower cost in men and money. . . .

No, Jim, I cannot bring myself to believe that we will gain lasting "prestige" by a shot we may make six to eight years from now. I don't think we should play the game according to the rules laid down by our adversary. I do believe that such prestige is apt to be less . . . enduring as compared to the respect and friendship we will gain from foreign aid programs, well administered over the same six or eight years.[51]

Years later Glennan was still sending letters to Webb, congratulating him on successes and commending him when he took positions that seemed broadly based and not simply oriented toward racing to the moon with the Soviet Union. He never wavered in his support for the accomplishments of NASA, even though he disagreed with the policies of the Kennedy administration, and often wrote to Webb or his former NASA colleagues thoughtful statements recognizing successful launches or other space-related activities. Like many others he was pleased with the successful lunar landing mission in 1969, although it was the capstone of a project he believed had been ill-advised.[52]

While he watched the direction of NASA and the space program during the 1960s, it did not consume his interests. He continued to direct the affairs of Case Institute of Technology, building its infrastructure, resources, and student body. When the opportunity came in the early 1960s to move toward an affiliation with CIT's sister institution, Western Reserve University, he and Western Reserve President John S. Millis took it and began a long process that resulted in federation of the two schools. Case's science and technology program coupled with Western Reserve's liberal arts, law, and medical schools ensured the success of both in the long term. The two presidents pushed for a linking of the two schools, finally achieving the goal with the beginning of the 1967-1968 academic year, after

---

[51] T. Keith Glennan to James E. Webb, 21 July 1961, Glennan Personal Papers, 19DD4.

[52] T. Keith Glennan to James E. Webb, 6 November 1962; T. Keith Glennan to Hugh Dryden, 20 November 1964; T. Keith Glennan to M.S. Rowan, Rand Corp., 26 March 1968; T. Keith Glennan to Richard E. Horner, Senior Vice President, Northrop Corp., 3 June 1968; all in Glennan Personal Papers, 19DD4.

Glennan's retirement.[53] For unfortunately, Glennan did not remain as Case president to see this federation to completion. He had developed health problems in the early 1960s, especially diabetes, and decided to retire after eighteen years of service in the summer of 1965. When it took the search committee longer than anticipated to find a suitable replacement, however, he remained another year and retired on 30 June 1966 at age 61.[54]

After leaving Case, Glennan spent two years as president of Associated Universities, Inc., a Washington-based institution involved in technical issues.[55] From his position in Washington he watched the student unrest on the nation's campuses during the latter part of the 1960s and was deeply troubled by what he saw. He could never fully understand what was taking place, although he made an effort to do so. He wrote to a friend, Polycarp Kusch, a Nobel Prize-winning physicist and alumnus of CIT, in May 1968 about the riots at Columbia University where Kusch was on the faculty. "The troublesome aspects of the Students for Democratic Society as an advocate of disruption and even of violence bother me very greatly," he confided, and asked for Kusch's analysis of why young people who were the most well-educated and provided for in the history of the United States could seek to destroy the society that had nurtured them so well.[56] Kusch was more sympathetic to the student demonstrators and tried to explain that oppression by his university had been allowed to go unchecked. "To my mind," he wrote, "we require a new mood or spirit or style; this is not going to be forthcoming through a readjustment of statutes. . . . What puzzles me now both personally and in more formal roles is how all the pieces can be put back together again into a university that may flourish in this age."[57] Glennan continued to watch the uproar on the campus but failed to

---

53 "Case-WRU Joint Calendar Committee," 4 January 1960-1967; "Appointment of Presidential Committee on Plans, Programs and Resources," 1 May 1964; "Joint Statement of Dr. T. Keith Glennan and Dr. John R. Millis," 1 November 1965; "Resolution Adopted by the Board of Trustees of Western Reserve University," 8 September 1966; Henry F. Heald, *et al.*, "Second interim Report of the Case-W.R.U. Study Commission," 22 November 1966; "Resolution on Federation," 10 January 1967, all in Case Central Files, 19DC, Case Western Reserve University Archives; Ruth Fischer, "Case Western Reserve: Federation Fever," *Change: The Magazine of Learning,* 10 (October 1978): 38-43; Lawrence S. Finkelstein, "Case-WRU Study Commission at Work," *Western Reserve University Outlook,* 4 (Winter 1967): 3-5.

54 T. Keith Glennan to Thomas P. Murtaugh, 16 March 1962, Glennan Personal Papers, 19DD4; "Glennans Cherish 18 Years Here," Cleveland *Plain Dealer,* 27 March 1965; T. Keith Glennan to Dr. Raymond L. Bisplinghoff, Alexandria, Virginia, 24 June 1965; Case Institute of Technology Faculty Announcement, 25 June 1965; Elmer L. Lindseth, Chair, Board of Trustees, to Willia C. Treuhaft, 13 August 1965; "Action of Trustees Without a Meeting," 20 August 1965, all in Central Files, 19DC.

55 T. Keith Glennan to Harry S. Rowen, Rand Corporation, 21 August 1967; T. Keith Glennan to Lewis L. Strauss, 1 October 1968; T. Keith Glennan to Franklin A. Long, Cornell University, 25 January 1968; T. Keith Glennan to Cong. Emilio Q. Daddario, 24 July 1967; T. Keith Glennan to James R. Schlesinger, 1 May 1968; T. Keith Glennan to Kingman Brewster, President, Yale University, 16 Oct 1968, all in Glennan Personal Papers, 19DD4.

56 T. Keith Glennan to Polycarp Kusch, Department of Physics, Columbia University, 13 May 1968, Glennan Personal Papers, 19DD4.

57 Polycarp Kusch to T. Keith Glennan, 17 May 1968, Glennan Personal Papers, 19DD4.

recognize until much later that a social transformation was underway that reshaped the structure of the U.S. in the latter 1960s and early 1970s.[58]

Since the early 1970s Glennan has been involved in a variety of public service and civic activities. As of this writing (August 1993) he is 87 years old but still available to lend advice and support on a broad front to those who ask. He presently lives in Mitchellville, Maryland.

The Glennans and their children in late 1992. Standing behind Dr. and Mrs. Glennan from left to right are Pauline (Polly) Watts, an educator who, at the time the photo was taken, lived in the Virgin Islands with her lawyer husband; Thomas K. Glennan, Jr. who lived in McLean, Virginia, with his wife Martha and was an economist with the Rand Corporation; Catherine (Kitty) Borchert, who was an ordained Presbyterian minister living in Cleveland, Ohio, with her husband Frank, a vice president for planning and budgets at Case Western Reserve University; and Sarah (Sally) Oldham, who lived in McLean, Virginia, with her architect husband, Ted, and was president of Scenic America, a Washington-based, non-profit organization dedicated to the beautification of the nation's major highways.

The document that follows was dictated by T. Keith Glennan to record for his own children his observations and priorities while head of NASA during the Eisenhower administration. The first part of it is actually a memoir, consisting of

---

[58] T. Keith Glennan to Polycarp Kusch, 23 May 1968, Glennan Personal Papers, 19DD4. For a discussion of this transformation see, Allen J. Matusow, *The Unraveling of America: A History of Liberalism in the 1960s* (New York: Harper & Row, 1984).

a summary of major events in which he participated during his early years at NASA in 1958 and 1959. Most of the rest is in a diary format, made up of daily summaries of events, including his personal reflections from 1 January 1960 through the end of the Eisenhower presidency over a year later. There also follows a postscript written by Glennan in 1963 to record his thoughts on the space program during the presidency of John F. Kennedy. The actual document was assembled by Glennan in the early 1960s, after his return to Case, typed up, and then bound in two volumes in 1964. He made copies of the diary for each of his children, kept one for himself, and placed one (restricted for a time) at the Eisenhower Presidential Library for research purposes. Later he made a copy available to the NASA History Office for incorporation into its reference collection. For many years historians have recognized the importance of this diary in charting the early history of NASA and the U.S. space program. It has been studied by historians and quoted at length in several historical texts, but until now the diary itself has been accessible to only a few people who could travel to Abilene or Washington to use it.

Since the diary was dictated into a tape recorder rather than written and was never prepared for formal publication, Dr. J. D. Hunley, who has done most of the editing for this version of the diary, has corrected spelling, punctuation, and grammar in some places, shortened many sentences without changing their sense or general flavor, and omitted certain sections of the diary that contained purely family-oriented comments not germane to understanding Glennan's role at NASA. Dr. Hunley took over the editing responsibilities when other demands of my job made it clear that I could not finish the project in a reasonable amount of time. I had done the preliminary editing on the first three chapters, but he has revised those and done all of the editing for the remaining chapters. In every case we have shown the results of the editing to the author before committing them to print. To a very real extent, therefore, this published version of Glennan's diary represents his statement. We have not changed any of the observations or conclusions he reached in the unpublished version, although Glennan suggested that perhaps some of them did not prove to be valid and that he might have changed his mind over the course of the last quarter century, especially in relation to some of his hard comments about individuals with whom he dealt. He agreed, however, that the diary accurately represents his position as it crystallized in the early 1960s and that in itself makes it an exceptionally valuable historical document.

Throughout the document, footnotes provide background information about matters that Glennan brings up in the diary but does not elaborate upon. The notes also add supplemental information about budgetary data, organizational matters, and the like, as well as occasional identifications of individuals mentioned in the diary but not fully identified. Where they are brief, sometimes we have provided identifications in brackets right in the text of the diary, and brackets sometimes also indicate where material from the original diary has been deleted. This latter practice is a rare one, however, used only when needed for transition. We have deleted a great deal of personal material and shortened some sentences without specific indication of the omissions because otherwise, the text would have been littered with ellipses. At the end of the diary, a biographical appendix provides sketches of the lives of the major figures Glennan talks about.

Throughout this project Keith and Ruth Glennan have been exceptionally supportive and helpful. Without their assistance the project could not have been completed. I also wish to acknowledge the help of R. Cargill Hall, who shared his wealth of knowledge about the space program of the Eisenhower administration; Lee D. Saegessor, who helped track down illustrations and sources for footnotes; Robert H. Ferrell, who generously read several chapters of the diary and offered us his excellent editorial advice; and J.D. Hunley, who besides doing most of the editing, read and edited various drafts of this introduction, and provided valuable advice. In addition to these individuals, I wish to acknowledge and thank the following people who aided us in a variety of ways to complete this editorial assignment: Donald R. Baucom, Virginia P. Dawson, Andrew J. Dunar, Aaron K. Gillette, Michael R. Gorn, Adam L. Gruen, Richard P. Hallion, Mark Hayes, Marianne Hosea, H. C. Hunley, Sylvia K. Kraemer, Theresa Kraus, Beverley Lehrer, Cathleen S. Lewis, John L. Loos, Howard E. McCurdy, F. Mark McKiernan, Jerry Meyer, H. C. Erik Midelfort, Jeffrey Michaels, Jennifer Mitchum, John E. Naugle, Allan A. Needell, Michael J. Neufeld, Stephanie Pair, Tricia Porth, Joseph N. Tatarewicz, Joyce Thompson-Stipe, Dennis Vetock, William C. Walter, Mary Weyant, and Bruce Wolf. My thanks also go to Ellwood Annaheim and Raymond A. Falvo for their persistence in preparing this manuscript for publication. Finally, I extend my thanks to the people at NASA's Printing and Graphics office for their help in bookmaking, and to the staff of the archives of Case Western Reserve University for their assistance with research. Needless to say, since we have not always followed all of the advice these people have kindly offered, Dr. Hunley and I retain responsibility for any errors in the book.

T. Keith Glennan was responsible, probably more than any other single person in the early history of the agency, for creating a National Aeronautics and Space Administration that could carry out a broad-based scientific and technological program. As such, he left an enduring legacy to the agency. His personality and beliefs, fiscally and socially conservative in some ways but progressive in others, also helped to shape NASA. His diary leaves a valuable legacy to historians, scientists, engineers, and public policy analysts seeking to understand the evolution of the U.S. venture into space. Glennan left a remarkable account of his work at NASA, and his record of meetings with President Eisenhower, the National Security Council, and other bodies as well as of the evolution of NASA as an institution provides important perspectives not only for those specifically interested in NASA but also for students of the recent history of science and technology in the United States.

Roger D. Launius
Washington, D.C.
11 August 1993

CHAPTER ONE

RECOLLECTIONS OF NASA'S EARLY YEARS

1958

In spite of my membership on the Board of the National Science Foundation, the agency providing the funding for the Vanguard Project, I had taken no more than casual interest in the efforts of this nation to develop a space program following the successful orbiting of Sputnik I by the Russians on 4 October 1957. The aftermath was marked by a continuous chorus of lament over the fact that the Soviet Union had stolen a march on the United States in fields that had seemed to be the special province of our own country. In reaction, President Eisenhower appointed Jim Killian of MIT as his Science Advisor. I thought this a most excellent appointment and sent a telegram to Jim congratulating him and stating that I would be happy to assist in any possible way.

In April 1958, the president sent to the Congress a bill calling for the establishment of an agency to develop and manage a national space program. Quite naturally, there was much debate about the actual management of this program—should it be handled by the military departments or by a civilian agency? The proponents of the civilian management won out, and the bill was passed and signed into law on 29 July by President Eisenhower. It called for the creation of the National Aeronautics and Space Administration using the then existing National Advisory Committee for Aeronautics as its foundation. That distinguished 43-year-old agency employed some 8,000 people, with major laboratories in Cleveland; Langley, Virginia; and Moffett Field, California. There were smaller field stations at Edwards Air

Official seal of the National Advisory Committee for Aeronautics (NACA), established by Congress in 1915. The seal depicts the first human, powered and controlled flight by the Wright brothers at Kitty Hawk, North Carolina, in December 1903.

Force Base in California and at Wallops Island, Virginia. Its budget for the 1959 fiscal year had been set at $101 million as I recall.[1]

The policy statement in the preface of the Act called for the establishment and prosecution of a program aimed at the development of useful knowledge of the space environment and the exploration and exploitation of that environment for peaceful purposes and for the benefit of all mankind. In recognition that space might well be used for military purposes, the law provided that any activities concerned principally with the defense of the nation were the responsibility of the Department of Defense.

As already stated, I paid about as much attention to all of these events as the ordinary citizen—not much more. Imagine my surprise when on 7 August 1958 I received a call from Jim Killian asking me to come immediately to Washington. I flew down on that same day and met with him at his apartment that evening. He said his purpose was to ask me, on behalf of President Eisenhower, to consider becoming administrator of the new agency, which of course was the National Aeronautics and Space Administration (NASA). He handed me a copy of the bill, which I had not previously seen. I read it through rather hurriedly and pointed out immediately the built-in conflict that seemed to me to be present whereby the Defense Department most certainly would dispute the claim of the civilian agency to important elements of any program that might be initiated. After some considerable discussion, I agreed to meet with the president the next morning.

The meeting with President Eisenhower was brief and very much to the point. He said he wanted to develop a program that would be sensibly paced and vigorously prosecuted. He made no mention of concern over accomplishments of the Soviet Union although it was clear he was concerned about the nature and quality of scientific and technological progress in this country. He seemed to rely on the advice of Jim Killian. I agreed that I would give the matter consideration and would give him a reply within a few days.

Discussions with Killian were followed by a visit to Don Quarles, deputy secretary of defense. I had known Quarles for years, since my stay in New London during World War II. It was apparent that few people had been asked to recommend a candidate for the NASA job, and I gained the impression that Quarles had only heard about the proposal that I be offered the post. He urged me to take it but expressed some unhappiness over the fact that he had not acted more promptly on a matter troubling him—head of the research and development activity in the Office of the Secretary of Defense. He stated that he had intended to offer that job to me. Although flattered, I assured him that I would not have been able to accept because of my conviction that only a scientist should handle that job.

---

[1] On the creation of NASA see, Robert L. Rosholt, *An Administrative History of NASA, 1958-1963* (Washington, DC: NASA SP-4101, 1966); Enid Curtis Bok Schoettle, "The Establishment of NASA," in Sanford A. Lakoff, ed., *Knowledge and Power* (New York: The Free Press, 1966), pp. 162-270; Alison Griffith, *The National Aeronautics and Space Act: A Study of the Development of Public Policy* (Washington, DC: Public Affairs Press, 1962), pp. 19-24. The history of the NACA is analyzed in Alex Roland, *Model Research: The National Advisory Committee for Aeronautics, 1915-1958* (2 vols.; Washington, DC: NASA SP-4103, 1985).

Returning to Cleveland, I discussed these matters with my wife Ruth, several of my associates on the campus, and members of the board of trustees. Frederick C. Crawford [chairman of the board of trustees at Case Institute of Technology] urged me to take the post, and after two or three days of soul searching I called Killian to say I would accept—but only if Hugh Dryden, the director of the NACA, would endorse the appointment and would agree to serve as my deputy. Events began to move rapidly. Fred and I agreed that it would be desirable to ask Kent Smith to serve as acting president during my absence since John Hrones [Case's vice president for academic affairs] had been with us only a year and was not acquainted with all facets of the campus. Fred and I talked with Kent and Thelma and, in spite of the fact that they had planned a year abroad, Kent agreed to take on the job.

The swearing-in was set for 19 August in Washington. Ruth, Polly, and Sally drove down with me and Ruthie and Jack Packard attended the ceremony in

President Eisenhower handing T. Keith Glennan (to his left) and Hugh Dryden their commissions as administrator and deputy administrator of NASA at their swearing-in ceremony on 19 August 1958.

the executive offices of the White House.[2] A crowd of friends attended the brief ceremony, and the family had a chance to speak with President Eisenhower who

---

[2] John S. Packard, and his wife Ruth, were friends of the Glennans. Packard was a businessman, a partner in Treadway Inns Corp. Polly and Sally are two of Glennan's daughters, formally named Pauline and Sarah.

presided and handed us our commissions, Hugh Dryden having been sworn in at the same time. Together with the Packards, we had lunch at LaSalle du Bois with everyone a bit punch drunk over events of the day. Ruth Packard and my Ruth immediately started a search for an apartment, and I returned to the NACA offices to become acquainted with members of the staff of that organization, soon to be absorbed by NASA.

Although my visit had been billed as casual, I found myself thrust into the problems of the new agency. Dryden called in Abe Silverstein and some of the top operating people who wanted to discuss budget. I will not try to describe the budget cycle in Washington agencies; suffice it to say that we were attempting to put together a budget that should have been initiated months before. Staff members were seeking my approval of a figure toward which they might work on the budget, which had to be submitted within weeks. Imagine my consternation when they proposed that we seek $615 million. The Case budget at that time was in the neighborhood of $6 or $7 million and I doubt that I had much feel for $615 million. Members of the staff made the point that when NASA was to be declared "ready for operations" we would be taking over from the Defense Department projects, together with manpower and funds already appropriated. It appeared that we would have about $300 million for FY 1959 (July 1958-June 1959). Their arguments must have been convincing, for I approved a budget for FY 1960 using the guideline figure of $615 million. This, then, was my introduction to what was to become one of the major activities of the federal government.

In accepting the appointment I had stipulated that I would take a vacation before reporting for duty and had set the reporting date at 9 September. In addition to taking a vacation, I had to complete my annual report for Case. We were able to find a cottage on Martha's Vineyard and after depositing the children in Cleveland we drove immediately to Wood's Hole and took the ferry to the Vineyard. This was a delightful spot, and I was able to complete my report even though I spent time on the telephone counseling with Dryden and others about additional top personnel.

I want to record my first brush with the inflexibility of bureaucratic procedure. The Case trustees had voted to continue my salary throughout the balance of 1958, paying me in a lump sum determined to be a legal procedure. I did not want to accept a check from the federal government until I was on the job, so I asked our financial officer at NASA to determine how this could be managed since my salary was supposed to begin when I was sworn in. He shook his head but agreed to make the attempt. When I returned to Washington on 9 September, I called him in. He stated that the only possible way to manage this affair was for me to accept the payment and to return it to the federal government as a gift. I would have to pay income tax. Since there seemed no way of circumventing these regulations, I decided to keep the salary although I suppose I could have paid the tax and returned the balance of the salary less the tax. The whole procedure seemed so unbusinesslike that I guess I acted as much in pique as from any sense of conviction.

Now my work began in earnest. Ruth was engaged, with the help of Mr. [W. Donald] Bacome, a Cleveland decorator, in making the apartment livable.

We bought drapes, a bookcase and a room divider, a daybed, and a rug and shipped furniture from Cleveland. As I look back over my appointment schedules for those days, I wonder how I kept anything straight. I was concerned with acquiring a number of good men to fill top positions in the agency and I seem to have spent a good bit of my time on this task. Hardly a day passed without a visit from the representatives of some industrial concern—usually the president—and meetings with top people in the Department of Defense and some of the other agencies with which we would be dealing.

Although NACA contained many fine technical people, it had been an agency protected from the usual in-fighting found on the Washington scene. Its staff, composed of able people, had little depth and little experience in the management of large projects. Considerable thought had been given by the staff to the organization that might develop, and these plans served to get us underway. It became apparent almost immediately that further studies would be needed and that some good people would have to be hired.

Let me discuss the philosophy with which I approached this job—a philosophy about which I had thought while vacationing at Martha's Vineyard. First, having the conviction that our government operations were growing too large, I determined to avoid excessive additions to the federal payroll. Since our organizational structure was to be erected on the NACA staff, and their operation had been conducted almost wholly "in-house," I knew I would face demands on the part of our technical staff to add to in-house capacity. Indeed, approval had been given in the budget to initiate construction of a so-called "space control center" laboratory at Beltsville, Maryland, an action I approved.[3] But I was convinced that the major portion of our funds must be spent with industry, education, and other institutions.

Second, it seemed to me that we were starting virtually from scratch and with little in the way of rocket-propelled launching systems. Thus it seemed to me that we should mount an aggressive program that would build on the advancing state of the art as we came to understand more about technologies with which we were dealing. Third, it seemed clear that we should not lose sight of the propaganda values residing in successful launches—yet we had to be aware of the limitations imposed upon us by the lack of availability of proven launch vehicle systems. This was because the military missile program was just reaching the testing stage and these same rocket-propelled units were going to have to serve as "booster systems" or, as we came to call them, launch vehicle systems for our space shots.

Fourth, in the nature of things it seemed necessary that we structure our program in accord with our own ideas of fields to be explored and the pace at which progress could and should be made. This meant we must avoid the undertaking of particular shots, the purpose of which would be propagandistic rather than directed toward solid accomplishment. Fifth, we faced the prospect of carrying to comple-

---

[3] This became the Goddard Space Flight Center, opened in 1961. See Alfred Rosenthal, *Venture into Space: Early Years of Goddard Space Flight Center* (Washington, DC: NASA SP-4301, 1968).

tion the projects started by the Advanced Research Projects Agency of the Defense Department, called into being by Secretary Neil H. McElroy during the period between 4 October 1957 and the operational beginnings of NASA. At the same time we must be planning our own broadly-based program of science and technology and organizing to accomplish all these tasks.

I was under pressure to appoint individuals to important jobs within the agency by members of the executive branch on only two occasions during my more than 29 months in Washington. The first was the more important and it occurred shortly after I had taken office. Sherman Adams was serving as President Eisenhower's chief of staff. I was searching for a general counsel and had sought nominations both within the government and from friends in the legal profession. Word reached me through General [Wilton B.] Persons [in the White House] that it would be pleasing to Governor Adams if I would appoint John Adams (I think that was the name) to this post. It will be recalled that he was the general counsel for Army Secretary Stevens during the McCarthy-Stevens trouble[4] and that he had been dropped from his position following that altercation. Whether his departure was requested or was voluntary is a matter about which I have no knowledge. My reaction was in the negative since I determined that our agency should not be involved in political machinations.

I did talk with John Adams and was not at all impressed by either his attitude or his ability. He seemed like a beaten man and I became even more convinced that the appointment would be viewed as a move to provide a haven for someone who had served the cause. I so informed Persons who asked me to speak with Governor Adams directly. This I did, and to my surprise, he listened without comment and then asked, "Is this your considered decision?" I replied in the affirmative and Adams stated that there was no need for further discussion.[5]

The second incident related to my appointment of several people to an advisory panel. Persons called to ask if I had knowledge of the political alignments of individuals I proposed to appoint. I responded that I neither had that information nor was I concerned about it. I wanted men whose scientific and technological competence and judgment were respected. Persons accepted that statement and throughout my term I had no further incidents.

This was not the case with respect to Congress. Quite a number of requests were made and I am certain that a few met with a favorable response. I recall one made by the majority leader of the House, John McCormack. He wanted a man

---

[4] This was a famous event in which Senator Joseph R. McCarthy (R-WI) accused the Army of harboring communists and homosexuals. On McCarthy see David Caute, *The Great Fear: The Anti-Communist Purge under Truman and Eisenhower* (New York: Simon and Schuster, 1978); Richard M. Freeland, *The Truman Doctrine and the Origins of McCarthyism: Foreign Policy, Domestic Politics, and Internal Security, 1946-1948* (New York: Alfred A. Knopf, 1972); Jeff Broadwater, *Eisenhower and the Anti-Communist Crusade* (Chapel Hill: University of North Carolina Press, 1992). Specific discussion of the Army hearings in 1954 can be found in Stephen E. Ambrose, *Eisenhower:* Vol. 2, *The President* (New York: Simon and Schuster, 1984), pp. 161-89.

[5] In the unedited diary Glennan commented in a footnote: "Governor Adams resigned under fire for violation of 'conflict of interests' reasons about two months after the incident I have described. Washington requires that a man's operations not only be conducted with propriety but that they have all the appearances of being conducted with propriety."

taken on in our contracting operations, and because of his great power over our legislative program we made the appointment. We were to regret it when we found the man was not only incompetent but required watching because of possible dishonesty. None of his problems were sufficiently flagrant to cause us to take action but we did resist demands of the congressman for advancement in position and salary for this individual.

It is quite an experience to be catapulted into a job of the kind I undertook. There exists no program that I know of for providing reasonable orientation. I did meet briefly with two members of President Eisenhower's staff and was told something about operations of the National Security Council (NSC) and the Operations Coordinating Board (OCB).[6] The latter attempts to monitor decisions by the National Security Council and assure that all those who should have knowledge do have such knowledge. Gordon Gray was the operating head of the NSC and Karl Harr the chairman of the OCB. Bob Merriam, another of the White House staff, briefed me on operations of the president's chief counsel. Except for these experiences I had to learn by doing.

As an example of the latter situation, let me describe my initiation into the handling of meetings of the [National Aeronautics and] Space Council, an advisory body or sort of board of directors established in law and having a membership consisting of the secretary of defense, the secretary of state, the chairman of the Atomic Energy Commission, three civilians, and one additional person from government, in this case, Waterman of the National Science Foundation. I had been on the job only two weeks when it seemed proper to Jim Killian and myself that the first meeting of the Space Council should be called. The civilians on the council were Lt. Gen. [James H.] Doolittle, U.S.A.F. (ret.), William A.M. Burden, and Jack Rettaliata of Illinois Institute of Technology. Together with members of my staff, I set about the preparation of the agenda for the meeting.

We had agreed that Killian would "present the agenda," which was the method followed in handling NSC meetings. The person presenting the agenda presumably had knowledge of attitudes of individuals who might wish to speak on any of the matters under discussion. He would call up an agenda item, discuss briefly the problem, and suggest comments. The president would listen to the discussion and, if a decision were to be taken, would normally make the first comments. Often it was necessary for the person presenting the agenda to suggest to the president that he might wish to comment. Having put together the agenda with Jim, we agreed that it would be well to brief the president that afternoon so things would go as smoothly as possible the next morning. This was 23 September.

---

[6] The National Security Council was established by the National Security Act of 1947 "to advise the President with respect to the integration of domestic, foreign, and military policies relating to the national security so as to enable the military services and the other departments and agencies of the Government to cooperate more effectively in matters involving the national security." It was formally located in the Executive Office of the President. The Operations Coordinating Board was established by Executive Order 10483 on 2 September 1953. The board—made up of representatives from the Departments of State, Defense, CIA, United States Information Agency, and the Executive Office of the President—was responsible for the effective coordination and implementation of national security policies (*United States Government Organization Manual, 1959-60* [Washington, DC: General Services Administration, 1960], pp. 63-65).

This was my first meeting with the president after taking office. He participated vigorously in the discussion and we agreed on the nature of decisions to be taken, barring arguments by members who would be present. Jim asked the president how he wanted the meeting conducted—expecting that he, Killian, would be requested to present the agenda. Instead President Eisenhower stated that he expected me to handle the meeting. What a situation! Jim had prepared his papers in his own style and I had added my comments. Now I faced the problem of handling the meeting in my own way.

I set up a session with Gordon Gray and learned about the way he handled NSC meetings. I went home and spent most of the night preparing my statements and notes I would use in calling for participation. Fortunately the meeting went off without incident and I was complimented by Gray and others. Sometimes there is no better method than to be thrown into the center of a situation.

With the help of Dryden we began to pull together our staff. Initially we used the organizational arrangements proposed by the NACA boys and they served well at this point. Because we were a new agency involved in an exciting new field, it was not too difficult to encourage top people from other agencies to join us. The secretary of the Air Force, Jim Douglas, assured me he would not countenance my approaching his general counsel, John Johnson. Apparently he had second thoughts and called me two days later to say that, to his surprise, Johnson had evinced an interest. The upshot was that Johnson joined us within two weeks and proved one of the stars on our team. No man in Washington was or is better equipped. John could make a great deal of money in industry but seems dedicated to government operations and serves his nation with distinction. Douglas became a warm and firm friend.

In like manner we acquired the services of Al Siepert as director of business administration. He had held a similar position with the National Institutes of Health and came to us with quite a background in research. A young man, Wesley Hjornevik, left the office of the deputy secretary of HEW to be my assistant. He turned out to be a good man and ultimately moved into business operations and is now occupying a responsible post in the Manned Spacecraft Center in Houston, Texas. We made but one appointment that turned out to be completely unsatisfactory—Henry Billingsley of Cleveland as director of our office of international affairs. His services were terminated at the end of a six-month period—not without unpleasant scenes.

It was clear that the NACA boys had done a good job in thinking through the organization they would like to see established, but I was convinced we needed independent counsel. Having had experience with John Corson of McKinsey and Company, I asked him to undertake an organizational study. The attitude of our staff was lukewarm but they did not interpose strong objections. This study confirmed much of the structure proposed by the NACA boys but I insisted, even against the recommendation of Corson, that we needed a "general manager" to handle day-to-day operations. This was not particularly palatable to the staff. Even Dryden was lukewarm. We compromised on calling the man an associate administrator rather than general manager and started a search for the right person—a search that was to take six months.

On 1 October, a month before the date set in the act, we declared NASA ready for operations. We took over several programs from the military—particularly from ARPA—together with men and the money set aside to underwrite these projects.[7] Our budget jumped overnight to more than $300 million.

I visited Huntsville, Alabama, 17 September—one of my first trips outside of Washington.[8] I became convinced that the talents of this group—so dedicated to space exploration and so hemmed-in by the fact that the Air Force had been given control of air and was intent on extending that control to space—would be a useful part of NASA. I had met General [John B.] Medaris and been treated in a somewhat cavalier fashion by him. He was a martinet, addicted to "spit and polish," never without a swagger stick, and determined to beat the Air Force. He simply did not have the cards. I was to learn a lot from this experience as evidenced in our acquisition of the Army Ballistic Missile Agency (ABMA) at the end of 1959.

NASA seal.

As a result of discussions with Roy Johnson and Herb York of ARPA, Dryden and some of our own people, and with Quarles, I had come to the conclusion that the nation's space program would advance most rapidly if we had working within our framework the so-called von Braun team at Huntsville, and the Jet Propulsion Laboratory (JPL) operated for the Army by Caltech. There was no warmth on the part of the staff for this proposal. Dryden, an old hand in government, recognized the values that could be had from the association but believed, I think, that we would have difficulty in convincing the Defense Department and particu-

---

[7] The Advanced Research Projects Agency (ARPA) had been established under the Department of Defense in 1958-1959 as a means of conducting high technology research and development in areas that were not service specific and that were not immediately identified with a specific military requirement. The initial project assignments were oriented toward military space technology, ballistic missile defense, and solid propellant chemistry.

[8] Huntsville, Alabama, was the site of the Army Ballistic Missile Agency, a part of the Redstone Arsenal, that transferred to NASA and was renamed the George C. Marshall Space Flight Center on 1 July 1960 (U.S. Senate Committee on Aeronautical and Space Sciences, NASA Authorization Subcommittee, *Transfer of Von Braun Team to NASA, 86th Cong., 2d Sess.* (Washington, DC: Government Printing Office, 1960); Robert L. Rosholt, *An Administrative History of NASA, 1958-1963* (Washington, DC: NASA SP-4101, 1966), pp. 46-47, 117-20. See also note 25.

larly the Army that such a move should be made. I was naive enough and stubborn enough to proceed with my plans, and Dryden gave his full support, once I had taken this decision. I discussed the matter with Killian and briefly with the president. In each instance there was agreement we should proceed. But I was not prepared for what then transpired.

Meetings were held with Secretaries McElroy and Quarles and with Roy Johnson and Herb York. I was encouraged to believe I would get full support in the move I wanted to make. I thought I was making a deal with the office of the secretary and that Quarles or McElroy would handle the matter with Wilber Brucker, secretary of the Army. But such was not the case.

It must have been late in October when I met with Don Quarles one morning to complete these negotiations. Much to my surprise, he suggested that I walk down the hall and discuss the matter with Brucker. As I recall it, Bill Holaday, serving in the secretary's office as a sort of coordinator of the missile program, was with us. I noted that he was perturbed at this turn but, having no experience in these matters and having no knowledge of Brucker, I said I would do so immediately. Quarles arranged a meeting and I went to Brucker's office. With him were two generals. I think they were Trudeau and Hinrichs. They sat there like wooden horses throughout the entire conversation, never uttering a word. I began, in a halting fashion, to discuss the situation and finally made the proposal that we take over a substantial portion of von Braun's operation and the Jet Propulsion Laboratory. It became apparent immediately that "fools rush in where angels fear to tread."

Brucker became irate and, while stating the desire of the Army to be helpful, said he could not countenance such a move. He termed it "breaking up the von Braun team." To an extent, this was the proper characterization. I had not realized how much of a pet of the Army's von Braun and his operation had become. He was its one avenue to fame in the space business. He was supported by General Medaris, the commander officer of ABMA. I attempted to point out to Secretary Brucker that we wanted the technical competence of ABMA and that we recognized the necessity for providing for capability to carry on the Army's work on the Pershing missile and other systems such as the Sergeant.[9] Brucker would have none of it and was emphatic in his statements. I finally left with my tail between my legs and called a session of our people to determine strategy.

I do not recall whether I went back that day to talk with Quarles and report to him the result of the discussion. That same evening, while at home, I received a call from Phil Goulding, the Washington correspondent of the Cleveland *Plain Dealer*. He asked me about the session and led me to believe that one of the Baltimore papers—the *Sun*—was going to carry a strong article by Mark Watson

---

[9] The Pershing and Sergeant were Army systems with a small payload capability. The Sergeant emerged from JPL at the end of World War II as a follow-on to its solid-rocket work on the Private and Corporal launch systems. The Pershing was seen as a medium-range ballistic missile follow-on to the Honest John and Sergeant systems (David Baker, *The Rocket: The History and Development of Rocket & Missile Technology* [New York: Crown Publishers, 1978], pp. 73, 106-107, 132, 141, 143, 145, 191-195).

the next morning. At first I denied any significant problem. I tried to get in touch with the assistant secretary for public affairs at the Pentagon, Murray Snyder—and when I was finally able to do so, found that he knew nothing of any leak at the Pentagon. I determined that I had better make my peace with Goulding and called him back to tell him more about the story.

The next morning's Baltimore *Sun* carried an account that could only have been given to Watson by the secretary or by one of the generals who sat with us in that office. Seldom have I been so angry. But I determined I was not going to argue in the press and gave orders that we would have no comment. Although I am sure I talked further with Quarles, I do not remember the nature of those discussions. But I found it necessary that I talk once more with Brucker. I had told him over the phone just how I felt about the way in which he or his aides handled the matter. He denied any knowledge of how the word had gotten to Watson.

About a week later I set up another meeting with Brucker but insisted that only himself be present. Again I attempted to explain how we proposed to "split" the ABMA group. He reiterated his former position that I was attempting to break up the team. I pointed out that it was necessary and proper to divide developmental activities and this had been done without injury to the organizations other than immediate inconveniences. He became irate and leaned across his desk to say, "You young whippersnapper, are you trying to tell me that I don't know anything about research and development organizations?" I kept my temper and merely smiled at him and stated I had no intention of telling him anything but that I had experience in this field and was attempting to give him the benefit of it. Needless to say, nothing productive came of that particular meeting—at least, immediately.[10]

I had many meetings with Killian, Quarles, and others at this time. The first week in November, I made a trip to the west coast to visit industrial organizations and our Ames Laboratory at Moffett Field.[11] While there, I learned that the Army had a proposition to make to us—proposing to turn over the Jet Propulsion Laboratory, provided we would make provision for that laboratory's mission on the Sergeant program. I assigned several people including [Al] Siepert and [Wesley] Hjornevik to negotiate.

Fortunately, the Army was represented by General Lyman Lemnitzer, one of the finest military men I have met. Quickly we came to terms and agreed that the transfer should be effective the first of the year. Dryden and I visited the west coast and had a talk with Lee DuBridge [president of Caltech] who seemed lukewarm about the transfer but agreed in principle.

---

10 In the unedited diary Glennan made the following comment in a footnote: "Wilbur Brucker was a man completely dedicated to the Army. In many ways, he was one of the most stupid persons I have ever met. But he fought for his people and they believed in him. In my opinion, their retention of the ABMA worked to the detriment of the modernization of other parts of the Army—those which were most particularly concerned with their principal mission."

11 The Ames Aeronautical Laboratory had been activated by the NACA near San Francisco in 1940. On its history see Elizabeth A. Muenger, *Searching the Horizon: A History of Ames Research Center, 1940-1976* (Washington, DC: National Aeronautics and Space Administration, SP-4304, 1985).

To finish up this narrative, McElroy and I visited the president to get his signature on the executive order making the transfer.[12] Ike felt we were making a mistake—that except for the fact that he respected our agreement to move as we proposed to move, he would prefer to make the ABMA shift right away. Be that as it may, we won half a loaf, and a lot of headaches. JPL was not a well-managed laboratory—I doubt that it ever has been. Possessed of a bright and aggressive staff of young men, it lacked any sense of management techniques necessary to handling programs such as ours. They had done a good job for the Army under supervision but possessed more technical competence than the Army ever thought of having and thus were able to get their way. Such was not to be the case when they ran up against the people in NASA. We continued employment of McKinsey and Company to assist in integrating the JPL organization into NASA. It was to take many months, hard feelings, and a great deal of work.

I have not mentioned any space vehicle launchings. I cannot resist, however, mentioning one incident that took place on 5 December 1958. Having lost the opening battle for ABMA but having won transfer of JPL, I decided to go to Cape Canaveral to witness the launching of an attempt on the part of JPL and ABMA to "hit the moon."[13] Subsequent to decisions relating to transfer of JPL, we determined that we must use launch vehicle systems produced by ABMA in conjunction with JPL. This meant we had to enter into an arrangement with Medaris to cover the costs. I think my trip must have been in the nature of a good will mission but it certainly had some of the elements of curiosity.

The shot was scheduled for late on the evening of 5 December or early the following morning. We were given a briefing by the Air Force relating the nature of the operations. Then we moved to the cape where we were met by General Medaris. I was taken to the blockhouse where I found a "hard hat" with my name painted on it. This was the first and only time I witnessed a launching from the blockhouse. It was a dramatic episode and I found myself much more interested in watching the faces of the men in the countdown than I was in watching the launch pad. The shot was successful but one or more of the upper stages failed, and the spacecraft traveled [about] 70,000 miles into space and then fell back to the earth. Medaris was much in command and conducted the post-launch press conference about four o'clock in the morning. Everyone was dejected.

---

[12] This was Executive Order 10793, signed by President Eisenhower on 3 December 1958. This order is attachment A to T. Keith Glennan, "Executive Order 10793, December 3, 1958," *Management Manual, General Management Instructions,* 13 October 1959, JPL Organizational Files, NASA Historical Reference Collection, NASA History Office, Washington, DC.

[13] The launch of Pioneer III, on a Juno II booster, took place on 6 December 1958 at Cape Canaveral. It was the third U.S.-IGY probe launched. Its goal was to send a payload in the proximity of the moon, but it failed in that mission. It reached an altitude of 63,580 miles, however, and found that the Van Allen Radiation Belt was comprised of at least two bands. (Eugene M. Emme, *Aeronautics and Aeronautics . . . 1915-1960* [Washington, DC: NASA, 1961], p. 104.)

Meanwhile, we determined to undertake Project Mercury on 5 October 1958—five days after we declared ourselves operational.[14] I am certain that the allocation of such a program to NASA had been agreed between Dryden, Killian, and DOD before NASA was born. The project was to be handled by a manned space flight task group resident at Langley under the direction of Robert Gilruth. The philosophy of the project was to use known technologies, extending the state of the art as little as necessary, and relying on the unproven Atlas. As one looks back, it is clear that we did not know much about what we were doing. Yet the Mercury program was one of the best organized and managed of any I have been associated with.

In the course of the first months, it became apparent that the rocket-powered launch vehicles available (Thor, Atlas, and Jupiter) had been seriously over-rated as to their payload.[15] The principal long-range problem was that of developing launch vehicles possessing adequate thrust. We decided to call a conference—the military, industry, and our own people—to settle upon a program. This was scheduled for 15 December 1958 and participation was excellent. But the military services were not as responsive as they might have been. We did establish a program to lead to a family of launch vehicles extending from the Scout (capable of launching 75 lbs.) to the as yet unspecified vehicle that would use the F-1 engine (1,500,000 lbs.) for which we were about to contract.[16] It made sense at the time, and focused attention on what was then and was to continue to be the limiting element in the space program.

---

[14] On Project Mercury see Loyd S. Swenson, Jr., James M. Grimwood, and Charles C. Alexander, *This New Ocean: A History of Project Mercury* (Washington, DC: NASA SP-4201, 1966).

[15] The Thor was a medium-sized liquid-fueled booster developed originally for the Air Force in the latter 1950s. The Atlas was the first true intercontinental ballistic missile, built by the Air Force in the late 1950s. The Jupiter had been developed by the Army Ballistic Missile Agency in the mid-1950s as an intermediate-range ballistic missile. For more on these boosters see Frank H. Winter, *Rockets into Space* (Cambridge, MA: Harvard University Press, 1990); John L. Sloop, *Liquid Hydrogen as a Propulsion Fuel, 1945-1959* (Washington, DC: NASA SP-4404, 1978); Richard E. Martin, *The Atlas and Centaur "Steel Balloon" Tanks: A Legacy of Karel Bossart* (San Diego: General Dynamics Corp., 1989); Edmund Beard, *Developing the ICBM: A Study in Bureaucratic Politics* (New York: Columbia University Press, 1976); Jacob Neufeld, *Ballistic Missiles in the United States Air Force, 1945-1960* (Washington, DC: Office of Air Force History, 1990).

[16] The Scout booster began in 1957 as an attempt by the NACA to build a solid-fuel rocket that could launch a small scientific payload into orbit. To achieve this end, researchers investigated various solid-rocket configurations and finally decided to combine a Jupiter Senior (100,000 pounds of thrust), built by the Aerojet Corporation, with a second stage composed of a Sergeant missile base and two new upper stages descended from the research effort that produced the Vanguard. The Scout's four-stage booster could eventually place a 330-pound satellite into orbit, and it quickly became a workhorse in orbiting scientific payloads during the 1960s and beyond. It was first launched on 1 July 1960, and despite some early deficiencies, by the end of 1968 had achieved an 85 percent launch success rate (Richard P. Hallion, "The Development of American Launch Vehicles Since 1945," in Paul A. Hanle and Von Del Chamberlain, eds., *Space Science Comes of Age: Perspectives in the History of the Space Sciences* [Washington, DC: Smithsonian Institution Press, 1981], pp. 115-34). The F-1 engine—first conceived for use in the Nova launch vehicle that was cancelled in 1962—was developed for the Saturn V booster that launched the Apollo spacecraft to the moon. Five of these engines in the first stage generated 7.5 million pounds of thrust. This engine represented one of the most significant engineering accomplishments of the program, requiring the development of new alloys and different construction techniques to withstand the extreme heat and shock of firing. On this subject see, Roger E. Bilstein, *Stages to Saturn: A Technological History of the Apollo/Saturn Launch Vehicles* (Washington, DC: NASA SP-4206, 1980).

I will relate only one other incident that had its beginnings shortly after I reached Washington and which, for the sake of completeness, will be related somewhat out of its proper place in the approximate chronology that I am following. I discuss it here because it provides a good example of the sort of problem that faces an agency executive in the Washington scene. The villain in the piece is my good friend, Congressman Albert Thomas, chairman of the subcommittee of the appropriations committee that handles the independent offices' appropriations bills. I had known him somewhat casually when I was in Washington on the AEC.[17] Shortly after my return to Washington in 1958, I had a call from him expressing great pleasure at my return to the Washington scene. It soon became apparent that he had an ulterior motive in the call—he was anxious that his district in Texas centering around Houston should benefit from the space program. He suggested that we were going to need additional laboratory facilities. He told me that Rice Institute—now called Rice University—would be quite willing to give NASA 1,000 acres of land about 30 miles from Houston. He stated that there were several good institutions of higher education in that general area that could support such an installation. I responded that, so far as I was concerned, we were not about to build any new laboratory facilities beyond the one already authorized and on which construction had begun in Beltsville. He said, somewhat peevishly, that authorization had gone through without his sanction since he had been absent at the time. I pointed out that this was no fault of ours and that construction had begun and we expected that the laboratory would have to be completed. (At the time, I had expected we might spend as much as $10 million at Beltsville—I suspect expenditures exceeded $75 million by June 1963). He indicated that he would be talking with me about this matter as time progressed.

Throughout the fall, I had several calls from Thomas—each having to do with this subject. I maintained my position and even called on him two or three times to discuss the matter more fully. Our fiscal 1960 budget was sent to the Appropriations Committees about two weeks before the president's message to Congress. A few days after he had received it, Thomas called and opened the discussion in a pleasant way. Soon, he got around to the subject of the laboratory at Houston. I repeated my former comments and thought I was getting away with it when he broke in: "Now look here, Dr., let's cut out the bull. Your budget calls for $14 million for Beltsville and I am telling you that you won't get a God-damned cent of it unless that laboratory is moved to Houston." Fortunately, I retained my sense of humor and simply responded, "I think it's about time I bought you a drink, Albert."

We did not build additional large laboratories while I was in Washington, but during the first year of President Kennedy's term, the decision was made to "race

---

[17] On the history of the Atomic Energy Commission see, Richard G. Hewlett and Francis Duncan, *Atomic Shield, 1947-1952:* Volume II, *A History of the United States Atomic Energy Commission* (University Park: Pennsylvania State University Press, 1969); Richard H. Hewlett and Jack M. Holl, *Atoms for Peace and War, 1953-1961: Eisenhower and the Atomic Energy Commission* (Berkeley: University of California Press, 1989).

the Russians to the moon." Obviously, additional facilities would be needed. The first to be proposed was a "Manned Space Flight Center." I think every state in the union must have made a strong pitch for this "plum." But I was convinced, and so advised the representatives of the governor of Ohio, that Texas, and specifically Houston would be the site chosen. After all, Lyndon Johnson was, by then, the Chairman of the Space Council and Thomas continued to be Chairman of the House subcommittee on appropriations concerned with NASA's budgetary requirements.[18]

But to return to our activities during the year 1958, we had processed our budget request through the Bureau of the Budget and the White House and come up with an approved asking figure in excess of $550 million for FY 1960. We had initiated Project Mercury within days after we became operational. We had initiated the 1.5-million-pound thrust rocket engine project to be known as the F-1 and had called for bids on its development. We had taken the first steps toward the establishment of a national launch vehicle program. We had attempted several shots with indifferent success—most of them previously planned by the military for which we had taken over responsibility. It was in the course of these activities that we became acutely aware of the limitations in thrust that faced us. We had attempted to secure the transfer of ABMA and JPL from the Army and had come a cropper on ABMA! As a consolation prize, the Army did transfer JPL. We had acquired a number of very fine people for top posts and had solidified our thinking about organization. We had begun a national space program, although not in a formal manner, using the advice of the Space Science Board of the National Academy of Sciences, people from industry and the universities and members of our own staff. I suspect we had made no lasting enemies (other than Brucker) and probably no steadfast friends. All in all, a great deal seemed to have been accomplished in those first four months.

1959

As I prepared to dictate these notes on the important or interesting aspects of the year 1959, I found myself up against a formidable problem. In the first place, I did not want to set down a play-by-play account of each day's activities, nor could I have done so had I wanted to. Consequently, I am attempting merely to give a summary of some of the important and, to me, interesting events of the year. Accordingly, to an even greater extent than was the case in the earlier portions of this summary, I will deal with incidents, impressions and problems as I can recall them without much concern for chronology.

---

[18] Lyndon B. Johnson (1908-1973), had been interested in the U.S. efforts in space since Sputnik I and as Vice President for John F. Kennedy between 1961 and 1963, he chaired the National Aeronautics and Space Council. See, Robert A. Divine, "Lyndon B. Johnson and the Politics of Space," in Robert A. Divine, ed., *The Johnson Years: Vietnam, the Environment, and Science* (Lawrence: University Press of Kansas, 1987), pp. 217-53.

One of the problems that faced us early in the game was the extent to which NASA should develop a program in the life sciences. This problem arose from Project Mercury, a perfect example of the man-machine system in which we knew a great deal more about the machine than the man. The Air Force and the Navy had developed strong and, at times, competing programs in bio-astronautics in connection with their need to understand man's ability to withstand the stresses in high-performance airplanes flying faster than the speed of sound and at exceedingly high altitudes. But there was little of fundamental science involved in their activities. We undertook to develop such a program and were fortunate to acquire the services of Dr. Clark Randt of the Western Reserve Medical School, an experimental psychologist interested in the field although having no real experience in it.

This was never a happy situation and I give great credit to Randt for his dogged pursuit of a sensible answer to the problems that faced us. All of our people were physical scientists or engineers and thus he had little in the way of understanding supervision. Furthermore, Project Mercury had been staffed with doctors from the Navy and the Air Force, and there was resentment against having Randt in that program, on the part of both the service people and the NASA staff. We called together a committee of eminent life scientists under the chairmanship of Dr. Seymour Kety of NIH and agreed upon a program of development. The cognizant committees in Congress viewed all of this with a certain amount of mistrust and I am sure they were aided in their suspicions by people in the military departments. Randt stuck it out for more than two years and accomplished a great deal, although I am sure he would be the first to say it was an uphill battle and productive of much less than satisfactory results. To have plunged ahead on a program of our own without concern for those already existing in the military services would have been possible. But to have done so would certainly have adversely affected our relationships with the military services. Thus I would have to admit to a reasonable sense of failure in attempting to deal with this most important problem.[19]

I spoke earlier of processing our budget request through the BOB and the Executive Office of the President. Maurice Stans was the Director of the Bureau of the Budget throughout my term in Washington and I came to know and admire him. NACA had established a reputation for forthrightness in dealing on budgetary matters. We continued this throughout my term and yet fought with the personnel of the BOB assigned to our agency. Usually, I took the initiative in stating our case to Stans before he could clue me on what he wanted on behalf of the president and indicated the figure I proposed to shoot for. Naturally, this was always cut back somewhat, but in no instance did I accept as final my discussions with Stans. I fought for the figure we wanted, always involving the president with Stans in a final determination of the figure to be included in the budget. On two or three occasions, others such as Bob Anderson, secretary of the treasury, were involved.

---

[19] On the life sciences program during the early years of NASA see, Mae Mills Link, *Space Medicine in Project Mercury* (Washington, DC: NASA SP-4003, 1965); John A. Pitts, *The Human Factor: Biomedicine in the Manned Space Program to 1980* (Washington, DC: NASA SP-4213, 1985), pp. 13-90.

Once the figure had been settled upon, I made it a rule never to undercut the president's budget in any discussions before congressional committees. This seemed to earn for me and for NASA the respect of General Persons and others on the White House staff who then carried my colors into battle with the president. I tried to operate as a member of the president's team and I think we were successful. Hugh Dryden's unquestioned sincerity, integrity and scientific competence provided the greatest support in these matters.

As the year began, I had my first introduction to the preparation necessary for defending our budget in Congress. It was quite apparent that my top staff was concerned about my ability to handle this matter and we set up, late in January, several sessions—each three or four hours—of what we termed the "murder court." Members of the staff questioned me about the budget as though they were members of the congressional committees. In most instances, they were asking questions about matters where their knowledge was significantly more complete than my own. This was a humbling and, at times, frustrating experience. I would attempt answers and could see by their faces that they were less than satisfied. In exasperation, I would ask, "Well, what in hell would you say in answer to that question?" This exercise proved excellent training for it was important that the administrator take the lead in answering questions posed by members of the congressional committees. Usually, I had Hugh Dryden with me and could turn to him for advice or even ask him to supplement my answer. This practice seemed to be accepted, and we managed to get through the hearings, which extended over many weeks, without loss of money. It should be remembered, however, that the climate in Washington and particularly on the Hill was such that only a blundering fool could have seriously harmed the program. We continued this practice of budgetary rehearsal throughout my time in Washington.

My initial appearance before a congressional committee—other than the brief hearing held at the time of our confirmation—was before the so-called Senate preparedness subcommittee. Lyndon Johnson was chairman but appeared seldom. [The subcommittee included] Senator [Stuart] Symington who was determined to prove that none of us knew much about what we were doing. In company with our assistant administrator for congressional liaison, Jim Gleason, I had paid a call on Symington on 16 January 1959. Gleason had formerly been the head of the office staff of Senator William Knowland of California and thus was known to Symington. The Senator lectured us on organizational practices and left no stone unturned in his determination to paint himself as the expert on industrial organization. I developed a dislike for him at this meeting that has continued to this day, and I must say that many incidents have occurred in the interim to support this feeling.

In this instance, Symington drove ahead in an attempt to extract from us the organizational relationship between NASA, the Department of Defense and the individual military services. He wanted a long-range plan. We had been in business for little more than four months and had been heavily occupied in doing a variety of things that taxed the capabilities of the people on our staff. Our dealings with the military services, though at arm's-length, were reasonably under control. But we had not begun to develop a long-range—say, a ten year—program. I tried to point out that time had not permitted such a program—that we needed to know more about our capabilities in rocket thrust, payload weights, etc. Symington did not shake our adherence to this stand but I must confess that these hearings had a great effect on our determination to develop such a program. It took almost a year before the program was available for discussion with the Congress.[20]

Symington also wanted to probe into the activities of the Space Council, by law, an advisory body to the president, who was its chairman. Under these circumstances, he could invoke the doctrine of "executive privilege" and refuse to reveal to Congress anything about the advice he was getting from the council. For several hours, I attempted courteously to point this out to Symington. We gave him dates of meetings, but stuck to our guns with the support of President Eisenhower. It was a most unpleasant experience, but I think it did me good in the sense that I learned there is no other approach to be taken in dealing with congressional committees than to be patient, courteous and forthright. In any event, we lived through the experience and gained by it. A year later, I was to have another brush on this matter with the House Committee on Science and Astronautics.[21]

As a result of the studies conducted by John Corson, we decided to employ an associate administrator (general manager). Dryden and I had been searching for such a man. We turned to industry and to some of the top people in government agencies, particularly the Department of Defense. I tried to interest Jim Dempsey of Convair and had a talk with Frank Pace of General Dynamics. Nothing came of this approach, although Frank was helpful. The job, as I recall, paid $21,000 and it was difficult to find a man in industry at that salary who would be capable of handling the job and willing to take the beating that was sure to be entailed.

I had developed a friendship with Dick Horner, then the assistant secretary of the Air Force for research and development. I tried to interest him in the job but he pointed out that he had been in government for almost twenty years and his children were getting to a point in life where he must think of college education and the attendant expenses. He planned to retire as of June 1959. My relations with

---

[20] This plan was completed late in 1959. See, Office of Program Planning and Evaluation, "The Long Range Plan of the National Aeronautics and Space Administration," 16 December 1959, NASA Historical Reference Collection. A less-detailed version called the 10-year plan was not completed until early in 1960.

[21] The National Aeronautics and Space Council had been created by law in 1958, at the same time that NASA was mandated in Public Law 85-568. The council was specified as advisory to the president in fulfilling his duty to survey all federal aerospace activities, "develop a comprehensive program" of federal aerospace activities, designate responsibility for direction of those activities, provide for cooperation between NASA and DOD in the conduct of aerospace activities, and resolve inter-governmental disputes regarding aerospace activities. The council, initially called a board, was the inspiration of Senator Lyndon B. Johnson, who sought a central focus for space planning so that "nothing vital to the national interest would be lost or overlooked by assigning responsibilities solely to NASA and DOD." Eisenhower had little evident enthusiasm for the Council, as did Glennan. The president did not appoint an executive secretary, and that function was filled by staff borrowed from NASA or the Office of the President's Science Advisor. The council also met infrequently. For more on the council see Edward C. Welsh, "Highlights of National Aeronautics and Space Council History," 1968, NASA Historical Reference Collection.

General Thomas D. White, chief of staff of the Air Force, had been good. I appealed to both of them to nominate one or more general officers who might be encouraged to retire and take on this post. I did the same with Admiral Roscoe Wilson of the Navy.[22] Several persons were proposed and some little time spent in interviewing. Finally, it appeared that General Monty Canterbury of the Air Force possessed the qualifications. There ensued a number of sessions with him spread over several weeks. Each time he came in, he had another set of conditions. At first, we acceded to his wishes but it became apparent that he was really a victim of a plot on the part of his superiors. I decided that we should break off these negotiations and discuss the matter with Dick Horner who agreed that Canterbury was becoming impossible.

T. Keith Glennan flanked (on his left) by Richard E. Horner and (on his right) by Hugh L. Dryden.

Much to my surprise and gratification, Dick Horner then said to me that he had led me down the primrose path for a long time and he felt honor bound to take the job for a period of one year if we still wanted him. The deal was made without delay, although the Air Force had not yet fully accepted the fact that Dick had made up his mind to retire. Horner turned out to be a good choice and, because of his past associations with the military, we were able before the end of 1959 to come into reasonable agreement with the Air Force on the division of effort between us. I shall always be grateful for Dick's willingness to put aside his interests and take on that difficult task. I treasured his friendship as much as I appreciated his effective efforts.

Another person with whom I developed a reasonable rapport was General Bernard Schriever, now commander of the Air Force Systems Command. Bennie had been in charge of the missile program for the Air Force and had carried that program forward with energy and effectiveness. He was determined to secure the largest possible role for the Air Force in the space program, but I found it possible to deal with him on a reasonable basis for the most part.

---

[22] There appears to have been no Admiral Roscoe Wilson of the Navy at that time. Glennan may have been thinking of Vice Adm. R.E. Wilson, deputy chief of naval operations (logistics), in Washington during this period.

President Eisenhower had the habit of asking the heads of independent agencies to sit in on cabinet meetings or meetings of the National Security Council from time to time. Usually, these occasions involved a discussion of overall budget policy or some element of the nation's involvement in science and technology. On several occasions, Dryden and I made presentations.

In a quite different aspect of my job, I must admit to an aversion, strongly held, to arguing with anyone in the press and, although I recognize the right of the public to full disclosure of the uses of its money, I find it difficult to think continuously in terms of the impact of the actions of an agency such as NASA on the public. I think I must be one of the few persons who has spent the better part of 30 months in an important job in Washington without indulging in a press conference called for the purpose of revealing my thoughts.

I did participate in several press conferences when it was necessary that we make some public pronouncement. One occurred on 9 April 1959 when I had the privilege of introducing our seven astronauts to a waiting and breathless world. Our press room was a little-used, unventilated, unattractive assembly room. On this day, of course, the room was packed with photographers and reporters. I did little more than open the meeting and introduce the men. I then turned the meeting over to one of our press officers.[23] One other press conference remains fixed in my mind. This was the one where I again had the privilege of introducing two astronauts—in this case, the monkeys, Able and Baker. Again, the room was crowded with photographers and newsmen. I suppose it was only my own obtuseness that prevented me from understanding the great interest the public would have in these animals.[24]

I developed a special antipathy toward the trade press during my stay in Washington. Two publications seemed always to get in our hair—*Aviation Daily* and *Missiles and Rockets*. These publications existed primarily on advertising, and their mission was advocacy rather than objective reporting. Their attacks on the

---

[23] The NASA astronaut selection process began early in January 1959 under the leadership of Robert Gilruth and the Space Task Group at Langley Research Center. Contrary to a NASA priority that these six astronauts be civilians, President Eisenhower directed that they come from the armed services' test pilot force. A grueling selection process had narrowed the candidates to seven by early April 1959. Unable to cut the last candidate, NASA decided to appoint seven rather than six men as astronauts. NASA publicly unveiled the astronauts in a circus-like press conference on 9 April 1958. The seven men—from the Marine Corps, Lt. Col. John H. Glenn, Jr. (1921-   ); from the Navy, Lt. Cdr. Walter M. Schirra, Jr. (1923-   ), Lt. Cdr. Alan B. Shepard, Jr. (1923-   ), and Lt. M. Scott Carpenter (1925-   ); and from the Air Force, Capt. L. Gordon Cooper, Jr. (1927-   ), Capt. Virgil I. "Gus" Grissom (1926-1967), and Capt. Donald K. Slayton (1924-1993)—became heroes in the eyes of the American public almost immediately, a development that surprised many people at NASA. The astronauts essentially became the personification of NASA to most Americans during the Mercury project. On this subject see Tom Wolfe, *The Right Stuff* (New York: Farrar, Strauss, and Giroux, 1979), pp. 109-20; Swenson, Grimwood, and Alexander, *This New Ocean*, pp. 159-65, among numerous other sources.

[24] This meeting took place on 30 May 1959 in Washington, DC. There were about 50 media people at the press conference, and Glennan told them only partly jokingly, "I'm not sure that the monkeys' ordeal in flight is comparable to the one that they have just been through [sic] here." Able, an American-born rhesus monkey, and Baker, a South American squirrel monkey, had flown in a life science experiment aboard a non-orbital mission on 28 May 1959 in a capsule atop a Jupiter booster. On this event see, Jack Raymond, "2 Monkeys' Pulses Steady in Space," *New York Times*, 31 May 1959; Emme, *Aeronautics and Astronautics...1915-1960*, pp. 109-10.

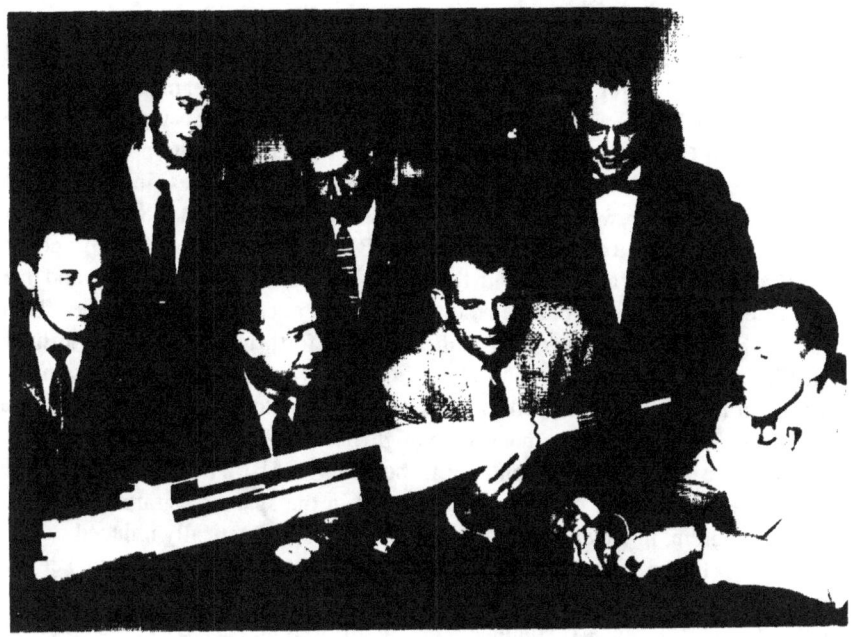

Mercury Astronauts examining a model of the Atlas booster and Mercury capsule. Seated, left to right: Gus Grissom; Scott Carpenter; Deke Slayton; Gordon Cooper. Standing, left to right, Alan Shepard, Walter Schirra, and John Glenn.

agency were hard to stomach. Similarly, the absolute authority with which people at *Time* magazine castigated our operations was also hard to live with.

However, I did find a few members of the press who seemed more interested in responsible reporting than in headlines. Among them were Phil Goulding of the Cleveland *Plain Dealer*, Courtney Sheldon of the *Christian Science Monitor*, and Richard Harkness of NBC News. Through the latter, I had two or three opportunities to speak to top columnists and news bureau chiefs in off-the-record sessions. Correspondents took the vow that they would not violate my confidence. In my experience, they always respected this vow. And their columns seemed more responsible than others. Roscoe Drummond was another columnist—at the New York *Herald Tribune*—for whom I developed considerable respect.

We always had a press conference after a launching and usually held another in Washington to fill in the information not readily available at the post-launch conference held at Canaveral. These sessions were usually chaired by a member of the public information staff with questions handled by our scientific and technical staff. We adopted the practice of holding an explanatory press conference at the time the budget went to Congress. This was usually handled by Dryden and by Associate Administrator Horner or, subsequently, by Robert Seamans.

Let me now talk a bit about the events that led up to the transfer, finally, of ABMA to NASA. Throughout the first seven or eight months of 1959, ABMA was under contract to NASA for launch services and for the provision of certain launch vehicle systems. We came to have considerable respect for Wernher von Braun and his "team." I had never lost interest in bringing them into the NASA fold and, with the help of Hugh Dryden, laid plans to do this in the latter months of 1959.[25] In the interval, Don Quarles had died suddenly—his passing was a real loss to the nation. Tom Gates, Secretary of the Navy, had moved up to the post of deputy secretary of defense. In conversations with McElroy and Gates, initiated about August or September, we reviewed the problems facing us in late 1958 when Wilbur Brucker had frustrated our efforts to acquire ABMA. I made it clear that I proposed to make a new deal, if any, only with the Office of the Secretary of Defense and that I expected Brucker to be told the results of the deal once it had been made. Tom Gates, who has remained a good friend, was a very forthright person and McElroy left the bulk of the discussion to him. For once, we were able to contain the information until our negotiations were essentially complete. By this time, I am sure Brucker had recognized that, much as he might desire it, he was not to have a major role in space and could not continue to hold on to ABMA much longer. The Huntsville people were much more interested in the space program than in the missile program, in which their participation had been drastically reduced.

As I recall it, the deal had been pretty well set and I think that Brucker had been told of the impending change when I made a visit to Huntsville to look at Saturn.[26] General Medaris conducted me on this tour and acted as if he had little knowledge of what was going on. During the course of our visit to several shops, I noted that he was called away to the telephone on several occasions. Finally, he called me to one side and asked me to come to his office. As I recall, there had been considerable discussion in Washington involving Herb York and ARPA with the Air Force and, to a limited extent, NASA people. The Defense Department had been able to find no real reason for developing a super-booster such as Saturn but wanted to upgrade Titan I to a more powerful configuration that would be able to launch their DynaSoar. Word had leaked that Saturn might be abandoned. Medaris revealed that he had some knowledge of the impending transfer and stated that if I would agree for Saturn to be retained, he would personally go to Washington and urge the transfer. As I recall, I was noncommittal—taking the position that all of the evidence was not in. However, I am sure I did indicate that we had no immediate reason for canceling Saturn and that we would deal with the problem on the basis of our requirement for a powerful first stage booster that could carry into orbit a spacecraft weighing many thousands of pounds.

I returned to Washington and shortly thereafter, the president signed an executive order transferring ABMA to NASA and instructed the Department of

---

[25] The Army Ballistic Missile Agency was transferred from the Department of the Army to NASA effective 1 July 1960. The agreement for this change, however, was executed on 16 November 1959. See, "Agreement Between the Department of the Army and NASA on the Objectives and Guidelines for the Implementation of the Presidential Decision to Transfer a Portion of ABMA to NASA," 16 November 1959, Glennan subsection, NASA Historical Reference Collection.

[26] On the development of the Saturn launch vehicle, which had first started as an Army project but transferred with the ABMA to NASA in 1960, see Bilstein, *Stages to Saturn*.

Defense, specifically Army, to negotiate the details. It was necessary that such an executive order be reported to the Congress for its consideration and debate, if any. Sixty days must elapse, by law, before the transfer would become final unless the Congress interposed objections. The executive order became public, as I recall, shortly before or after the end of 1959. We had finally achieved our end, although the transfer did not occur until 1 July 1960 so the problems of transferring such a large group of people—5,500—and a considerable amount of property could be worked out.

Toward the end of 1959, Wernher von Braun asked to see me and I asked him to come to my apartment one evening. He tried to secure something of the plans we had in store for ABMA and to negotiate an operating position for himself and his team. I refused to be drawn into this discussion and simply said we had great need of their talents and he could rest assured that they would have every freedom possible to pursue the programs we proposed to assign to them, involving the development and operation of launch vehicle systems. Wernher finally ended the conversation by saying, "Look, all we want is a very rich and very benevolent uncle." What a personality!

Throughout my years in Washington I developed an affection and a great deal of respect for President Eisenhower. I suppose I saw him in his office on the average of once every four weeks and at intervals at meetings of the cabinet or the NSC. Prior to Sherman Adams' departure from the government, I had often thought that Ike was being shielded from some of the nasty problems of organizational relationships and that he exhibited less than the force one might expect from a man in his position. All of this changed drastically when Adams left. Ike seemed much more on top of his job, more decisive when matters were brought to his attention. I tended to work closely with his staff and particularly with Stans, Persons, Killian, and later George Kistiakowsky, his general counsel Gerald Morgan, and certain members of the cabinet. Parenthetically, I may say that it seems my successor, [James] Webb, places more reliance on his relations with Congress and less on his relations within the executive branch than I ever did.

In late November 1959, we were preparing our budget for the 1961 fiscal year, which would be proposed to Congress in January 1960. We had started out with a request in excess of $1.250 billion and had been whittled back to a little more than $900 million. This was later increased by a supplemental to about $965 million.[27] The trade papers had been taking NASA and the administration to task for their apparent unwillingness to throw caution to the winds and place really large sums on programs that I was certain could not move much faster than plans called for.[28] In truth, we lacked a rocket-powered launch vehicle that could come anywhere near the one possessed by the Soviets. And it would take years to achieve such a system, no matter how much money we spent.

---

[27] The NASA budget for fiscal year 1961 turned out to be $964 million. (*Aeronautics and Space Report of the President, Fiscal Year 1991 Activities* [Washington, DC: Government Printing Office, 1991], p. 180.)

The Air Force had finally come to the point of firing several Atlases downrange from Cape Canaveral with amazingly good results.[29] Nevertheless, the trade press seemed to be saying that the ability of the Soviets to orbit very heavy payloads was to be equated with their ability to place nuclear warheads on any spot in the United States. This was a new twist—one not understood easily by the man in the street.

It seemed to me that it was time for the president to state publicly the policy of his administration—that we were pursuing an orderly but aggressive policy in space and were proceeding as rapidly as our technologies would permit. The recent successes of the Atlas launches would permit him to portray dramatically— possibly by television or a demonstration for the press—the effectiveness of the Atlas as a nuclear warhead carrier. I decided I would broach this matter to the president and, accordingly, set up a date to talk with him at Augusta where he was vacationing in November 1959. I wrote a lengthy memorandum stating my position, which wound up with a request that he make a public statement on these problems. I reproduce here, in full, this letter that I read to him at Augusta. I stated that I understood he did not like to have lengthy memoranda presented to him; therefore, I proposed to read it if he would listen. He smiled and agreed to hold his peace until I had finished. The letter read as follows:

16 November 1959

The President
The White House

Dear Mr. President:

On many occasions over the past several months we have talked about management and the objectives of the nation's space program. There is no question that there exists in the mind of the public today substantial and disturbing confusion on these points. A difficult condition has been brought about in the main by our basic space law being hastily and, in some of its provisions, wrongly enacted with the result that it became a creature of compromise and a ready instrument for political

---

[28] On this controversy see, "Glennan Looks to Moon, But With Purpose in Mind," *Times Herald,* 4 February 1960; "Space Death Wouldn't Halt U.S. Effort, Glennan Says," *Baltimore Sun,* 11 April 1960; "Glennan Has Goal in Space," *New York World Telegram,* 5 February 1960; "Capital Circus," *New York Times,* 30 December 1959; "An Exclusive Interview, T. Keith Glennan," *Boston Sunday Globe,* 23 October 1960.

[29] Of three launches of the Atlas in 1957, only the last, taking place on 17 December was successful. In 1958, 8 of 14 launches were successful to some degree, and in 1959, 14 of 23 launches were successful. See, General Dynamics, "Atlas Flight Program Summary," 27 March 1969, Atlas Launch Vehicle file, NASA Historical Reference Collection.

exploitation. Extensive attention has been given to this subject by the editorial writers and the reporters of the major newspapers and trade magazines of the country, and an increasingly alarming proportion of political speeches are being devoted to this subject. Extensive hearings by Congressional Committees have tended further to add to the confusion, and these hearings doubtless will be resumed in January.

It has not been possible wholly to reassure the public or others regarding the objectives of our national program in space because of the existence of conditions which have tended to support, to a degree, the statements made by critics of the present situation. My own observations and discussions with others lead me to the conclusion that a program of corrective action is both possible and desirable. If you agree with my conclusions after you have heard me out, I can assure you that your leadership in this situation will make more certain a belief on the part of all concerned—the public, the Congress and those engaged in the prosecution of the program—that we have well defined aims and are sensibly organized to move firmly and at reasonable speed toward those goals.

There are several elements in this picture, each of which requires attention. The principal ones may be described in the following fashion:

a.   The National Aeronautics and Space Act of 1958, after declaring the policy of the United States to be that of devoting its activities in space to peaceful purposes for the benefit of all mankind, places upon the President the responsibility for managing the nation's space program [Sec. 201(e)]. The Act is somewhat unusual in making it incumbent on the President to become intimately involved in the management of an activity of this kind. This requirement arises out of two facts. First, that the Congress desired the space program to be at a level in Government as close to the President as it was possible to provide; and second, that both the DOD and NASA are enjoined in the Act to be active in the field of space research and development with a consequent need for coordination of their individual efforts.

There is provided to "advise the President with respect to the performance of his duties" The National Aeronautics and Space Council. Further, to assist in coordination of the work of the DOD and NASA in the space field, there is established a Civilian Military Liaison Committee. Neither of these activities has been particularly useful or effective. Under present circumstances, and in the light of past experience, it is doubtful that either of these agencies can usefully be employed in the management of the nation's space program.

b.   In the hearings conducted in the Congress last spring, the Committee members insisted upon establishing, for the record and in the minds of the public, the concept that our national program is the sum of the military and the civilian programs in space research. To an extent, our testimony aided in establishing this concept. This situation, together with other testimony taken at that time, has given rise to the "multiple and confused" management argument. Certain administration officials testified that an insufficient amount of money was being requested for the nation's space effort and that there was inadequate coordination of programs. The doctrine of executive privilege was attacked strongly as a device to hide the lack of action on the part of the President and the Space Council in cleaning up the management picture.

It is important to note, in this connection, that the DOD has adopted and supported the position that "space is a place—not a program." This philosophy supports the concept that space projects in the DOD are undertaken only to meet military requirements. The reasoning goes that these projects must compete, therefore, with more conventional projects intended to accomplish the same or similar military objectives. Carried to its logical conclusion and coupled with the stated policy calling for the exploration of space for peaceful purposes, one would be led to the conclusion that there is no need to provide specifically for military space activities in the Space Act of 1958! What the military needs to do in whatever medium—on land, on sea, in the air or in space—they can and should do under their statutory responsibilities for defending the nation.

Recent actions taken within the DOD would tend to support this viewpoint. Secretary McElroy's recent directive allocated to the individual military services those satellite programs that are consistent, in each case, with their assigned roles and missions in the more conventional fields of weapons development.

It might now be held that the "nation's space program" is, indeed, a non-military program—the responsibility of NASA. If the law were modified to reflect this changed concept, a much more believable organization would be possible with management responsibilities placed directly on the agency concerned. Provision should be made, of course, to ensure the close cooperation which has always existed between the military services and the former NACA.

c.   It is probable that a substantial proportion of the nation's citizens equate Soviet space accomplishments—especially those indicating the effective use of guidance systems in their shots at the moon—with the accuracy and effectiveness of our IRBMs and ICBMs. Clearly, it is necessary that the U.S. nuclear warhead carrying missile capability—both as to thrust and guidance—be separated in people's minds from the present shortcomings of our space vehicles which now use those same missiles as base boosters for the space vehicles. The President's word on this has the best chance of being believed. It may be desirable to accompany such a Presidential statement by a dramatic demonstration of the accuracy of our missiles when used as missiles. More of this later!

d.   There continues to be, both in the public expressions of certain members of the press and of members of the Congress, a question regarding the objectives of the U.S. in space. The phrase "competition with the Soviets" recurs constantly and I am not aware of any authoritative public pronouncement that has dealt effectively with this question.

Personally, I do not believe we can avoid competition with the Soviets in this field. I do not believe we should want to avoid that competition. But I do believe that we can and should establish the terms on which we are competing. We could thus place the "Space race" in proper perspective with all the other activities in the competition between the US and USSR. In doing this, thoughtful consideration should be given to the particular and dramatic role occupied today by space activities in the whole gamut of international competition. Here again, the word of the President will gain the most credence in any such statement of national policy.

e.   Accompanying the actions that might be taken under the immediately preceding paragraph, it would be desirable to delineate, publicly, a program of non-military space research, development and exploration. Such a program should set forth both short term and long term objectives and the planning that has been done to move toward those objectives.

f.   Again, in the context of paragraph (d) above, firm indications of intent to propose budgetary support consistent with the statement of objectives and policy should be possible. Actions to be taken with respect to a supplemental request for FY 1960 as well as FY 1961 regular budget would be involved, of course.

Mr. President, if my views as given in this memorandum coincide in any significant way with your own thinking, I would suggest that legislative proposals be introduced to correct the present difficulties arising out of the present Act. Further, I suggest that you may wish to seize the initiative in announcing a single national space program, soundly organized with a single management responsibility clearly designated—one which could be supported before the Congress and the public.

Such a program of action might include the following:

1.   A proposal to amend the Space Act of 1958 so as to (a) remove Department of Defense activities in space from the context of the Act; (b) make NASA responsible for the nation's space program including the development of all *new space vehicles systems* whether for use by NASA or the military services; (c) abolish the Space Council—replacing it with a committee advisory to the Administrator; and (d) reconstitute the Civilian Military Liaison Committee as a Military Applications Committee with responsibility for keeping NASA informed of military interests and requirements. These changes, I believe, are generally consistent with your own thinking on this matter.

2.   A request for restoration of the $30,000,000 cut in the NASA FY 60 and supplemental FY 59 appropriation requests. Authorization legislation already exists for this approximate amount.

3.   Institute a demonstration of the effectiveness of the Jupiter, Thor and Atlas missiles—all on one day—with competent and reliable observers at the "splash net" to witness the successful landings of simulated warheads. (It is recognized that this suggestion would involve the DOD in an exercise which might not be to their liking. It may be that other methods of highlighting the point you would be making can be suggested. The important element in this situation is the dramatic impact that should be involved.)

4.   On the evening following the demonstration proposed in (3) above—on the same day, if possible—a major address by the President would effectively deal with two troublesome aspects of the nation's missile and space business. First, the adequacy and effectiveness of the missiles we now have in operational status would be discussed. Second, the President could make abundantly clear the policy of his Administration with respect to the competitive aspects of space research and development. He could briefly describe the program we propose to follow for the next several years and indicate proposed administrative and budgetary actions.

5.   Finally, the President would submit to the Congress a budget believed by him and his advisors to be necessary to implement fully the program described above.

It is my opinion that each of these elements is an essential part of a program designed to clarify the present somewhat confused and controversial situation. It should reassure the nation's citizens as to our objectives in space. It should provide an opportunity for the President to take the initiative in setting forth a philosophy of urgent concern with the development of strength in the science and technology that must undergird the future well being and leadership capability of this nation. It should be done in the space field because activities in that field represent an attack on the real frontiers of science today. It can be done with complete and honest reassurance to the nation that our defense needs are being cared for effectively. It is probable that the most serious obstacle to the successful carrying out of the suggested plan of action will be that of securing DOD agreement on certain of the proposed changes in the law—particularly that change which would place sole responsibility for the development of new space vehicles on NASA.

I am not sufficiently aware of the procedures which would assure the best possible reception of a legislative proposal such as the one outlined. Perhaps consideration should be given to the development of the suggested legislative changes with the leadership of the cognizant committees of the Congress. In any event, a first draft of the language believed by my staff to be necessary to accomplish the corrective actions desired has been prepared. It is my belief that the language of this draft is consistent with your own views on this subject. Indeed, it approximates in many respects the original proposal made by you to the Congress eighteen months ago.

It has been discussed in a preliminary way with the staff of the Bureau of the Budget and with members of the White House Staff—but not with representatives of the DOD.

I strongly urge thoughtful and favorable consideration of the proposals made in this paper. Naturally, I will be happy to discuss them with you at any time.

Sincerely,
T. Keith Glennan
Administrator[30]

Ike seemed to accept my proposal and I did attempt, with the help of his staff, to prepare a speech for him. Events overtook us, and it seemed that transfer of ABMA and an increase in the budget for Saturn provided a full answer to critics of the U.S. program. At least, the speech never was given.

Throughout my term in Washington, there was clamor on the part of the so-called scientific community and to a lesser extent, on the part of some industrialists expressing their strong convictions, that the space program was being developed without sufficient input from persons outside NASA. As early as 1959, there was a good bit of debate about continuing with Project Mercury—the nation's first manned space flight project—when so much needed to be known about the space environment before any really deep penetration of space could be undertaken by a manned spacecraft. Partly to answer these charges and partly because it is a good thing to seek outside advice, we did call into being, from time to time, advisory groups. One such was the Greenewalt Committee since Crawford Greenewalt of DuPont was its chairman. I recall that Paul Nitze, Frank Stanton, Jim Perkins, Mervin Kelly, Ed Purcell, Walt Rostow, and several others were among those present. Greenewalt turned out to be an excellent chairman although he was rather strongly biased against any attempts to put man into space. Mervin Kelley thought the entire space program much overrated. But out of those sessions there came a reasonable endorsement of our activities, with the usual caveats about overdoing any part of the program.

I never failed to find interesting and cooperative people to serve on committees of this sort. The Kimpton Committee on organization, was another good example.[31] In all these operations, we tried to be as objective as possible in presenting the entire picture to the committee, although I am certain there was little

---

[30] This letter is also available in the Glennan subseries, NASA Historical Reference Collection.

[31] This committee, formally called the "Advisory Committee on Organization"—under the chairmanship of Dr. Lawrence A. Kimpton who had just resigned as Chancellor of the University of Chicago in favor of a job with Standard Oil of Indiana—was formed in April 1960 to review the overall organizational structure of NASA and to make recommendations for streamlining and other improvements. It held eight 2-day meetings over a 6-month period between April and September 1960. It used the studies that had been completed by McKinsey & Co., and submitted a final report on 12 October 1960, the Kimpton Report. It's findings revolved around the "basic concept" guiding NASA, the overall organizational structure flowing from the concept, the relations with other governmental agencies, headquarters organization, and several miscellaneous organizational issues. For more on this report see, Robert L. Rosholt, *An Administrative History of NASA, 1958-1963* (Washington, DC: NASA SP-4101, 1966), pp. 162-70.

question in the minds of committee members as to the beliefs of our people in the program. Our relations with the Space Science Board of the National Academy of Sciences were not nearly so cordial. Composed almost entirely of scientists, this board had a strong urge to run the program—not just be an advisory group. We did pay attention to its recommendations but treated it more routinely as a group comparable to our own scientific staff.[32]

[32] On this subject see, Norriss S. Hetherington, "Winning the Initiative: NASA and the U.S. Space Science Program," *Prologue: The Journal of the National Archives,* 7 (Summer 1975): 99-108; John E. Naugle, *First Among Equals: The Selection of NASA Space Science Experiments* (Washington, DC: NASA SP-4215, 1991), pp. 29-40, 71-74.

CHAPTER TWO

JANUARY 1960

First, I must record my grave apprehensions about the activities in which I will be engaged during the next three or four months. For the past three or four months, it has become increasingly apparent that Congress, the newspapers and the magazine writers, and the radio and TV commentators have been sharpening their knives for the nation's space program. I think I have been able to arouse some of the people in the White House and the Bureau of the Budget to the concern of Congress and outsiders over the "space race." I hope that the actions we can take in the month of January will regain the initiative for the president and the administration as well as my organization. Much good work has been done by the staff during the past month or so in drafting a revision of the basic statute under which we operate, in negotiating an agreement with the Army on the transfer of von Braun's group to NASA from ABMA and on preparations for the presentation of our budget to Congress.

Friday, January 1: Ruth is in Cleveland with the children but will return on Monday, 4 January. Last evening was spent at Abe Silverstein's. A large group of NASA people gathered during the evening and drank the New Year in. I left about 12:30 and retired to my bed—none the worse for wear. Most of the morning was spent in working over a draft of a speech for President Eisenhower. Ed McCabe of the White House staff had been assigned to prepare a draft of this speech, which we hope to persuade the president to give at an early date. It is intended to clarify the misconceptions held by many people about the effectiveness of our missiles. The critics' argument goes somewhat as follows: the Russians make difficult shots with their space rockets. This must mean that they can hit a target here on Earth—in the Washington area—with pinpoint accuracy. We are not able to match their space shots with our rockets, and thus we are thought to be unable to hit a target in Russia with a nuclear warhead. This translation of our inability to match the Russians in space shots into an inability to hit a target on earth with extreme accuracy is a very annoying situation. The facts are that we have great accuracy and increasingly good reliability in our intercontinental ballistic missiles and intermediate range ballistic missiles. The speech, in addition, is intended to state that we are not going to attempt to compete with the Russians on a shot-for-shot basis in attempts to achieve space spectaculars. In fact, we cannot match them at the present time; it will be several years before we will have rockets of sufficiently high thrust to do this sort of thing. In the meantime, it is necessary to recognize that we are competing with the

Russians on many fronts, and that space is one of the most glamorous and important of these. It is the area in which the Russians clearly have a lead at the present time. Our strategy must be to develop a program on our own terms that will allow us to progress sensibly toward ultimate leadership in this competition. To do this we will use the rocket systems available to us while we continue to develop systems with higher thrust such as Saturn and ultimately, Nova.[1]

I think we have the words set down for a fifteen minute speech that will convince the country of the rightness of the president's thinking in this matter. Importantly, too, a statement from the president will gain the initiative in the forthcoming struggle with Congress. After a somewhat skimpy lunch, I took a walk for an hour and then returned to the apartment to watch football and then go out to dinner with a friend named Millard Richmond.

Saturday, January 2: Another look at the speech for the president and some rewriting. I took it down to the White House where I was able to get it retyped and given to Ed McCabe, for further editing. The rest of the day was spent in thinking over the kinds of questions that Congress is apt to ask of us. It will be very important to have all the senior members of the staff speak with a single voice. I began to set down a host of questions divided up into the principal areas of activity, and I am going to ask the staff to add to them. We will then have a session at which we will answer the questions—some quite factually, others in terms of policy. The day ended with about nine pages of questions set down.

Sunday, January 3: Our new director of launch vehicle systems had written out a number of questions regarding NASA policy or my own convictions on subjects he felt might be discussed outside the office. Once again, I found myself writing out longhand about 10 or 12 pages of answers. I am going to have them typed and distributed to the senior members of the staff. This effort, together with the questions written down yesterday, should give us a good base for our policy framework for the coming year. In the middle of the afternoon, I started out for a walk and must have walked for two hours. I did enjoy listening to Senator Kennedy on one of the news programs. He answered questions about his candidacy for the presidency with candor and much good sense.

Monday, January 4: During the night I was awakened with a familiar pain in the right foot. At first I thought it must have been a strain from the strenuous exercise of the day before, but I guess it is the old malady, the gout. At four in the morning, I got up and took one of the new pills, Decadron, which had been recommended to me by George Rincliffe several weeks ago. During the course of the day I took three more of them and the pain seemed to disappear. I hope that I won't have to use the pills very often, but I am glad to know that they can be

---

[1] Nova was a proposed super rocket several times the size and thrust of Saturn. It was projected to use a first stage of four 1.5-million-pound engines under development by the Air Force, a second stage of one such engine, plus third and fourth stages incorporating liquid-hydrogen-fueled engines (four in the third stage and one in the fourth stage). Anticipated for use in ambitious lunar flights projected for the seventies, Nova was discarded by 1962 because of its technical demands and expected funding problems. See Bilstein, *Stages to Saturn*, pp. 37, 39, 50-53, 57-60, 63, 65, 67.

effective. Colchicine has so many bad side effects with me that I dread taking the drug. Off to the office at the usual hour, a staff meeting awaiting me. We debated the issue of opening one of the next Little Joe shots at Wallops Island to the press.[2] Apparently we had half-promised the press we would open one of our shots at the Island for them. Normally, we maintain a strict control over the press at the Island believing that our people ought to be able to do their research and development tasks without the press breathing down their necks. I finally gave in; the next shot will contain a biomedical specimen, which should please the press very much.

Following the staff meeting, Dryden and Horner sat with me for some time as we worked over the draft of the president's speech and discussed several other matters having to do with the timing we should follow regarding the supplemental budget, the new statute, the ABMA transfer and the delivery of the president's speech. We also discussed the distribution and use of the 10-year plan now nearing completion. Calling in Stewart, Gleason and Johnson, we finally decided to develop an unclassified version of the document and provide an appendix containing the schedules and the estimates of total program cost.

At 12 o'clock, Commander Williams of the White House staff came over and sat with Horner and myself reviewing the sections of the president's State of the Union message of concern to us in NASA.[3] It appears to be a fairly good statement. We were able to make several useful suggestions that I believe improved the document materially. Lunch at the desk in order to get more work completed. Then a short discussion with Frank Phillips about the proposed changes in the Space Act. He is secretary of the Space Council and will be out of a job if the law is changed, since the Space Council will be eliminated under the new statute. I asked him to consider whether or not he was interested in working directly with me. I also asked him to review the report from our bio-science advisory committee so that we can take action on it later in the week.

Gleason and Nunn came in to talk about the preparations for congressional hearings on various matters. I gave them the series of questions I had prepared over the weekend and asked them to fill them out and to put them into such shape with respect to the policy and operating matters involved that members of the staff could be briefed on them. At 3 o'clock, I moved over to the White House to meet with Secretary Tom Gates of the Department of Defense and Gerry Morgan, chief counsel to the president. Tom has now agreed that the proposed changes in the law

---

[2] Little Joe was a relatively small and simple, solid-propellant rocket with a quarter of a million pounds of thrust. It was used as an inexpensive booster, with about the same performance as the Army's Redstone, for testing various solutions to the many problems associated with manned space flight in the Mercury program, notably the problem of escaping from an explosion during takeoff. For details, see Loyd S. Swenson, Jr., James M. Grimwood, and Charles C. Alexander, *This New Ocean: A History of Project Mercury* (Washington, DC: NASA SP-4201, 1966), passim.

[3] Possibly Ross Norman Williams (1927-), who became a rear admiral in 1975 and had received his commission in 1951. He was head of the Trident program coordinating branch, Office of the Chief of Naval Operations, 1969-1973 and then commanded the U.S.S. L. Y. Spear, 1973-1975. In 1975 he became the military assistant to the deputy undersecretary of defense for research and engineering, Strategic and Space Systems Office, Secretary of Defense.

make a good deal of sense from a management standpoint. He says the law should have been written this way in the first place, but he doubts very much whether it is politically expedient to do anything about it now. His attitude is one of letting Congress initiate the laws we know it is planning and letting the president veto them if we can't change the minds of Congress in debate on the Hill. I represent the opposite view. I believe the president must take the initiative in this situation. Tom was very decent about the whole matter. He said that the Defense Department would go along with whatever decision was reached; I urged that he speak to the president personally. He did not want to do this, and I therefore urged Gerry to represent to the president Tom's attitude. It is really important that we be together on this.

At 4 o'clock, I sat down with Ed McCabe and worked over the draft of the speech again, giving him the result of the efforts Dryden, Horner and I had made in the morning. Back to the office about 5:15. A call to the apartment elicited no response. About 15 minutes later, a call came from Ruth saying that she had arrived. I moved into high gear and we managed to get the station wagon unpacked and everything in good order by about 6:30. At this juncture, George Burgess [an engineer and fellow Yale graduate] called to say that he was in town. Ruth agreed and immediately we asked him to come out for a drink and potluck supper with us. He seems just the same as always, and it turned out to be quite an enjoyable evening.

January 5: This is Tuesday and the president will make his State of the Union message to Congress on Thursday. The White House is completely occupied with preparing the speech. It seems that every member of the staff has a chance at it, and even some of us on the outside get our 10 cents worth. I am told that this will go on right up to the few minutes before the speech.

At NASA, we are preoccupied with the preparation of the defense of our budget and, more importantly, the defense of our past activities before the committees of Congress who have set hearings for the latter part of January. This is a time of soul-searching and attempting to rationalize some of the actions we have taken in the past and to decide on policy regarding certain of the things we should be doing in the future. It is necessary to determine who is to say what before which committee. All in all, it is a strange and wonderful climate. It is democracy in action and seems completely inefficient. But this kind of operation in a goldfish bowl is probably necessary because: it is easy to be complacent; the atmosphere that is developing at this time of year requires one to support his thinking and to rethink attitudes toward Congress and plans for the coming year; and while the political overtones make all of this somewhat unpalatable, in a nation where the individual is king, there must be a constant check on that individual. We are all human and have our normal complement of human frailties.

The day started out early in the morning with a discussion of the speech Ed McCabe and I are trying to put together for the president and then a session with Maurice Stans, director of the Bureau of the Budget. Meeting with several of his staff, we discussed schedules for sending material to Congress on behalf of NASA. It became apparent there were political problems involved, and we determined that

I should see Bryce Harlow on these matters later in the day. I did have a call from Gerry Morgan asking for the papers on the proposed new statute and for copies of the proposed speech. These will get to him later in the day.

Shortly after lunch (at the desk) Bill Holaday, chairman of the Civilian-Military Liaison Committee (CLMC), came in to talk about the timing of his resignation.[4] The law we are proposing will eliminate the CMLC, and Bill wants to resign before the law is submitted to Congress. I cannot disagree with him. Incidentally, Holaday is one of those persons retired from industry and attempting seriously to assist here in Washington. He has difficulty operating in the jobs he has had recently, but he has tried diligently to make decisions and make them stick. Unfortunately, he is in the position of advising without any responsibility for carrying out the advice he is giving.

A little later in the afternoon, a long session with a variety of our top people to discuss budget defense and the scheduling of the papers that are to go to the Hill. It is interesting to note the extent to which one must be on one's toes to see to it that all of the top people know what is going on at all times. At 4:30, I had a visit with Bryce Harlow at the White House. Bob Merriam and Ed McCabe participated. We discussed the timing on all the events in the next week and had quite a little debate on the timing for the proposed speech. I am not sure just where we will get off on this, but there will certainly be more discussion later in the week. Back home at about 6:30 and a good dinner with some time left after dinner for starting a speech for 27 January.

Wednesday, January 6: A brief meeting first thing in the morning with Hjornevik about some recruiting he wants to do. We still have the problems of working with other people and like to tell the boss of the man we are trying to hire what we are attempting to do. A little later in the morning we heard a briefing by the General Dynamics people, including Fred de Hoffmann and Ed Creutz, on Project Orion.[5] This is really quite a project. It is the propulsion of a heavy space ship into orbit through successive explosions of atomic devices. It is an expensive way of doing the job, but it may be the only way to put a very heavy object into orbit.

A quick lunch with Phil Farley at the Statler. Phil is with the State Department and has concern for atomic and space matters. We discussed the desirability of cooperative programs with the Russians and how these might be

---

[4] His resignation became effective 30 April 1960 and he became a consultant.

[5] De Hoffmann and Creutz were apparently General Dynamics personnel. Project Orion was an effort begun by the Advanced Research Projects Agency (ARPA) to employ low-yield nuclear bombs as propulsion devices for rockets. It began seriously in 1958 with ARPA's award of a $1 million contract to General Dynamics to explore and develop the idea. In the next few years, there were additional contract awards from ARPA and NASA in the attempt to achieve a high-performance rocket for rapid manned trips to other planets, but the nuclear test ban treaty of 1962 prohibiting nuclear testing in the atmosphere, in the oceans, and in space prevented development of such a propulsion system. The Air Force finally terminated the program in 1965. See Kenneth W. Gatland, "Project Orion," *Spaceflight* (Dec. 1974): 454-455 and Bill Wagstaff, "A Spaceship Named Orion," *Air & Space* (Oct./Nov. 1988): 70-75.

brought about. It appears there is no reason why we should not write directly to our opposite numbers in Russia so long as we keep the State Department advised. I talked with him somewhat about the speech for the president and found him interested and hopeful that this speech could be made. More work during the afternoon on budget and the speech for the president and then, at 5:30, cocktails with Bob McKinney and his wife. Bob is doing another study for the Joint Congressional Committee on Atomic Energy and will be spending time abroad. He is looking into the international aspects of atomic energy and the AEC's international program.

Back home about 6:30 and into a dinner jacket to attend the annual congressional dinner of the Women's National Press Club. It was a really noisy cocktail hour, and finally we were able to sit down about 8:15. Having finished the main course, we came to the magic hour of 10 o'clock, which was important to me. I had recorded last week a television interview with Howard K. Smith of Columbia Broadcasting System. The subject of the hour-long discussion in which many other people participated was, "Can democracy cope with the problems of management in the missile and space programs?" We wanted to see the show and managed to get home about 20 minutes late but saw the last 40 minutes. My own little piece came off reasonably well, but I suspect the nation is no more able to understand the management problems than it was before the film was made.

January 7: The regular Thursday morning meeting with the staff at 8:30 prompted discussion of the CBS program the night before. Almost everyone agreed that our part in it had gone fairly well but that the management of the missile and space program must seem a mess to the general public. Gleason felt that we should have stayed out of it. Certainly, I would prefer to stay out of this sort of thing, but I doubt very much whether a person in a responsible position in the management of the nation's space program can avoid being drawn into discussions of this kind on TV.

At 10:30 with Dr. Kistiakowsky to rehearse a discussion with the president scheduled for 4:30 this afternoon. Back to the office in a hurry to get together with Dryden, Harry Goett, Horner, Silverstein, Wyatt and George Low on the program to follow the present Mercury. This was a very short discussion because I wanted to get up on the Hill to hear the State of the Union message. We came to no conclusion except that we would have to get together after lunch. Over to the White House in a hurry and into one of the White House cars that took me to the Hill. Hustling through the corridors, we were able to find standing room behind the railings on the House floor. [Discusses the ceremonies that followed.]

The president started off reading his speech at about 12:35 pm. He spoke vigorously and with great clarity for about 20 minutes. It seemed then that he became rather obviously tired and seemed to fumble a little bit with the pages of his speech as he turned them. This situation continued for five or six minutes when, all of a sudden, he seemed to get his second wind and finished the 45 minute talk with great vigor and clarity. There was a significant amount of applause when he announced that the budget for 1961 shows a surplus of something more than $4 billion. Only two or three people in the official family knew that he was going to give this figure. It was not in the printed text, and most of the people on the White House staff had no knowledge of the figure prior to the president's announcement. All in all, the ceremony, the occasion and the participants made a deep impression on me. The members of the diplomatic corps, including Mr. Menshikov of Russia, must have acquired a new respect for the democracy that is the United States.[6]

A hurried lunch at the White House mess and back to the meeting we chopped off just before lunch. We finally decided that there should be a follow-on project, and that it would have, as its objective, manned flight to the moon and back. In those few simple words, I am describing a project that may well cost between $5 and $10 billion and probably will occupy the attention of very good men for a period of ten to fifteen years.[7] What an age in which to live!

Earlier arrangements for an appointment with Senator Dodd of Connecticut were called off because of the caucus of the Democratic party. I'll see him tomorrow for lunch. The meeting with the president at 4:30 was called off because of a long session of the National Security Council, but I am to be at the White House at 8:30 in the morning. I was able to keep a date with Bob Anderson, secretary of the treasury, who always gives me a great lift. I was rehearsing with him some of the discussion I was to have with the president and urging upon him the support of my desire to have the president make a speech clarifying our efforts in the space program. I intended to ask Bob to be the commencement speaker at Case but couldn't bring myself to do it after taking three-quarters of an hour of his time to seek advice. I'll get to him at some other time on the Case commencement.

Friday, January 8: I had an 8:30 appointment with the president this morning to check over the matters undergoing debate and preparation in our shop and at the hands of the White House staff these last several weeks. I had scheduled a visit to the dentist for 9 o'clock but had been able to delay this by 15 minutes, thinking that I would be through with my discussion at the White House in 30 minutes. Unfortunately, it soon became evident that there had been a complete lack of any briefing by the White House staff, so the president was completely "cold" on all matters I wanted to talk about. Fortunately, I had made an outline of what I had intended to discuss, and it started with a briefing on our discussion in Augusta, Georgia. With me on this occasion were Dryden, General Goodpaster, General

---

[6] Mikhail Menshikov was then Soviet Ambassador to the U.S

[7] Actually, Apollo lasted 11 1/2 years and cost about $23.5 billion. It ended in December 1972, having landed 12 men on the moon, and yielded an incalculable fund of information and technology, the latter including hardware systems enormously more capable than their predecessors, such as the Saturn launch vehicles, the command and service module, the lunar module, and the lunar roving vehicle, to mention only the obvious. For further details, see Bilstein, *Stages to Saturn*; Charles D. Benson and William B. Faherty, *Moonport: A History of Apollo Launch Facilities and Operations* (Washington, DC: NASA SP-4204, 1978); Courtney G. Brooks, James M. Grimwood, and Loyd S. Swenson, Jr., *Chariots for Apollo: A History of Manned Lunar Spacecraft* (Washington, DC: NASA SP-4205); and William D. Compton, *Where No Man Has Gone Before: A History of Apollo Lunar Exploration Missions* (Washington, DC: NASA SP-4214, 1989).

Persons, Harlow and Kistiakowsky. I was attempting to put the situation on the Hill into perspective and to suggest the course of action to be followed. The president, who had discussed with me the proposed changes in the statute, seemed to have forgotten our earlier conversation. Almost immediately, we branched off into a discussion of doing away with the Space Council. At this late date, with the law completely redrafted and ready for submission to Congress, I was flabbergasted. We did debate these issues for 15 or 20 minutes, and it appears clear that a re-examination of certain elements of the proposed changes is now necessary. I doubt that anything will come of this re-examination, however.

We went on to discuss the supplemental budget request and the transfer of ABMA, and on these there was no difficulty. I then brought up the matter of the speech already discussed with the president on several occasions, most recently at Augusta. I had sent him a confirmatory letter following my visit to Augusta restating my conviction that a statement by him was absolutely necessary. Another 10 or 15 minutes was taken up on this matter, and the president did read the draft speech. It was very much a rough draft, but it attempted to clarify the missile-space situation and to state the objectives of the United States' space exploration program. Finally, the president seemed to agree that he would make such a television speech and that it should be done with proper advance billing.

It became evident that discussions between the president and the legislative leaders (especially Lyndon Johnson) as well as with the members of the Space Council would be necessary before the new statute could be proposed to Congress. On this note, we ended the meeting. At least I thought we had ended the meeting, but as we turned to go, the president said the really important thing is to get that big booster out as soon as possible. He asked me when I thought this could be done, and I told him it would not be operational before 1965 and would not be reliable even then. He expressed great concern and said something about his willingness to put additional money into this if much time could be saved. I immediately suggested that as much as a year might be saved if we spent an additional $50 to 100 million in the coming year. I did not press the point, but I intend to do so later. Following this meeting—by now I had lost my chance to go to the dentist—we talked a little bit with the White House staff about the next steps. It was agreed that we would review the message to Congress on the proposed statute after we considered changes that would make it more palatable. I was also to see Senator Styles Bridges, Republican minority leader of the Senate, with Bryce Harlow to discuss strategy with respect to submission of the bill. I was somewhat discouraged as I left the White House.

I had made an engagement with Senator Dodd of Connecticut for lunch. I was not in a very good mood but found that the lunch was an exceedingly pleasant one, for Dodd was only too happy to see me and expressed himself completely convinced of the sincerity with which I was attacking my job. He offered to help in any way he could. This is very important to me because he is a Democrat and, as such, might be expected to be exceedingly critical of anything we are doing.

Following that discussion, Jim Gleason, Harlow and I visited with Senator Bridges for an hour. I pointed out the proposed changes in the law and found that he was in complete agreement with them. Bridges is an old hand in the Senate and knows his way about. He has been less than completely cooperative with the White House, but he has never voted against the presidential position without first telling the president that he was going to do so. I found it a very pleasant occasion and enjoyed debating strategy. Senator Bridges suggested that Lyndon Johnson and he visit with the president to talk about the elimination of the Space Council from the Space Act. He also suggested the desirability of visiting with Representatives John McCormack and Joe Martin.

This discussion had somewhat restored my spirits, and I was anxious to get back to the office where we were having a discussion about the Atomic Energy Commission and its part in our program. The AEC is developing a nuclear reactor for use in rocket propulsion. It depends upon us to show interest and to support this program by providing certain facilities and non-nuclear developments. This is a long-term proposition, and it is highly improbable that we will have a nuclear rocket before 1966. Even then, it will be necessary that we have remote launching sites because of the probability of danger from a nuclear accident.[8] The immediate reason for this conference was to prepare ourselves (Dick Horner and myself) for possible discussions with Chairman McCone of the Atomic Energy Commission, who is going to Huntsville, Alabama, to visit the von Braun operation tomorrow.

In the middle of these discussions, we received a proposed press release from the public information officer of the Jet Propulsion Laboratory. It was to cover a scientific paper being given in Nice, France, by one of the laboratory personnel, including a proposal for a cooperative program with the Russians that had no official standing. There were statements of reservation throughout the release, but it was aimed principally at getting the headlines on its being a proposal for a cooperative program with the Soviet Union. A quick call to the West Coast determined that we had not had a chance to see the document, although we had approved an abstract in which there was no mention of international cooperation. When we did get a copy, it appeared that there were only three sentences in the entire paper having anything to do with international cooperation! We had to ask that the paper be altered, and thus ended what might have been a minor crisis if we had not caught it in time.

This was the day the Russians announced that they were setting up a range in the Pacific to test more powerful space boosters. Two or three of their telemetry ships are already on station as we know, and the exercises are expected to take place

---

[8] Studies of nuclear propulsion began shortly after World War II, leading to the Rover program for the development of a nuclear rocket in 1955, renamed NERVA (Nuclear Engine for Rocket Vehicle Applications) in 1961. Budgetary considerations led NASA and the AEC to announce the termination of the program in 1972 after some 20 reactors had been built and tested. In 1983 the DOD, the Department of Energy, and NASA began to build upon the technological legacy left behind by Rover and NERVA, beginning the SP-100 project to develop a nuclear power source for use in space. See various documents in the NASA Historical Reference Collection dealing with propulsion, Rover, NERVA, and the SP-100.

sometime between January 15 and February 15. This should make the cheese a little more binding as we attempt to maintain our own program on a sensible basis. About this time I should be attending a reception for the foreign news correspondents for the NBC, but I had to say I would be unable to attend. There is a limit!

Saturday, January 9: Today, we took off at 7:30 in the NASA Convair for a visit to our new group under von Braun. John McCone was with us for a visit to ABMA. John is a member of the Space Council and much concerned about our lag in the competition with the Russians. I have had some difficulty in attempting to convince him recently that there really is no way of bridging the very deep and broad gap between our capabilities in rocket thrust and the Russians'. That gap will be closed within two to four years, but I doubt that there is anything that can be done to speed it up more than perhaps six months to a year even if we spend $100 million more a year extra.

Von Braun met us at the plane with General John Barclay, the Commanding General of the Army Ballistic Missile Agency. They had laid on a briefing with their staff in attendance. The briefing covered the Saturn Project very well. Von Braun carried the whole discussion and, as usual, spent about half his time describing exotic trips into outer space. The Saturn booster project covering some 22 boosters looks to cost us about $1.2 billion.[9] It is one of the most amazing combinations of engineering, plumbing and plain hope that anyone could imagine.

After lunch we toured the facilities. I am always amazed at the quality of workmanship necessary to make these beasts fly and certainly, this German-American team has a great deal of knowledge in this field. It really is a superb group. Now that it is a part of our NASA family, it is hard to go down there without being beset by any of a dozen supervisors who ask for additional funds and men, etc. We took off for home about 5 o'clock our time and made it back in two and a half hours. We had dinner on the plane, and I was able to get back to the house at 8 o'clock.

Sunday, January 10: This morning we were able to take it a bit leisurely, because I did not have to go to the office until about 10:30. I met with Dr. Kistiakowsky to brief him on the meeting of the Space Council to be held on Tuesday morning with the council members and with the president in the afternoon. This will be an interesting meeting; the president will have to tell the members that he is proposing to eliminate the council. Back home to work over the various papers I have to understand for Tuesday and to use in preparing notes for a discussion with the president tomorrow morning. I need to go over with him the Space Council agenda and to talk with him about the possibility of increasing funds for Saturn. At 6 o'clock, we watched Stuart Symington on the "Meet the Press" show. Here is a man who is a positive menace on the national scene, in my opinion. He uses

---

9 The actual cost was $9.284 billion, according to Bilstein, *Stages to Saturn*, p. 422. (I have subtracted the figure for the lunar roving vehicle, which seems more appropriately an Apollo than a Saturn expenditure.) See this book for details of the Saturn development and the various boosters and stages that were involved.

innuendo in ways that are as serious as were some of the statements dropped loosely by McCarthy several years ago. I know that I am prejudiced because of my activities last spring when Symington was chairman of the committee to study the organization of the space and missile business. I gained no respect for him then, and I have even less today. It would be a great calamity for this country were he to be a serious contender for the presidential position.

Monday, January 11: A short staff meeting brought forth no very real problems. I reminded the staff that the Monday meetings were supposed to deal with problems for resolution by the administrator, and that the Thursday morning staff meetings were intended to provide an opportunity for bringing to the staff items about which the staff ought to be thinking. A little sternly, I asked them to be better prepared for each of these meetings. Over to the White House at 10:30 for the meeting with the president. Prior to this I had discussed at some length with Horner, Dryden and Johnson the proposed changes in the statute. It has been possible to give effect to the president's suggestions of last Friday and add two clauses to the law that I think will improve it. Again, I was glad that I had spent a good many hours on Sunday preparing for the meeting with the president. I had been warned by General Persons that he was not in a very good mood. This did not appear to be the case throughout our conversation, which lasted about 70 minutes. I began by briefing the president again on the changes in the law and then listed the things that were accomplished by the revised law and pointed out that it did not provide for any acceleration in the program or any promise of acceleration in the program. Further, it did not provide for advisory committees or coordinating committees. We agreed that in my letter of transmittal to Congress, and in the president's message to Congress, we must cover these items thoughtfully and fully.

I then went on to discuss with him the other items on the agenda of the Space Council, which meets tomorrow, and he gave thoughtful attention to the items in dispute on space policy. He made useful suggestions, and I was given the assignment of having available for the meeting tomorrow suggested changes for resolving the dispute between the Budget Bureau, the Joint Chiefs of Staff, and the rest of the participants. We then got to the matter of the highest national priority for the Saturn project. It was obvious this was going to be a shoo-in. I was able to talk with him further about the possibility of accelerating this program by providing an additional $100 million. This would have to be requested through the supplemental appropriation route. This money would provide us with an operational Saturn system 12 months before it would be available under the present funding plan. The president reiterated his oft-repeated statement that a very powerful booster system was of the utmost importance in the space business. I got him to agree to discussions between Persons, Stans and myself. Hopefully, I can get Stans to go along with this proposal. Actually, it will make the changes in the law, the transfer of Saturn and ABMA to NASA, and the president's statements relating to the accelerations of the big booster program much more believable, and it really will speed up the availability of Saturn.

This news was gleefully received by Dryden and Horner. A meeting was set up with Staats for 5 o'clock, since Stans was not available. A quick lunch with Wallace Brode at the Cosmos Club was not very inspiring.[10] I wanted to talk with him about his concern over the state of science and the support of science in the federal government. He really hasn't any very good ideas. I am in the same boat. A meeting at 4 o'clock with Dr. Kety to discuss the report of the bioscience advisory committee. This was a pleasant discussion, and we came to substantial agreement on the form of the report.[11] The meeting with Staats at 5 o'clock went well. I am optimistic about the result and will push very hard tomorrow to secure authority to put in the additional supplemental.

Plaque on the Little White House, which served as headquarters for NASA from its birth until 1961. Built by Benjamin Tayloe, it became the home of Dolly Madison and was later part of the Cosmos Club until after World War II, when the club moved to its present location on Massachusetts Avenue.

Off to the Cosmos Club at 6:15 for a quick martini with Alan Waterman of the National Science Foundation. We reviewed the agenda for the Space Council

---

10 The Cosmos Club was an exclusive gentlemen's club in Washington, DC, founded in 1878 for 200 "persons interested in science or literature," although the membership grew much larger thereafter. For its history to 1949, see Thomas M. Spaulding, *The Cosmos Club on Lafayette Square* (Washington, DC: Banta, 1949).

11 Copy of the report in a biographical file on Kety in the NASA Historical Reference Collection.

meeting of tomorrow, and I was able to get home at 7 o'clock for a pleasant dinner and about two hours of reading before going to bed.

Tuesday, January 12: This is the day when the Space Council meets—probably for the last time. Arriving at the office at the usual time, I called in Horner and Dryden to tell them about my discussion last evening with Elmer Staats. I was mildly optimistic in asking that papers be drawn for a second supplemental covering the $100 million we hoped to get for extension of the work on Saturn. I then hastily put together the words I was going to use in describing the long-range plan to the Space Council. Fortunately, the morning meeting is informal, offering a chance to sharpen up a presentation of this kind. The meeting started out with Kistiakowsky in the chair and the initial presentation by Dr. Ling on the comparative United States-Soviet standing in the space business. Ling, an associate of Hendrik Bode of the Bell Telephone Laboratories, had worked with Bode and several others in preparing this evaluation. He made a very able presentation which occasioned a good deal of comment, question and debate. In particular, John McCone took off on a 10- or 15-minute cold-sober indictment of the present administration policy. He called for a single space program to be managed by either the civilian agency or by the military services, a totally new space vehicle booster system as a backup to Saturn, more money for Saturn, and a straightforward determination to push ahead in pursuit of the single objective of beating the Soviets in this field. More sober counsel prevailed and indeed, John had to take back some of his rather broad statements. Apparently, it is his way to attack with the idea that he will be quite willing to withdraw if he is found to be seriously in error. In any event, he did us a good service in getting these things off his chest and letting them be debated in the open prior to the meeting with the president. The second item on the agenda was discussion of the National Security Council paper on space policy. No progress was made on the matter of resolving the differences between the Budget Bureau and the rest of the participating agencies, so the entire paper with its differences will have to go to the president for final adjudication. I made a presentation of our 10-year plan. It took about 15 minutes and provoked one or two useful comments.

At this point I had a call from the White House (General Persons) asking me to be in the president's office at quarter of 12 for further discussion with Stans on the additional money we want for Saturn. I called Stans immediately and found that he wasn't, at the moment, debating the issue of whether or not we should have the additional money; it was really just a question of the proper tactics to use in getting it.

Hugh Dryden and I repaired to the White House and talked briefly with Bob Anderson, Persons and Stans before going in to see the president. I waxed a little bit eloquent in attempting to put in a straightforward fashion the problem we saw in delaying over-long the submission of the supplemental. Anderson was concerned that the president not send up the budget one day and amend it the next. This sort of indecisiveness is political fodder of the most explosive type. On the other hand, I think that the president stands to lose a good deal more by not taking this step immediately than he would in admitting that a second look had inclined him to request more money for Saturn. In any event, we didn't solve it in the president's outer office.

The president started off by saying he was pretty well fed up with people coming in and asking for more money. He says here you come and bother about $100 million while I'm trying to solve the problems of the world with $50 billion. He said he was quite certain that we were going to have to spend an extra $100 million on Saturn during the course of the spring, and he thought it ought to be settled at once. Stans put in his concern over amending the budget almost before it had been presented to Congress. Bob Anderson spoke in the same vein although, in the most gentlemanly way, he appealed to the president's quieter nature in attempting to convince him that delay would not be serious. Both tried the gambit of using the $23 million supplemental we have already asked for Project Mercury—using it for Saturn with the intention to replace it with a second supplemental that would be requested at a later date—Stans [suggested] some months later. I immediately spoke up and said I couldn't understand the delay. We would take a look at the possibility of diverting some of our present funds or asking for an additional $7 million that could be covered by the authorization legislation we already have. The meeting wound up with the president saying he felt that a quick study should be made to be certain of the amounts we would request.

A quick lunch with Hugh and back to the office to get the staff working on this problem. Then a return to the White House to go over the mechanics for the afternoon meeting with General Goodpaster. It is always necessary to see that the arrangements made on the day before the meeting haven't changed overnight.

This was an unusual meeting of the Space Council since the National Security Council members had been asked to attend because of the attention to be given to the NSC paper on space policy. Some really important people were in the room, which was quite crowded. Among them were President Eisenhower; Vice President Nixon; Ambassador Henry Cabot Lodge; Secretary of the Treasury Robert Anderson; Under Secretary of State Livingston Merchant; Director of the CIA Allen Dulles; Special Assistant to the President for National Security Affairs Gordon Gray; Chairman of the Atomic Energy Commission John McCone; Director of the Bureau of the Budget Maury Stans; Deputy Secretary of Defense James Douglas; Chairman of the Joint Chiefs of Staff, General Twining; Dr. Bronk, Dr. Waterman, and a good many others.

The first presentation was again given by Dr. Ling, and it was a beauty. He spoke for 45 minutes, using beautiful English and giving a detailed but dispassionate picture of the relative standing of the two countries (the U.S. and the Soviet Union). The president was interested throughout and asked several questions following the presentation. The upshot of this discussion was agreement that the Soviet Union, on balance, is now ahead of the United States in space technology (not necessarily in space science), and that this lead was probably to continue for several years—a period of two years at the minimum. We then moved on to the National Security Council paper, and after some discussion, due to the homework I had done with the president yesterday, we were able to come to a conclusion on the language. This was the first time that I have seen Stans bested in a discussion of this kind.

We moved on to the discussion of our long-range plan. The president was interested and asked several questions. We were able to answer them all satisfactorily, although McCone again brought up some of the points he had discussed during the morning. It was obvious that the president was not going to buy additional and costly programs but that he did believe the program we had described represented a satisfactory level of effort for the present. He did not comment unfavorably on the fact that our extrapolation on costs took us to some $ 1.6 billion annually by 1968.[12]

Dr. York then presented the same sort of a picture for the Department of Defense. He did it with his customary good humor and disarming offhand manner. He is a master at the discussion of complicated matters in a very understandable way. By this time we had been in the room for over two hours, and it was obvious that the president wanted to end the meeting. I did call for the last item on the agenda, a request for the highest national priority for Project Saturn. The president disposed of this in very short order, at least partially due to the discussion I had had with him yesterday. Following that item the president spoke up and said that he felt the Space Council had probably served its purpose. He indicated he had ideas about changes in the law that I was familiar with and that I was to talk with each of the members of the Space Council about these changes. I had thought he was going to tell them what the changes were, and that the law was going to be sent to Congress but apparently, I'll have to do this interim chore before we can get on with the real move.

Back to the office and off to the house for dinner preparations. Ruth was preparing dinner for eight of us—four from NASA and four from the Department of Defense. Herb York, Joe Charyk, Dick Morse, Jim Wakelin, assistant secretaries of the Air Force, Navy and Army respectively. Dick Horner, Homer Joe Stewart, Hugh Dryden and I made up the balance of the party. We were presenting to this group our 10-year plan, and it seemed that it would be more pleasant to do this at home over a drink and a good dinner. The evening went very well, and the dinner was good and well received. We finished up about 10:30.

Wednesday, January 13: Wednesday and Thursday are really significant days in my stay here in Washington. We started out Wednesday morning with a staff meeting on policy matters, an expansion of the series of questions I had written down some two weeks ago, and we were trying to deal with the problems that have arisen as a result of those questions, calling for a good bit of discussion and some beating out of new policy positions. Involved in the discussion, in addition to Hugh and Dick, were Johnson, Gleason, Nunn, Sohier, Rosen, and Golovin. I must say that at the end of three hours I was really done in. We made some progress, but it is clear that another day and a half will be required to finish with this task.

A hasty review of the letter prepared for my signature addressed to the vice president and to the speaker of the House transmitting the proposed legislation was followed by a pleasant lunch. At 2:15, Dr. Bronk came in and we discussed the proposed changes in the law. He questioned me quite a bit, but seemed reasonably happy with the decisions that had been taken. As a matter of fact, he said that he had

---

[12] In fact, the NASA budget peaked at $5.25 billion in 1965 and was $4.587 billion in 1968. (*Aeronautics and Space Report of the President, FY 1991 Activities,* p. 180.)

been planning to talk with the president about the desirability of abolishing the Space Council.

The day had been broken into a little earlier by a call from Bryce Harlow asking me to be at the White House at 5 o'clock to meet with the president, Lyndon Johnson and Senator Bridges. I managed to get over there about 4:45 and we went over to the Executive Mansion. We were shown into the Oval Room, and the president joined us at 5 o'clock precisely. A minute or two later, Senator Johnson and Senator Bridges arrived. Drinks were served and the somewhat strained atmosphere was relieved a little bit. The president launched into a discussion of the space business and the reasons for his desiring to change the Space Act. Lyndon listened attentively without saying a word. It was obvious that the president was finding the going a little bit rough. I was called upon to make a presentation, and I am sure that I made it about three times over. Johnson just kept leaning on his hand and looking at me somewhat quizzically. Finally, Johnson said, "Well, Mr. President, you will remember that you were the one who really wanted this Space Council, and if you want to do away with it now, I'm certain it will be all right with me."[13] Styles Bridges asked two or three questions about the protection afforded the Defense Department in its use of space and about the intention of NASA to continue to use outside contracting for the major portions of its expenditures. Satisfactory answers were given, and I think Styles registered his points with Johnson.

The president turned the conversation aside to talk about Sid Richardson, late multimillionaire of Texas. Richardson had been a great friend of both the president and Senator Johnson. It was obvious that the president wanted to bring the conversation to an end but Lyndon called for another drink and kept the president talking for another 20 minutes. At about 10 minutes of six, we departed feeling that we had done a reasonably good job with the leaders of the Senate on the changes we were proposing in the Space Act. When I reached home about 6:15, I was so thoroughly tired that I'm afraid I was little or no company for Ruth. I went off to bed at 8:30, watched the television for a bit and gave up for the night.

Thursday, January 14: Up at 6:45 and off to the White House for breakfast with John McCormack, Joe Martin and Overton Brooks, the president and Bryce Harlow. The breakfast was pleasant with the president leading the conversation. Incidentally, he suggested to me that if I had to discuss the proposed changes in the law with these guests, I might sell it only once, not two or three times. The dining room is a very pretty room. Breakfast consisted of pink grapefruit, oatmeal and cream, scrambled eggs and sausage and bacon, toast, and several kinds of jam and coffee. The president ate only the grapefruit and the cereal. He remarked that

---

13 In his message to Congress of 2 Apr. 1958, the president had requested a national aeronautics and space board appointed by him. The Senate version of the bill that created NASA added the word "policy" to the board's title and located it in the Executive Office of the President. Eisenhower did not like this change but agreed with Lyndon Johnson to accept the board, renamed The National Aeronautics and Space Council, if it were similar to the National Security Council with the president as its chair. (White House press release, 2 Apr. 1958; Confidential Committee Print, Senate Special Committee on Space and Astronautics, S. 3609, 31 May 1958, pp. 4-5; Public Law 85-568, National Aeronautics and Space Act of 1958, 29 Jul. 1958, pp. 2-3, all in box marked "White House, . . . Eisenhower, National Aeronautics and Space Act of 1958," NASA Historical Reference Collection; Robert A. Divine, *The Sputnik Challenge: Eisenhower's Response to the Soviet Satellite* [Oxford: Oxford University Press, 1993], pp. 146-148.) For the outcome of Glennan and Eisenhower's effort to revise the initial law, see chapter seven, note 14.

he had to watch his intake of any foods containing cholesterol.[14] We launched into a discussion of the changes in the law. I felt very much more at ease than the evening before, perhaps because I had a night's rest. In any event, both McCormack and Brooks questioned me on several points. Later in the day, Brooks issued a statement saying that he did not know of anybody who would oppose the changes in the law. I doubt that this will be true to the same extent of Congressman McCormack. During the conversation, Martin sat absolutely still, said nothing and did not crack a smile. It is obvious that he is not well—a lonely man whose future is behind him.

Back to the office for some discussions with the staff and then, Hugh Dryden and I drafted a letter from the president addressed to me authorizing me to use initial overtime on the Saturn project and requesting the completion of a study of any funds necessary to support an accelerated program in the super booster field. I took this over to the Bureau of the Budget and was able to get agreement on it and to work out a program of action to send up an amendment to our 1960 budget within the next two or three weeks. This is a real triumph.

Lunch at the Hay-Adams [hotel] with Admiral Hayward. I gave him a complete briefing on the changes in the law and secured his agreement to support our proposals before Congress. At 1:30, I was back at the White House and managed to catch General Persons. We made some minor changes in the letter the president was going to be asked to sign, and the president made subsequent changes in three or four words; we were finally able to get it signed at 2:30—just before the cabinet meeting. I was able to give a copy of the letter to Anne Wheaton, assistant press secretary to the president. Thus, it got on the wires at 4 o'clock and we were in business again. I count this one of the best days I have had since coming to Washington.

At 2:30, I sat with the cabinet and listened to briefings on the budget, on the missile program and on plans for "Operation Alert."[15] Budget Director Stans displayed a chart showing that we have a national debt of $280 billion and that we have fixed obligations in the future of some $790 billion, most of these arising from costs of past wars. Back to the office where I signed the documents for the submission of our budget. All of our staff are very happy about the accomplishments of this day.

Off to the apartment and to dinner with the Richmonds after which we will go to the Russian Symphony. The concert took place at Constitution Hall which is hardly suitable for a fine symphony orchestra. The audience was very generous in its applause as indeed, it should have been. [Prize-winning Russian pianist Emil] Gilels was superb. The orchestra gave two encores and Gilels, one. We couldn't help remarking that just this morning Khrushchev was making loud noises in Moscow about the capabilities of his rockets and nuclear weapons while here tonight in Washington, a large and enthusiastic audience was applauding the Moscow State Symphony Orchestra and its guest performer, Emil Gilels.

Friday, January 15: This was not one of my better days. Going to the Statler Hotel for a haircut at 8 o'clock and then to the dentist was not the best way

---

[14] In 1955 Eisenhower had suffered a coronary thrombosis. For a recent account of the heart attack and his recovery from it, see Chester J. Pach, Jr. and Elmo Richardson, *The Presidency of Dwight D. Eisenhower* (rev. ed.; Lawrence: University Press of Kansas, 1991), pp. 113-114, 117-118.

[15] Operation Alert was an annual civil defense exercise in which people practiced what they would do in the event of a foreign attack.

to start out the day. At long last, I am having a replacement tooth put in my upper left jaw. In order to accomplish this improvement on nature, it is necessary for the inlays in the teeth on either side of the blank be removed. New inlays will then be put in that will form the bridge to which will be attached the replacement tooth. It was a rough morning, believe me. I really did not have much to do today. Discussions with several members of the staff, lunch with Hugh Dryden, and a three o'clock appointment with Jim Gleason and John Johnson to discuss my testimony before the House Science and Astronautics Committee on Monday. I think it must have been the letdown from the excitement of the week plus the distress over the drilling in my mouth this morning, but, I count this as lost.

As an addendum to yesterday's discussion, I want to record here an anecdote about the president. As we were at breakfast, somewhat sheepishly, he said he had learned from his mother never to speak when angry. He was somewhat ashamed of having spoken out the day before when a woman reporter asked him in an accusing manner why he was being partisan in his support of the defense program. He really lashed out at her. I saw it on television and could tell that he was quite angry. His worry about that sort of an incident is all to his credit. Many times in the past, I have said that I think his real problem is that he cannot avoid doing in a gentlemanly fashion the things that he thinks ought to be done. I think he must worry about every decision he makes. I don't mean this in a derogatory sense; I think he is just a sensitive human being—much too much so to be President of the United States. Somehow or other, the men who hold this office must play God, and this is not in Mr. Eisenhower's make-up.

Saturday, January 16: [Relates that Ruth was off to Cleveland for an operation to correct varicose veins.] Most of this day was taken up with writing a speech I have to give on 27 January at Jackson, Michigan. The occasion is the Dinner for Ike celebration which will take place in some 75 cities over the country. Apparently, some one person will give the speech at each of the dinners to be followed by a closed circuit television appearance by the president. There has been some untoward comment in the newspapers concerning the heads of two independent agencies—NASA and the Federal Aviation Agency—taking part in this activity.[16] The statement is that these are and of right should be independent and free of all politics. So far as I'm concerned, that is the case, but I still work for Mr. Eisenhower and am glad to speak on this occasion. I will make clear to the people concerned that my speech is one dealing with our program, not with politics.

A call from Dick Horner, who had flown in from Salt Lake City overnight, suggested the desirability of a brief meeting with him and others before the press conference we are holding this afternoon at 4:30. This is to explain the budget to

---

[16] Like NASA, the FAA was created by Congress in 1958. It had the duties of regulating air commerce so as to promote its development and safety as well as fulfill the requirements of national defense; promoting and developing civil aeronautics; controlling and regulating navigable airspace in the U.S.; consolidating related research and development; as well as developing and operating a system of air traffic control and navigation for both civil and military aircraft. (*United States Government Organization Manual 1960-1961*, pp. 385-386.) Without change in mission, the FAA became part of the Department of Transportation in 1967 and was redesignated the Federal Aviation Administration. (*United States Government Organization Manual 1968-69*, p. 407.)

be submitted by the president to Congress on Monday, noon. This discussion was worthwhile, and I returned home without waiting for the conference to take place. [After dinner with his daughter, Polly,] I finished up some more of the speech. It is almost completed.

Sunday, January 17: Polly and I made breakfast about 9:30 and enjoyed it very much indeed. I then finished my speech and am reasonably well satisfied with it. It is over-long, but it is much easier to cut than to add to a speech when it is finished. Polly cooked chicken breasts in white wine for lunch. Corn and a salad with Roquefort dressing completed the menu. One of mother's baked apples sufficed for desert, and it was really a pleasant experience.

I forgot to mention an incident that took place yesterday. At 5:35, we launched from Wallops Island a test rocket, which carried aloft a 100 ft. mylar balloon. Everything seemed to work perfectly, and the report was that it had gone some 250 miles above the surface of the earth and had come down some 400 miles from the launching point. Polly and I watched it from the roof and were able to see it quite clearly.[17] The balance of the day was given over to preparations for the

The 100-foot-diameter Echo satellite during inflation tests. Undergoing development flights early in 1960, the satellite achieved successful launch in August. It was made of a micro-thin film of plastic coated with a film of aluminum.

[17] This mylar balloon launch was a development flight in Project Echo, a program to develop a passive communications satellite, which did have a successful launch on 12 August 1960; Echo I, likewise a 100-foot aluminized plastic sphere, reflected a radio message from the president across the nation and was the largest and most visible satellite launched to date, remaining in orbit for almost eight years. (Emme, *Aeronautics and Astronautics . . . 1915-1960*, pp. 118, 126. For background on Project Echo, see Joseph A. Shortal, *A New Dimension: Wallops Island Flight Test Range: The First Fifteen Years* [Washington, DC: NASA Reference Publication Publication 1028, 1978], pp. 686-695. In 1994 the American Astronomical Society expects to publish Donald C. Elder's study, "A History of Project Echo," in its history series published by Univelt, Inc. in San Diego, filling a gap in the literature.)

coming week. A call from Ruth indicates that she is now in the hospital awaiting the tender ministrations of the doctor tomorrow morning.

Monday, January 18: This was another of those days. I had a call from Ruth about 10:30. She sounded a bit groggy and yet she had all of the laughter and good humor in her voice that is so characteristic of her. I called her back at 7:30 p.m. and found that she was doing well. She said that her circulation had seemed to be quite satisfactory, and she would not know until tomorrow how long she would be in the hospital. Apparently Sally had been out to see her during the afternoon.

My day at the office started out with a staff meeting in which we reviewed the schedule for the month ahead. I said we seemed to be worrying a great deal about money, but it was time we started to worry about results. I recognized the difficulties with the R&D program, that is Research and Development, but it was time we made a more determined attempt to meet the schedules we had set.

At 4 o'clock, we went over to the Defense Department to sit in on a meeting with Secretary Gates and perhaps 30 of his staff, both military and civilian. The purpose was to determine whether or not we had arguments that would compromise the position of either agency in testimony before Congress. We had a long discussion of the changes to be made in the law and had an opportunity to answer some of the questions I'm sure had been bothering a good many of the people at the Pentagon the past several weeks. It must be remembered that the negotiations leading to the proposal that the law be changed were carried on somewhat behind closed doors. Discussion of the Cisler Report led to the conclusion that we should deal with this as a matter still under study.[18]

Back to the office—one or two more appointments of no considerable importance and home for dinner with Polly. Her sense of humor—her ability to take criticism and to recover from the immediate pangs of remorse—these are something to watch and be thankful for. A sweet child in every respect.

Tuesday, January 19: Down to the dentist at 9 o'clock. This was not such a bad session as the one last Friday. Impressions were taken for an inlay in

---

[18] Walker L. Cisler was a special consultant to the secretary of defense who had submitted a letter to him on 30 Nov. 1959 proposing the establishment of a Central Scheduling and Control Office under the Secretary of Defense with broad authority and responsibility over both DOD and NASA ranges and space flight ground stations. He had sent Dr. Glennan a copy of the letter, and Glennan replied on 17 December that the director of this office should be jointly appointed by NASA and the DOD. Glennan further suggested the "office" be called the "Space Flight Ground Facilities Board" and that it concern itself primarily with problems common to both DOD and NASA, allowing each agency to conduct its own operations separately. This suggested, Glennan said, that there needed to be a single office in DOD with centralized control of ranges and stations as there was in NASA already. (T. Keith Glennan to Cisler, 17 Dec. 1959, Glennan subsection, NASA Historical Reference Collection, miscellaneous, C-Ci.) Dryden had sent a copy of the letter to Dr. Herbert F. York in the DOD, and the DOD apparently responded to Glennan's suggestion for a single office by giving York control over its missile ranges and ground stations. (Mark S. Watson, "Dr. York Is Given Control of Services' Space Work," *Baltimore Sun*, 8 Apr. 1960 in York biographical file, NASA Historical Reference Collection.) The proposal for a central space flight ground facilities board seems not to have reached fruition, except perhaps in a different and expanded form in September 1960 when NASA and the DOD established a broader Aeronautics and Astronautics Coordinating Board with Dryden and York as co-chairmen. On the latter board but none of the preceding negotiations, see Rosholt, *Administrative History of NASA*, pp. 172-173. On Cisler, see also Chapter III of the diary.

preparation for the removal of the second inlay. Back to the office for budget rehearsals for the balance of the morning. These left a great deal to be desired, and I was quite worried about the program as laid out. It is clear, once again, that engineers are not the most able and effective salesmen. Lunch with Del Morris, a candidate for the job of deputy director for business administration at Huntsville. He turned out to be an attractive fellow, and we were able to close a deal with him before the day was over. His present occupation is that of deputy operations manager for the Atomic Energy Commission in San Francisco.

At 5 o'clock, Dr. Hagen came in with the document that had been sanitized by a group of representatives of the various agencies involved in evaluating the present status of the space business in Russia and the United States. At 5:15, Dryden, Horner, Gleason and I made preparation for the congressional hearings. We agreed that a complete change in the schedule would be desirable. By this time, I had seen Polly to the train and sent her back to Swarthmore for her exams.

Wednesday, January 20: A full day of questions and answers in an attempt to solidify our position on a variety of subjects. I was in better spirits than I was at the last session, and we managed to get through with credit to all concerned. Lunch at Duke Zeibert's with Admiral Bennett. He has proven to be a very good friend here in Washington. Much discussion of the Navy's position with respect to the space business. Back to the office for a brief meeting with Professor Massey of London and several of his colleagues.[19] They are over here to discuss the proposed cooperation between the United Kingdom and the USA. We are going to fly a payload for them in one of our first Scout vehicles.[20] Everything seemed to be going well. Back to the question and answer session to be interrupted at 4:30 by a gentleman from the Central Intelligence Agency. Several of us were briefed on the Russian shot into the Pacific.[21] The amount of information available is really something to comprehend. Six-thirty and a dinner at the Cosmos Club with Professor Massey and his group. I brought it to a close at about 8:30 because they were tired, and so was I. And so to bed to do a little bit of writing on the statement for the congressional committee next week.

Thursday, January 21: A staff meeting at 8:30 and then a variety of activities of no particular consequence. A lunch at 12:30 with John Oakes, editorial writer for the *New York Times*. This was a delightful affair with plenty of give-and-take in conversation and some real straight talk. Back to the office to receive a

---

[19] Probably Sir H. S. W. Massey, Professor of Physics, University College, London, 1950-1975.

[20] The Scout was a low-budget, solid-propellant booster used to launch small payloads into orbit. Its development began at Langley Research Center in 1957. The first four-stage Scout arrived at Wallops Island in 1960, and NASA used it to launch many small satellites and probes in the years that followed. The Scout evolved with time, its payload more than doubling by 1965, for example. For its early history, see Ezell, *NASA Historical Data Book*, vol. II, pp. 61-62.

[21] This was a long-range ballistic missile fired by the USSR. According to Soviet statements, apparently accepted by the Pentagon, the missile traveled 7,762 miles and landed within 1.24 miles of its intended target. ("Shot in the Pacific," *Chicago Tribune*, 24 Jan. 1960, reprinted in NASA *Current News*, 25 Jan. 1960, p. 13.)

Mr. William Seaver at the request of Congressman McCormack.[22] Seaver was interceding for a friend of his who needed a job. Damn these congressmen!

At 3 o'clock, Jules Whitcover, Washington correspondent for the *Huntsville Times*, came in for an interview. He is a youngster and was very easily dealt with. Actually, one wants to help a person in this situation, and I think he was able to get a good story. Up to the House Office Building at 4 o'clock to record a radio interview with Congressman Silvio Conte of Massachusetts. Back to the office for further conferences with Dryden and Horner. We settled two or three things, and I was able to get away at 6:30 for dinner with Warren Morris at the Colony. A pleasant evening and now I am back home attempting to finish up the statement for Congress.

Friday, January 22: Another visit to the dentist. We are making progress but the mining operation takes time. Back to the office and a briefing by the United Research Corporation people on a new configuration for solid rockets. It was an interesting one with one stage nested on top of another. I doubt that we will be able to put the money into it; they want $100 million and three years to bring it to operational condition.[23] Luncheon with the Swedish Ambassador at the Swedish Embassy. It was a really pleasant visit.

Further discussions with Dryden, Horner and Gleason about the order of battle for the congressional hearings to begin next week. It appears that I am to be clobbered with the "executive privilege" problem. In the selection of contractors, we use a very involved, thoroughly organized and honest system of evaluation. A technical team and a business administration team report to the selection board, which finally makes a report to me. We give to the congressional committee all the bids, all the specifications, a statement of the reasons for my final decision, but we refuse to give to the committee any papers on the advice given to me by my subordinates on these evaluation teams. If I were to do this, it would mean that very soon the objectivity of the advice given to me would disappear. This is a strongly held belief throughout the executive branch.

Over to the White House with Johnny Johnson to talk with Gerry Morgan and Roemer McPhee about the executive privilege matter. It is apparent that they want this executive privilege on matters such as the one I have described to continue. I was able to make a date with the president on Tuesday to talk further with him about this since I am going to quote him in this matter. The White House staff are not allowed to testify before the Congress. At least, they never do, and as a result of this "walling off" of these people from the political facts of life, I think they often take viewpoints on operating and policy matters that are a little less than realistic. Those of us who have to appear before the public and congressional committees must stand up and be counted on matters with which we are concerned. This is not always easy,

---

[22] Seaver appears not to have been prominent in his own right.

[23] United Research does not appear as a NASA contractor in either volume I or II of the *NASA Historical Data Book*.

but it is a part of the job. The White House, when asked for opinions, often holds to a line of argument that might be somewhat modified were it to face the public as we do. This is just one of the facts of life in Washington.

Saturday, January 23: Up at 7:30 and off to the office after doing my washing for the week. Von Braun, Rees and a couple of other people are to be in to discuss the additional money they hope to get. It was a good session. However, it was obvious that von Braun and Co. expect to get the entire $100 million, whereas some of us feel that the large, single chamber engine ought to be given additional support, as well. Finally, I asked von Braun to go back and tell me what had to be done to hold the additional money to $75 million for fiscal year (FY) 1961. We are to get the answer next week, Wednesday.

A pleasant lunch with [advertising executive] Ward Canaday at the Metropolitan Club. Then, home for a brief rest before going to the Alfalfa dinner with John Parker, formerly of Remington Rand. The Alfalfa Club is just one of those things. Apparently its principal function is to have a very elaborate dinner once a year—black tie, wines and champagne, etc. Much of the evening is given over to the induction of new members, the installation of a new president and the nomination of a candidate for the presidency of the United States. The brass of the town were in evidence throughout the evening.

Sunday, January 24: A leisurely day. I managed to stay in bed until 9:30, although the last hour was given over to reading the papers in bed. Then I made myself a generous breakfast and did a stint of housekeeping in anticipation of Ruth's return next Tuesday. I cleared up some of the material that I have to provide to various congressional committees, dinner meetings, etc., next week.

Monday, January 25: The staff meeting at 8:30 concerned our public information problems. Walt Bonney presented one of his usual round statements that really got us no place. It is becoming increasingly clear that his abilities are limited and that his field is that of public information in the most restricted sense. The planning of well-thought-out developmental programs in public information is not his ball of wax. There followed a lengthy discussion on the classification of some of the materials we want to give to congressional committees. The compilation of charts and diagrams indicating launching dates for a considerable number of months or years has been giving us a great deal of difficulty. While there is nothing classified in the strict sense, it has been the conviction of most of us that to provide this kind of information to the Russians is not in the best interests of the nation. There is a great deal of strongly held opinion on this kind of a problem, and the answers are not straightforward or easily come by. At 10:30, Frank Stanton, president of Columbia Broadcasting System, came to see me at my request. I talked with him about the problems of properly informing the public on the nation's space program. We had talked of this briefly in New York some weeks before. He expressed great interest in finding a solution to this problem. We finally agreed on a strong story line that might serve as a syllabus does in a college course. I think we will undertake this.

A call from the White House (Kistiakowsky) about the story in the Baltimore *Sun*. This is a banner headline indicating that the Democratic Science Advisory Council has tagged the president as being bewildered. The article apparently attacks the management of the space administration and the priority being given to Project Mercury.[24] A look at the roster of the committee members indicates that Dr. Frank Goddard, Jet Propulsion Laboratory liaison man here at headquarters, is a member. The familiar "viper in the house." I called him to determine whether or not he was a member of the committee, and he readily admitted as much. I thanked him and hung up, but he called me back a few minutes later to ask if he could explain himself. I said I thought he could, after lunch. I then called Pickering, his boss on the West Coast, and found that Pickering had expected that Goddard would have resigned long before this. How strange are the ways of men when politics are involved!

Lunch with Kistiakowsky and a discussion that resulted in our agreeing that the president should be advised not to answer this particular charge but to make a strong speech at a later date. Back to the office for a discussion with Goddard— one of the most unhappy ones I have had. He is a brilliant but devious young man who looks you straight in the eye with a powerful gaze. Apparently, he expects thus to convince you of his sincerity. He attempted to engage me in a philosophical discussion about the propriety of his working for an agency of the government while serving as an advisory committee member in the Democratic Science Advisory Council. I told him he would have to make up his own mind about matters such as this—that I thought professional ethics entered into this, and I reminded him of the tenets of academic freedom, which I thought applied similarly in this instance. He was becoming more confused and flustered by the minute and finally suggested that maybe the thing for him to do was to resign. I told him I thought this would be exactly the wrong thing to do because it would be construed as resignation under duress. Finally, I told him that the best thing he could do is to go back to his desk and work diligently at doing the things he was being paid to do. He thanked me and apologized for not calling me before I called him, but then finally said that he had left the initiation of the call to me thinking that this was my right. I immediately put him straight on that by telling him that whenever he felt that he was in a compromised position or he had been in error in some action, he'd better make the call immediately. What an ordeal!

At 4 o'clock, several of us got together to discuss the proposed U.N. conference on space activities. Whereas the Russians first proposed this conference, we now find it difficult to get them to sign up and say what they want to do.

---

[24] The Democratic Science Advisory Council was a 17-person committee of scientists who formed an adjunct to the Democratic Advisory Council. It had prepared a lengthy report critical of the Eisenhower administration's space program. One section of the report, allegedly not intended for publication, characterized Eisenhower as "bewildered by the problems of space technology and the threat Russia has posed by its sputniks." The council members urged that the Mercury program be deemphasized in favor of communications and weather satellites. ("Space Survey is Admitted," Baltimore *Sun*, 25 Jan. 1960, reprinted in NASA *Current News*, 25 Jan. 1960, p. 2.)

Nevertheless, we cannot wait longer, and I authorized the expenditure of up to $2 million to underwrite our participation in the conference. At 5 o'clock, we had a briefing from the CIA, and at 5:15, George Feldman, formerly counsel and now consultant to the House Space Committee, came to see me at my request. I was really fishing for information and was not very successful in my quest. I did tell him that I thought it was nonsensical to go into the matter of executive privilege as was now being threatened—that Congress couldn't win and that I proposed to take a very firm stand even though I could be held in contempt. This was just a thrust in the dark. He seemed to take the matter seriously and said he would stay over tomorrow to discuss it with Congressman McCormack. On to the DuPont Plaza at 7 o'clock for dinner with Kistiakowsky and several members of the Federal Council on Science and Technology.

Tuesday, January 26: Eight thirty in the morning at the dentist's for an hour and one-half of drilling—a fine way to start the day. At 11 o'clock, I saw the president with Gerry Morgan and Romer McPhee. I wanted to discuss the executive privilege matter as it related to our contract negotiations. I was now insisting that the president know what I was going to say, because I wanted to use his name. Much to the surprise of everybody, he questioned whether we might not be stretching the doctrine a little too far in cases of this kind. He was so darn human in his discussion of this matter that I once again found myself very much lost in admiration for the high ideals he seems to maintain in this difficult office. It was a long session extending for almost an hour. It wound up with a decision that discussions with the attorney general should determine whether the policies should change. In the meantime, I'll get out of the mess tomorrow as best I can. I think this is just one more instance of the difficulties that face a president in the use of his staff. They can become overly protective all too easily—usually, of course, without intending to. It seems clear to me that the job of a cabinet officer or the administrator of an agency such as NASA includes frequent discussions with the president on policy matters.

I dashed off for a lunch with Polly at 12 o'clock and left her to do some wandering around in the downtown area while I went back to work. Another series of conferences on the classification of materials we are using for the discussion on the Hill. Later, I found that we were not going to be able to see the attorney general; he is deep in the preparations for an appearance before Congress. How familiar that sounds! A call from George Feldman to say that I should not worry about the hearings on the morrow; they will be conducted in a gentlemanly fashion. There would be no thought of contempt. How literal can you get when you want to be devious?

Back home at about 6:30 to find Ruth, Fred [Watts, later Polly's husband] and Polly there. Since Ruth is in the room with me, I will not say how much it means to have her back in the community. She might become a little too conceited.

Wednesday, January 27: This is the day on which our bouts with congressional committees begin. Discussion on last minute strategy until we were to go to the hearing room of the House Committee on Science and Astronautics. The hearings started off with the matter of executive privilege being posed by the general counsel, Mr. Keller, of the General Accounting Office. His prepared statement certainly made it appear as though we were withholding very important information from the GAO and from the House committee. Actually, the material withheld is the advisory notes and reports from my own top staff of their evaluation of the respective merits of bidders on the big engine and Mercury capsule contracts. These were Rocketdyne and McDonnell.[25] The hearing was punctuated by much discussion from both sides of the aisle with Mr. Keller being questioned sharply by both Republicans and Democrats. It seemed that he was not as well prepared as he might have been.

Since it was known to Chairman Brooks that I was to travel to Jackson, Michigan, in the afternoon, he finally excused the witness and asked me to take the stand. It was apparent that he wanted me to testify in rebuttal to Mr. Keller. I said that I was not prepared—that I had heard of this particular hearing through the newspapers and wanted time to prepare a reply. There was a bit of an exchange with the final decision being that I could make the statement I had prepared to start off our defense of our 1961 budget. This went rather well, and the newspapers commented favorably on the whole affair.

With Congressman George Meader of Jackson, Michigan, I took off from Butler in the Lubrizol Learstar at 1:30.[26] It was a nasty day and we found we couldn't get into Detroit or Jackson and finally came down at Cleveland. A hurried reconnaissance in the Federal Aviation's office led to the suggestion from the pilot that we go to Battle Creek, Michigan, which is some 60 miles west of Jackson. It turned out to be a wise idea, and we landed at Battle Creek with an 800 ft. ceiling at 5:15. A car met us there and took us to Jackson just in time for the reception. The meeting was a good one, and I was given a check for $11,250 to take back to Washington to the Republican National Committee. I spent almost an hour with the press and radio in Jackson. It is pleasant to be with a group that is not attempting to find fault or drive a wedge between you and some other agency of the government. Off to bed about 1 o'clock hoping that we will be able to get off tomorrow morning from the Detroit airport.

Thursday, January 28: There is not much to tell about today. We drove to Detroit and found that it was highly improbable that we would be able to fly back to Washington at all during the day. After waiting two or three hours, George Meader called a Ford official and made arrangements for us to have a car to drive

---

[25] On 31 May 1960 NASA selected Rocketdyne Division of North American Aviation to develop a 200,000-pound-thrust engine using hydrogen and oxygen propellants for use in the second stage of the Saturn program. On 12 January 1959, NASA had selected McDonnell Aircraft Corp. as the source for the design, development, and construction of the Mercury capsule, which McDonnell delivered to the agency on 12 April 1960. (Emme, *Aeronautics and Astronautics . . . 1915-1960*, pp. 106, 121, 123.)

[26] "Butler" was Butler Aviation at Washington National Airport, Butler having the contract with NASA to provide maintenance of NASA aircraft. Apparently this aircraft belonged to the Lubrizol firm, however.

to Washington. It turned out that we had a driver with the car. We drove in foggy weather almost the entire way, arriving in Washington at 11:30 at night. Just a day wasted!

Friday, January 29: I canceled my dentist appointment and huddled with the staff to prepare for the continuation of the hearing on executive privilege. There has been so much interest in these hearings that the committee moved to the House Caucus Room, which will hold three or four hundred people. This was a rough day. I read my statement, which was rather strong. The members on the Democratic side started questioning me, and it was 12:15 before I got off the stand. Each asked essentially the same questions and made the same accusations. The simple fact that I wanted to withhold the memoranda of advice given me by my top staff was blown up as a device for thwarting the proper interests of the GAO and the Congress. My plea was that these were privileged communications and if my advisors were to be subject to cross-examination after the fact by Congress, I would soon have little objectivity on the part of the staff. This is an argument that has gone on with Congress since George Washington's day, and I doubt it is going to be settled in this instance.

John Kusik [vice president of the C&O Railroad] and I had lunch at the Mayflower, and I returned immediately. I was on the stand from 2 o'clock through 4:15, and I was very weary before the day was over. Toward the end of the day, a Democratic congressman began to threaten me with dire consequences of this effort on my part to thwart the purposes of the committee. I simply said that I was sorry that they couldn't understand my point of view. I must say that Congressman Fulton, normally a maverick, did a fine job in supporting my position.

Returning to the office after a really tough day of testimony, I moved immediately to the Bureau of the Budget to discuss with Staats our request for support of the Saturn. We had requested $125 million, and the budget examiners were trying to hold us to $100 million. I finally suggested a compromise at $113 million, which would make our total budget request for 1961, $915 million.[27] This seemed a reasonably satisfactory compromise, but we were held off for a decision until Saturday morning. Off to the apartment and a little work before taking Fred and Polly and Ruth to dinner at Blackie's House of Beef. It was a pleasant dinner and we returned immediately to the apartment where I worked until about 10:30 before starting to watch the fights on television.

Saturday, January 30: To the office at 10 o'clock to discuss the hearings set for Monday morning in defense of our $23 million FY 1960 supplemental. This gets to be quite an operation, but I think we are in fair shape for this hearing. Lunch with John Corson of McKinsey and Company to discuss my request to propose a mechanism for studying our organization and our contracting procedures. Back to the office for an hour's dictation and then home for a bit of relaxation for the rest of the day. Correction—I forgot to report a call from Staats this morning to say that

---

[27] The actual budget for FY 1961 was $964 million. (*Aeronautics and Space Report of the President, FY 1991 Activities*, p. 180.)

the $113 million figure would be satisfactory. We are now set up for a meeting at the White House with General Persons at 8:45 on Monday morning and expect to see the president later in the day.

Sunday, January 31: This has been a day of worrying about the next several appearances before Congress, but it has also provided an opportunity for some contemplation and review of the situation facing the nation in carrying on its urgent and important business. We have now spent three days before the House Committee on Science and Astronautics. Only one of those days has been given over to the real business at hand. The balance of the time has been taken up with an argument over "executive privilege". We have at least three more days of testimony before this committee, which is presumably appraising our progress and will then deal with our request for authorization to obligate as much as $915 million in new funds in FY 1961. We then have the same sort of an appearance to make before the Senate committee. Appearances will be required before both of these committees on the proposed new legislation and another series of hearings on the von Braun transfer from ABMA to NASA. We will then have hearings before the House and Senate Appropriations Committees.

With respect to "executive privilege," it seems that this has been a dispute of long standing between the executive and legislative branches of government. Actually, George Washington first raised the issue when, I believe, he refused to give to Congress information it demanded. He stood on the right of the executive to declare "privileged" certain documents or conversations that if made public might endanger the welfare and security of the nation. For a good many years past, the Defense Department and certain other agencies of government have withheld certain confidential memoranda of an advisory nature prepared by subordinates for the head of an agency that dealt with contract matters. This is exactly what I have done in this instance, after advice by our own general counsel and by the general counsel of the White House. The General Accounting Office demands these documents as being necessary to its review of our contracting practices. Actually, they have had all of the documents on which we base our decisions except this particular document. In place of it, I have given them a statement on each contract indicating the reasons for accepting one contractor and rejecting the others.

It is very hard to have a rational approach to a matter of this kind. Politics and personal interests enter the picture very importantly. For instance, it is my considered opinion that Rep. John McCormack, majority leader, has a personal interest in determining why Avco was not given the contract in the Project Mercury competition. It so happens that the Avco proposal was one of the least desirable of the entire group. Politics being what they are, it would be very difficult to give this sort of information to Mr. McCormack and have him believe it.

The large companies that spend as much as $200,000 in making a proposal in a competition of this kind are understandably concerned over the loss of a contract. Some of them may even stoop to putting pressure on their congressmen to determine the cause for their elimination. This is not unusual; it is done every day.

John McCormack has put very heavy pressure on me to deal liberally with a particular friend of his—[William] Willner—whom we hired, unfortunately, at his urging a year ago. I think every call that I have had from McCormack's office has had to do with Willner and the manner in which we were treating him. Willner is a contract negotiator and McCormack wants to see him the deputy head of our contract negotiating group. When we proposed to move Wilner from headquarters, to a field station, McCormack called me from Boston saying that he would view such a move with grave concern. I suppose one calls this "good, clean politics"!

In the debate that lasted throughout the day on Friday, there were many moments of good humor. As a matter of fact, it would have been very funny if it had not been quite so serious. Rep. Fulton of Pittsburgh did a very good job in debating with the Democrats and stated our case as clearly and succinctly as it could be. Rep. Bass attempted to stop the nonsensical proceedings but was overruled by the chairman.[28] At several points in the debate, it was noted that the administrator didn't appear to be about to change his mind but they must make the record on this matter. Freshmen congressmen were very strong in their statements about the dire consequences of my failure to give in. Finally, at the end of a long and weary day, Congressman McCormack suggested that I speak personally to Joseph Campbell, comptroller general of the United States. Knowing Joe, this is apt to be a waste of time. However, any port in a storm. I believe thoroughly in the necessity for the conduct of public business in the complete view of the representatives of the people and of the people themselves. There does occur an instance, now and then, when effective administration is seriously jeopardized by the release of information such as what I have been attempting to withhold. With the pressures of the type just described being ever present, it seems reasonable to me to believe that some of my subordinates might well lose some of their objectivity and willingness to speak their minds frankly and forcefully if there were the certainty of congressional inquiry as to the advice they had given to their superior. Thus I find myself in sympathy with the democratic process but cautious about extending it to the point where it tends to destroy the initiative, competence and independence of the people on whom we have to rely for getting the job done.

---

[28] Probably Perkins Bass (R-NH), elected from the second district of that state since 1954, but possibly Ross Bass (D-TN), elected to every Congress since 1942. (*Congressional Directory*, 87th Congress, 1st Session [Apr. 1961], pp. 91, 154.)

CHAPTER THREE

FEBRUARY 1960

Monday, February 1: The day started off at 8:45 with a meeting at the White House in which Dryden, Horner and I participated with General Persons, General Goodpaster and Staats. We came to an immediate agreement on the $113 million as an added amount for our budget for FY 1961. This amount is to be spent on the super booster program. In order to make the most of this before von Braun testifies on the morrow, I suggested that we get a press release out immediately. The president being at Palm Springs about to take off for Denver, General Persons called him at 6:15 California time and secured his approval to the amount requested and made arrangements for Jim Hagerty to release the information at Denver. This was done.

Hurrying to the Capitol, we appeared before the House subcommittee on deficiency appropriations chaired by Albert Thomas. A 2-hour hearing on our $23 million supplemental appropriation for FY 1960 went fairly well. Thomas questioned us very closely on personnel and listened to the rest of the presentation. I doubt we will have very much trouble although he may try to do something with the personnel ceilings.

Back to the office and a hurried lunch at the White House mess with Dryden. Then a meeting with Wernher von Braun and Dr. Ernst Stuhlinger of the Huntsville Center. General Ostrander, von Braun's superior, was in attendance. I hope we answered their questions about additional money for the support of research in a satisfactory manner. It is truly amazing how easily one can throw millions and millions of dollars around in this business.

A visit with Julian Bartolini of the "American Forum of the Air" resulted in my agreement to go on that medium with one of the members of Congress on 11 March.[1] A little later, Mr. Strong of the CIA briefed Dryden and myself on some of the recent Russian shots. The rest of the day was taken up with visits to Congressmen Tiger Teague of Texas and Bernard Sisk of California. Both had given me a going over in the "executive privilege" discussions and I wanted to check out with them the seriousness of their views in this matter. Both had been most helpful to me in the past. Teague stated that I was wrong in continuing to withhold the statements of opinion by my subordinates. On the other hand, Sisk, who had started the whole darned thing, stated that he was a little sorry he had started it. He

---

[1] A search of standard reference sources turned up no reference to Julian Bartolini, but for the radio and TV program, "American Forum of the Air," see the entry in the biographical appendix under Theodore Granik.

pledged his undying support for me and the program but hoped that we could find a way around the "executive privilege" problem. Such is life!

A black tie dinner at the Norwegian Embassy with Ambassador Koht and his wife at eight o'clock in honor of Senator and Mrs. Fulbright turned out to be a very pleasant affair. Walter Lippmann and his wife were there as were a few others I had known. The only other congressman was Senator Karl Mundt and his wife. It turned out to be a very pleasant evening; I would like to know the ambassador better.

Tuesday, February 2: The day started with an hour and a half in the dentist's chair. The two inlays are now prepared and the bridge is about to be made. Otherwise, this was one of the better days. I saw Senator Sparkman and Congressman Jim Quigley—the former about his attitude on the transfer of ABMA, which I found to be favorable, and the latter about the executive privilege matter on which we had an interesting conversation. Quigley proposes that I offer to give these two reports to the committee without admitting that they should have them and without admitting that they will get any more of them. The idea is simply to clear the record for the present. An interesting idea; I wonder who put it in his mind?

I had a lunch brought into the office for John Johnson, Dryden, Horner, Siepert and myself. We discussed the desirability of two studies, of our contracting procedures in an attempt to develop a really useful improvement in the government's contracting procedures with industry and of our organizational structure. All concerned contributed well to the discussion and we agreed to go ahead with the studies. Later in the day, I talked with John Corson of McKinsey and Company to get these projects underway.

Wednesday, February 3: This was another of those days. Dryden, Gleason, Horner and I gathered together at 8:15 to discuss the order of battle for the hearings, now being held by the Senate Preparedness Committee combined with the Senate Committee on Aeronautics and Science. The upshot was that Glennan would "be on the pan" for both the Senate Preparedness Committee and for the Senate subcommittee on space when we present our 1961 budget request. Gleason suggested an executive session for briefing the appropriations committees of both the House and the Senate on the Kisti Report on U.S.-U.S.S.R. standings.[2] Believe me, the long shadow of the Pope is cast in each of these sessions.

At 10:00, a large group gathered in my office to discuss the executive privilege matter. I was to see the comptroller general, Joe Campbell, at 11:00. The discussion makes one wonder how long a government of this sort can keep its head above water. I have followed the recommendations of my advisors throughout in this case until, indeed, 10 days ago when they wanted to throw in the sponge. If we were going to do that we should have started that way—not carried out a gambit that must ultimately lose for us. I have stated that I am going to fight this out or give in completely—not adopt any halfway measures.

---

[2] This refers to a report by Dr. George Kistiakowsky, then Eisenhower's science advisor.

The visit with Joe Campbell and his general counsel, Mr. Keller, was indecisive. Campbell wants to think the matter over—he seemed to be cognizant of the problems that face anybody in an organizational sense but, of course, his own problems of serving the Congress in an auditing sense must come first. I showed him all of the documents, which he, through his organization, had stated he must have in order to assure Congress that our contracting efforts were sound and reasonable. He showed some understanding of our problem regarding possible injury to companies if frank comments were allowed to be published without good reason. We discussed methods of evasion of the GAO regulations or audits and all agreed that these could be devised. None of this appeals to me however. At the end of an hour, Joe asked for a further discussion of the matter.

I had a lunch engagement with Senator Clinton Anderson of New Mexico. A mixup found me waiting in the lunchroom for an hour while he waited in his office for an hour. We managed to get together for a few minutes and I was able to tell him a little bit about our proposed changes in the statute. On the other hand, and before I could get my oar in, he had told me about his concern over Project Rover.[3] He asked that I get together with John McCone to see if we couldn't get the money back into the project.

I then made a beeline for the old House office building to see Representative Ben Jensen of Iowa. I got there just in time to flag down his administrative assistant and return to the Capitol where I had 10 minutes with Jensen. My purpose was to try and get him to understand why we needed additional personnel to manage this larger responsibility we have taken on in the last several months. As the author of the Jensen rider, which is intended to prevent undue enlargement of the federal payroll, he saw this as "carrying coals to Newcastle." However, because I had taken the trouble to come up and see him, he promised to look with favor upon our request. Inevitably, he asked for my consideration in providing a job for the son of one of his constituents who wanted to take a law degree at George Washington. It seems almost to be a quid pro quo.

Back to the White House for a meeting with the attorney general, representatives of the Department of Defense, the president's staff, the Budget Bureau and NASA on this matter of executive privilege. I was the only nonlawyer among a group of 11 people. Everybody applauded my stalwart stand and said "keep it up." This sounds a lot easier than it really is. However, there is a real principle involved and I think I will play out the string. When I returned to the office I found that Joe Campbell had called me back and was attempting to find a way to forestall further unpleasantry. He had suggested that his organization did not want to keep our papers but simply to see them. Maybe there is a way out of this thing yet.

---

[3] Project Rover was a nuclear rocket program. Senator Anderson had a special interest in it, doubtless in part because of the involvement of Los Alamos Laboratory from his state in its development. On his involvement, see James A. Dewar, "Project Rover: A Preliminary History of the U.S. Nuclear Rocket Program—1906-1963," 26 Jan. 1971 (MS. in propulsion subsection of NASA Historical Reference Collection), esp. pp. 96-98 (to be published by Smithsonian Institution Press in an expanded version currently under revision).

Back up to the Capitol again to see Senator Clifford Case and Senator Cannon. I waited for half an hour to see Senator Case, who is really a fine person. He is a Republican who must run for office this fall again. He is most sympathetic to our problems and has always been willing to help. He is not really a leader, however. Senator Cannon was still in an executive session of the Senate Committee on Astronomics and Science, so I did not get a chance to see him. I left about 6:15 with the understanding that I would get back to him another day.

Thursday, February 4: The staff meeting this morning went well. There were many items to be discussed and the participation was exceptionally good. At 10:00 we started in on a discussion of the planning of spacecraft or payloads for the very large boosters we are undertaking to develop. The addition of $113 million to our FY 1961 budget request means that we must take a look at the requirements for additional monies for FY 1962 and later so we will have adequate payloads ready. The lead time on these payloads is anywhere from 2.5 to 3.5 years. This means that money must be available a long time in advance.

We stayed in for lunch with Dryden, Horner, Johnson and Frutkin. We were discussing proposed disarmament maneuvers in which the space business might well be involved. Again, it was a useful discussion. At 2:00, Dr. Kety, Dr. Morison, Dryden, Horner and Phillips sat with me to discuss the recommendations of our bioscience [advisory] committee on the selection of a director for the proposed bioscience division. It was a frank, dispassionate discussion and I think a very helpful one. We decided to move forward immediately with the letting of certain research grants and move as fast as possible in the search for the director of the proposed division.[4]

Friday, February 5: This was not one of the better days. I didn't seem to be able to get up much steam. A 10:30 meeting with Joe Campbell, comptroller general and his general counsel, Keller, found John Johnson and myself repeating for an hour the absurdity of the report the GAO had rendered to the House Committee on Science and Astronomics. Unfortunately, the report has been in the hands of the House committee for a long time and it is very hard for the GAO to retract or to say that it really didn't mean what it said. In the meantime, NASA suffers a blot on its escutcheon that it really doesn't deserve. The matter was left for Campbell to worry about and I don't know what he will do about it.

Tom Morrow of Chrysler Corporation came in to say how much his company wanted to win the competition for the S-4 stage for Saturn.[5] He is a very smooth article. He did report that the ABMA team in Huntsville seems much more

---

[4] On the Kety commission report, see Mae Mills Link, *Space Medicine in Project Mercury* (Washington, DC: NASA SP-4003, 1965), pp. 34-38. As Link reports on p. 38, "In line with the Kety committee recommendations, an Office of Life Sciences was established on March 1, 1960, with Clark T. Randt, M.D., a member of the Bioscience Advisory Committee, as Director." See also the entry under Randt in the Appendix.

[5] In fact, on 26 April 1960 NASA awarded a contract for development of the Saturn I second stage (S-4) to Douglas Aircraft Co. Its powerplant consisted of six Pratt & Whitney engines propelled by liquid hydrogen and liquid oxygen. See Ezell, *NASA Historical Data Book*, vol. II, pp. 56-57.

relaxed than it has been during the past seven years when he has had knowledge of them. If this is true, it is good news, indeed.

Saturday, February 6: The entire day was spent in attempting to prepare a statement for the hearings before the Senate committee. I have not been notified of the time of my interrogation but I suspect it must be within the next few days.

Sunday, February 7: Again, the entire day at home working on the statement. It has finally turned out to be 22 pages, handwritten and covering the gamut of our activities and plans. I hope the staff feels it a good statement; I find I lose my critical powers as I keep working on these papers.

Monday, February 8: The staff session at 8:30 had nothing unusual to offer. Peter Chew came in at 11:00 to talk over the statement I wrote over the weekend. I had read the statement to Dryden and Horner and had received general approbation on the tone and content. Chew will put it into final shape and check out the technical and statistical details with the proper people. More discussion of the ABMA situation by General Ostrander resulted in the report that they really needed more manpower in Huntsville. It is quite a task to stick to one's guns on matters of this kind. I don't believe that the von Braun group has ever had to face up to top-quality management. I am going to stick to my guns until I am proven wrong.

Up to the Hill at 4:30 to see Congressman Karth. He could not say enough in praise of my activities and NASA's program. He regretted exceedingly the inquisition on Friday last but said he guessed these things had to happen. No question about his support. Over to the office of Congressman Daddario, another Democrat. This was a more profound discussion. Daddario feels that NASA should have the entire responsibility for both military and civilian space. I hope I convinced him that we had all we could handle at the present time and that it was unlikely the military would agree to such a drastic change. I pointed out that one moved slowly in these matters if one wanted to move with assurance. It wound up with a protestation on the part of the congressman indicating his approval of our activities.

Tuesday, February 9: This was the morning! Dr. Brushart finally installed the new bridge. It is really a work of art, a true engineering masterpiece. I must say that I was nervous and fidgety before the two hours had passed. Today marks a high-water mark in the matter of supporting the congressmen in their home districts. I recorded three 15-minute television interviews—one with Congressman Westland of Washington state, a second with Congressman Osmers of New Jersey and the third with Congressman Belcher of Oklahoma. The same old questions deserved and got the same old answers.

Lunch with Senator Stephen Young of Ohio was a rather strange experience. Senator Young is a dedicated and restless man of 69 years. He seems ill at ease and a bit flighty to me. It is plain that he knows his own mind although I think he has difficulty expressing himself sensibly at times. He expressed great interest in our program—he is on the Senate committee—and I feel rather certain that we can count on him for further support. This might be called the day of "politicking" by the administrator.

Back to the office for a discussion with Frank Phillips. I am asking him to come into the office with me as my assistant. He is an able fellow and I am sure will do an even better job than did Hjornevik.

A meeting at 3:00 on the nuclear rocket—Rover—program was attended by Dryden, Horner, Ostrander, Finger and Rosen. The problem is one of making certain we know where we stand with respect to the Atomic Energy Commission. The AEC is developing the reactor while NASA is responsible for the engine, pump and other flight articles. We decided we must meet with the Commission to clarify our position before we appear before the Joint Congressional Committee on Atomic Energy on Monday next.

Wednesday, February 10: At 9:00, Dr. Clark Randt came in for an hour's discussion of his interest in heading the proposed NASA program in biosciences. Randt is a rather attractive and very positive young man who seems to know what he wants and how to go about getting it. I am of the opinion that he would make a good head for this activity, and it is clear that he wants to have the opportunity. He has had a year of exposure to the space business and, if he wants to stay with it as he says he does, I am going to recommend that he get the job.

Lunch with Admiral Rawson Bennett, which is always a pleasant affair. We discussed the report on the biosciences—I guess I should call it the life sciences. Rawson advised care in breaking the news to the service laboratories concerned with biomedical work. The rest of the afternoon was spent in the usual discussions and wound up at 4:00 with a visit to the AEC for a discussion with Commissioner John Floberg and the general manager, General Luedecke. We came to an agreement that we could indicate our support of a technologically sound program that matched up fairly well with the one originally proposed to the Bureau of the Budget. The Bureau of the Budget had cut the request and the AEC is now attempting to have the cut restored. This promises to be an interesting exercise, for Senator Clinton Anderson is calling the shots!

Thursday, February 11: A pleasant breakfast with John [Hrones, who had arrived the previous night on business for Case] and Ruth and much useful discussion. The staff meeting brought nothing exceptionally new and no real problems. I had a discussion with Dr. Robert Morison who is another possible candidate for the directorship of the division of life sciences. He is a rather quiet man—somewhat different from the picture given to me by Dr. Kety when he described Morison to me. We spent almost an hour together. I doubt that he has the leadership and drive to do the job. It may be that I was conditioned against him by the discussions with Kety and his committee, but Morison does not appear to me to have the quality required for this job.

A lunch with Roger Jones, chairman of the Civil Service Commission. We discussed all sorts of matters relating to the activities of the Civil Service Commission and some of the nonsensical routines and requirements that guide and guard the employees of the federal government. The highlight of the day was a visit extending over a period of an hour and a half with John McCormack of Massachusetts. John was a particularly sharp questioner in the hearings last Friday. We got off to a good start by discussing our wives and families. I described some of the activities of my children, all of which pleased him very much. We got into a discussion of the law changes and clarified a few points. We moved over to a discussion of executive privilege. I told him that I had been in sessions with Joe Campbell, the comptroller general, and had come to no satisfactory conclusion. I did say that we had worked out a method I felt sure would avoid the issue in the future. McCormack, much to my surprise, said that he would have done exactly as I had. I must not, said he, worry about the sharpness of the questioning. It reminded me of the announcer on "Meet the Press," who says "the questions do not necessarily reveal the interests or convictions of the questioner—it is just their way of getting a story." I wound up the session by suggesting that this was a hell of a time to give me this advice. In any event, before I left McCormack's office I was able to get him to call Congressman Albert Thomas in an attempt to speed up consideration by the full Appropriations Committee of our $23 million, FY 1960 supplemental bill. What will come of this, I have no idea. It was an interesting and instructive afternoon.

A call from General Persons conveyed to me the concern of the White House staff over the letter I had sent to the Bureau of the Budget with a copy to the Atomic Energy Commission. This letter simply said that we believed that the Rover project should be prosecuted at a technologically sound level and that we felt the program originally submitted by the AEC to the Bureau of the Budget was such a program. We very carefully avoided saying that the Bureau of the Budget should supply the extra funds needed. It is not our business to do so and, besides, the AEC has plenty of money—it should just learn how to reprogram as we have to do. I hope to get out of this argument with a whole skin, but it is interesting that we are arrayed against the AEC, the Bureau of the Budget, the White House, and Senator Anderson at one time or another. It will be a real feat if we manage to escape some kind of injury.

Friday, February 12: This day started with an interesting discussion on NASA's activities in the atomic field. We have a test reactor—a rather powerful one—at the Plum Brook Ordnance site near Sandusky, Ohio. It was built by the NACA some years ago for research and development work in connection with the Aircraft Nuclear Propulsion Project. A good many millions of dollars have gone into this facility, and it is a good one but it is not clear that there is much point in pursuing further the aircraft nuclear propulsion research—enough of that is going on elsewhere. In any event, Dryden, Horner, Golovin, Abbott, Conlon and one or two others had debated for an hour the nature and extent of the work that should be undertaken by NASA in support of nuclear propulsion for rockets. In this situation, it is clear that we can come in conflict with the AEC, which has tended to have a monopoly on research work in the reactor field. We did not settle the question but have set up a meeting on 24 February for a full dress debate with the Atomic Energy Commission. My contribution this morning was to suggest that we form a joint steering committee to determine how to divide the work in accordance with the capabilities of each.

Lunch at 12:30 with about twenty top correspondents and columnists. I discussed principally the differences between the adequacy of our missiles when used as missiles and their effectiveness when used as part of a launch vehicle system. There is the ever-present conviction on the part of a great many people that the Russians' great ability in hitting the moon, taking a picture of the back side of the moon, etc, is a measure of the effectiveness and accuracy of their missiles when used to attack a target thousands of miles away with a nuclear warhead. I think this is a reasonable extrapolation of potential effectiveness. The reverse is not true as many people would like to think it is. In other words, the fact that we have not been able to hit the moon or do some of the very spectacular stunts with our space launch vehicles is not to be taken as an indication that our missiles, when used as missiles carrying nuclear warheads, are not equally effective as the Russians' and able to do the damage we have estimated needs to be done in order to win in any serious altercation. It took me quite a while to get this point across. Actually, just before we broke up, one member of the group came back with a question that indicated he obviously had not understood. I went through it once again and was gratified to find that they were happy at having the review.

In addition, I discussed some aspects of our long-range program, the proposed changes in the law and the reasons for those changes, the transfer of von Braun and his team and the progress to be made on Saturn. As one might expect, the questioning, which was very sharp and at times difficult to handle, ran to the comparisons to be made between the Russian program and progress and ours. Obviously, 90 percent of the top correspondents and columnists in this town—and maybe the country—are solidly in the camp of spending more money, entering upon crash programs, and beating the Russians in space. They are not willing to recognize the fact that new technologies take time, which cannot be shortened merely by the application of money. I pointed out that our acceleration of Saturn by reason of the $100 million added to the budget for FY 1961 would be by about one year in four and a half. I said that the expenditure of an additional $1 billion as of the present time would probably not gain an additional four months although it obviously would provide additional assurance that the device would work at the appointed time for final test. This would come about because of the ability to make many more tests and to plan alternative methods of accomplishing tasks within certain of the sub-systems in the device.

Back to the office at 2:30 and busy for the balance of the afternoon—until 6:00 with Dryden, Horner and Frank Phillips. We discussed a variety of matters. Among these were the guidelines to be sent to our field stations in order that they may prepare their FY 1962 budgets.

Saturday, February 13: A day at home, most of it spent working on a variety of papers for both Case and NASA. We wound up the day with a visit for dinner with Senator Hartke of Indiana. He is a freshman Senator and had as guests at the dinner Senators Tom Dodd of Connecticut, Howard Cannon of Nevada, and Ted Moss of Utah. Apparently they are all freshman senators. It was not a highly spirited or enlightening evening. I managed to divert the attention for one hour for a viewing of the NBC television show—"A Missile Mess." It was a good show.

Back home at midnight in the first snowfall of the season. It looks like six or eight inches, and that means trouble in Washington.

Sunday, February 14: Valentine's Day. Ruth gave me half dozen very nice linen handkerchiefs, which I very much need, and I managed to find that she wanted the Victor recordings of the play "J. B."[6] I was able to get these and put them under her bed for Valentine's morning, which pleased her very much indeed. As a gag, I had purchased a snow scraper for the car and as luck would have it, we needed to use it last night so I gave it to her as a Valentine's present last night. This further confused her when she found the records this morning.

The snow was coming down this morning still and a call to the weather bureau indicated that it was highly probable that the New York weather would not be particularly good in the evening and that I might have difficulty in landing, although there was no question that I would land. Under the circumstances, I decided to take the train.[7] I talked with Elmer Lindseth today asking him to serve as a member of the committee that is going to oversee the making of a thoroughgoing study of our management at NASA. In his usual thorough manner, he wants to have a full background story on it and I am going to see that he has it. I'm off to New York. It is now Wednesday night and I can relate that the trip to New York was pleasant. Arriving early allowed me to get to bed early, although I did manage to read a mystery thriller before turning off my light.

Monday, February 15: Up at 7 o'clock and a haircut when the barbershop opened at 8. My first appointment was with Charlie Stauffacher, vice president of Continental Can and an old hand in Washington, although a relatively young man. I had asked him to be a member of the organizational study group we are presently putting together. Charlie agreed to serve, although he is going to be very much tied up in an antitrust suit the government has brought against his company.

I took a flight to Boston at 1:15 and was met there by the director of the development office of Worcester Polytechnic Institute. He drove me to Worcester where I arrived at 4:00 sharp. A cocktail party was scheduled for 4:30 so I had only a few minutes to get my bearings, change my shirt and shave. It was a pleasant group of Worcester businessmen and we managed some interesting conversation, although I, as usual, got into a bit of an argument with a dedicated Republican. We managed to get away from the cocktail party in time to change into dinner jackets and repair to the hotel. A very fine group of men—500 to 600—was assembled for dinner and they paid me the great honor of listening very attentively through a

---

[6] *J.B.* was Archibald MacLeish's verse drama based on the biblical story of Job. It opened on Broadway 11 December 1958 and did not close until 24 October 1959 after 363 performances.

[7] The purpose of the trip to New York appears to have been to speak with Charles Stauffacher, executive vice president of the Continental Can Co., about serving on the Kimpton committee, on which see note 31 of Chapter I. Stauffacher did agree to serve, as did five other business executives including Elmer Lindseth, mentioned below in the diary. He was president of the Cleveland Electric Illuminating Co. (Rosholt, *Administrative History*, p. 161n. See also the Glennan subsection of the NASA Historical Reference Collection; they contain correspondence with Stauffacher and others about the Kimpton committee.)

discourse of over 45 minutes. I think it was one of the better speeches I have made—at least the audience made me think so. Back to the reception room of the hotel for a drink and conversation for three quarters of an hour with the guests at the head table and then on to the home of [Arthur] Bronwell [Worcester's president] for a brief discussion of his desire that NASA support the research of some of his faculty members. Thus it is that one finds a type of pay-off expected in almost every walk of life.

Tuesday, February 16: Up at 5 o'clock for a very pleasant breakfast with Mr. and Mrs. Bronwell. Then, one of the guests at the dinner last night picked me up and took me in a Continental—no less—to the Boston Airport where I was able to get an 8 o'clock plane to Washington. This was not a particularly productive day but it was marked by one significant event—I went home for lunch!

Frank Stanton, president of CBS, called on me at 2 o'clock to counsel with me about my interest in having a television series produced and televised as a means of informing the people of the nation about the reasons for our being in the space business. Stanton is a really fine person and his advice is worth having. He agreed that we had moved as far as we could without professional advice and suggested that we now invite the presidents of the three large television networks to indicate their interest or lack thereof.

At 4 o'clock, we had a review of the Scout rocket launching system program, which was altogether discouraging. It is obvious that there is not sufficient attention being paid to the project by its sponsor, Langley Research Center. As a result, dates are slipping and the job is not getting done. Corrective action must be taken without delay and I have asked Dick Horner to get on the job.

Wednesday, February 17: At 9 o'clock, Admiral Monroe, commander of the Pacific Missile Range, called on me. He simply wanted to offer again the services of his command for whatever purposes NASA might have in mind. The balance of the morning was spent in preparing for the Stennis hearings, which are to take place tomorrow. At 12:45, off to the Metropolitan Club for lunch with Donald Power, president of General Telephone. Robert Fleming of the Riggs National Bank was our host.[8] Ruth was at lunch with the wife of Mr. Power at some other place in the city. At 2:00, the *National Geographic* came in to take a photograph of Dr. Dryden and myself and at 3:30 we moved into a discussion of the proposal for a study of our organization.

Thursday, February 18: This is the big day. If all goes well—and the weather this morning looks pretty fair—all of the children should be here by 9 o'clock tonight and we will start on our celebration of Ruth's fiftieth birthday. This is also the day on which the Naval Research Advisory Committee meets, so off to the Naval Research Laboratory at 8:45 to spend the rest of the morning listening

---

[8] Robert V. Fleming (1890-1967), who began with Riggs National Bank in Washington in 1907. He became its president in 1925 and its president and chairman of the board from 1935-1955. In the latter year he became its chairman of the board and chief executive officer.

to some very interesting developments. Lunch at the Sheraton-Park in honor of one of our men, Max Faget, who had been named one of the ten outstanding young men of the year in federal government service. Unfortunately, I had to leave the luncheon before the presentations were made because of the impending Senate hearing. When I looked for my car, who was in it but my good wife, who had decided that she wanted to go up on the Hill to hear the Senate committee's interrogation of her good or not-so-good husband.

A long afternoon—my testimony didn't take very long but I had to sit through testimony by General Medaris and General Schomburg. Some day, a general is going to resign from the service and not write a book and keep his mouth shut. Medaris is not in that category. Congressman Stratton of New York spoke against the House joint resolution calling for the early approval of the transfer of von Braun's group to NASA and Senator Sparkman spoke for it. At the end of three and a half hours, it appeared that we had won our point and I took Ruth back to the office—by this time it was raining cats and dogs—where she took a bus back home. Ostensibly, I was going to the dinner for the NRAC. Actually, . . . [describes Ruth's pre-birthday celebration, which their daughter Sally missed because her flight from Cleveland was diverted for bad weather. He finally picked her up at 3:00 a.m.].

Friday, February 19: I had planned to take Friday off but was unable to keep the faith. I went in at 1:30, somewhat the worse for wear. A visit with Stans and Staats of the Bureau of the Budget brought them up to date on our plans for the organization and contracting studies. They were sympathetic and helpful. I picked up the presents Polly and Tommy had sent in and repaired to the apartment. The Richmonds arrived at 6:30 and the fun began. The children brought out a big laundry bag full of fifty odd presents, singing "Happy Birthday." Each of them had written verses to accompany the presents and it was really quite good fun.

Saturday, February 20: Ruth and I made breakfast as quietly as possible so as not to disturb Polly and Sally who were sleeping on the daybed in the living room. I had an engagement at the office, which I had to keep. We were discussing budgetary matters and a variety of other problems that Dryden, Horner and I seemed to find it impossible to discuss in depth during the week.

Sunday, February 21: This is the day of departure [for Ruth's 10-day trip to the West Coast]. Ruth and I arose about 8:30 to prepare a corn pancakes and bacon breakfast for the kids. They managed to appear about 10:00 in various states of clothing, although I think all of them had washed their faces. It was decided that we should have dinner by 3:00 because of the necessity for preparing for the departure of the children, one by one.[9] In any event, Frank and I had a few good games of backgammon and we managed to get Polly to the station for her train. The place was a little like a morgue as we prepared for the departure of the children. It started to

---

[9] Glennan's children were Thomas (called Tom), who was unable to attend the birthday celebration, Catherine (Kitty, married to Frank R. Borchert, Jr.), Pauline (Polly), and Sarah (Sally). See his letters to Polly and Kitty in his correspondence file for Feb. 1960, as well as the deleted portions of the diary in the T. Keith Glennan subsection, NASA Historical Reference Collection.

snow and there was doubt about the flying weather. However, we did take Kitty and Frank to the airport at 10:00 and got them aboard their plane. Later advice tells us that they sat there for an hour and a half while one engine was being repaired and then had to be taken to a hanger for the removal of the accumulation of snow. They arrived home at 3:00 on Monday morning and Kitty had to be at work at 7:30. Such is life.

Monday, February 22: Washington's Birthday! It turned out to be a day of work for me along with many of the top staff at the office. Sally was to take a 4:45 plane to Cleveland but had hoped to have an opportunity to visit the Senate during the course of the day. Some shopping had to be done and in one of those peculiar quirks of timing, I missed them throughout the day. I went up to the Senate to try to find them but was unable to do so. We finally got together at 4:00 and put Sally on the plane. These partings, although seeming to be a routine affair, always provide for me a bit of a tug at the heartstrings. Sally's brave smiles and obvious regret at leaving are compelling.

Ruth and I decided we would have a drink and so went to the Admiral's Club where we spent three quarters of an hour leisurely reading the magazines before returning home. The house seemed a bit empty, although there was plenty left to do. Thus ends the saga of Ruth's fiftieth birthday and one of the very best weekends any of us has experienced. I should say that Ruth's birthday actually falls on February 26, but Sally had an extra holiday [from school] that made it best to get together on the 18th.

Tuesday, February 23: Another bout with the dentist at 8:00. This time the torture was not so bad. He cleaned my teeth and prepared a gum line cavity for a small inlay. It appears that this will end the dental program for the present. I hope it does because it begins to look to me as though a bill of $400 is in the offing. I had a 10:00 date with the Senate Appropriations Committee to defend our request for $23 million in supplemental 1960 funds. Only three members of the committee were present, although Senator Stennis came in for a few minutes and made a strong statement in support of our bill. Senator Hayden, one of the oldest men in the Senate and the chairman of the committee, conducts these hearings with dispatch. There seemed to be no trouble on the money, although we were able to place some statements in the record that ought to support the senators when they present the bill on the floor.

There was some considerable discussion about the request we have been making for 150 additional positions in our headquarters office. It may be remembered that Thomas, chairman of the subcommittee on appropriations in the House, had reduced that 150 to 75. His statement was that we just had too many good men in the headquarters—they should be in the field and, in any event, we didn't need them. He was particularly vitriolic about lawyers and public information men. Senator Dworshak questioned me on this matter and then said, "If you can convince the Senate so readily on this matter, why haven't you been able to convince the House?" Anticipating just such a question, I had determined to say that it was impossible to convince any man who didn't want to be convinced. Instead, wisdom got the better of valor and I said simply, "That fact has been troubling me a great deal these past few days." I doubt that the conference committee will change Rep. Thomas' actions.

This afternoon, I have spent some time with Walter Bonney discussing the problems I believe are present in our public information organization. It is composed largely of men trained in newspaper offices, whose normal interests are handling the news. Actually, we do a good bit of radio, television, and motion picture work, and it is very essential that long-range planning be had in each of these fields. This, Walter seems to have difficulty doing. Further, there seems to be very little in the way of organizational management on his part. He is sort of a big teddy bear who, quite sincerely, substitutes flattery of the boss for hard, driving direction of his own staff. This has got to come to an end and I am slowly bringing him to that realization. He is paid as well as anybody else in government for this activity and stands very well in the aviation and space communities. It is the old question of a person who ingratiates himself by a certain sort of subservience but who has less than the desirable best "on the ball" when the chips are down.

A last-minute meeting with Siepert on matters relating to the management studies we are proposing to undertake and on the office space situation. All of this was followed by a quiet evening at home with me toddling off to bed at an early hour.

Wednesday, February 24, 1960: Our early morning meeting resulted in a determination to go ahead with the appointment of Clark Randt, using the last of our available $21,000 spaces for this appointment. Immediately, we are behind the eight ball because we must have one of these for von Braun. Perhaps we can move the Bureau of the Budget and satisfy Congress by requesting additional spaces. Discussion of this is going on at the present time.

At 11 o'clock, Phil Farley and John McSweeney of the State Department came in to talk with me about the proposals we were drafting for possible cooperation on a space project with the USSR. We believe that such a project makes a lot of sense and are working up a proposal relating to the meteorological satellite field. We found that Farley and McSweeney were favorably inclined as I had expected. Apparently, Secretary Herter is working up the president's agenda for the proposed trip to Russia, and we laid out a program such that this matter could be brought to his attention as soon as he returned from the South American trip.[10] In the meantime, we will provide Farley with draft copies so that staff work may be attempted in the State Department.

---

[10] Eisenhower visited South America in late February and early March 1960. (*Public Papers of . . . Dwight D. Eisenhower, 1960-1961* [Washington, D.C.: GPO, 1961], pp. 202, 282-283.) His proposed visit to Russia in June was cancelled by Khrushchev at the aborted summit meeting in Paris in May (see entry under Khrushchev in Appendix and the sources cited there). Consequently, nothing appears to have come of the proposed cooperation in the area of meteorological satellites, although later in the Kennedy administration there was an agreement with the USSR in this area. (See Edward C. and Linda N. Ezell, *The Partnership: A History of the Apollo-Soyuz Test Project* [Washington, DC: NASA SP-4209, 1978], pp. 46, 47, 56, 128-129 and Arnold W. Frutkin, *International Cooperation in Space* [Englewood Cliffs, NJ: Prentice-Hall, 1965], pp. 4, 89.)

Dryden, Horner and two or three others met with me at one o'clock to discuss again our posture with the Atomic Energy Commission in matters relating to cooperative work between our two agencies. We are to meet with the AEC at 2:30. As we sat down, Chairman McCone called to ask if I could come by a little bit early. Our problem here arises out of the fact that the Bureau of the Budget reduced the AEC's budget for Rover, a nuclear rocket propulsion system, by some $12.7 million. Senator Clinton Anderson of New Mexico, chairman of the Joint Atomic Energy Committee, is determined that not only will this action be reversed, but additional sums will be provided. The AEC certainly has the ability to reprogram this amount of money within an operating budget of $2.1 billion and a construction budget of $224 million. We do this frequently throughout the year and I think it is highly probably that the Bureau of the Budget is more right than wrong in this particular situation. However, one does not stick his neck out on matters of this kind and we have avoided being called to task for approving or disapproving the budgets of other agencies.

I did meet with John McCone and found that his concern was principally about the rate at which we wanted to have them fund the Rover program. I made a point of saying to John that I thought they could reprogram their monies as we had and stop all the argument. If he really believed that Rover should go ahead as fast as Anderson wants it to, this sort of action could be forced within his own organization.

We went into the commission meeting expecting to visit with four or five of the commissioners and top staff. Instead of this, we found a full commission meeting with three commissioners and several members of the staff present. Obviously, the commission had not been apprised of the work that had been going on at the staff level between our two agencies, and it was a pretty sorry meeting. We did manage to salvage out of it some important elements. For instance, I think it became plain to the commissioners for the first time that it wasn't enough to go ahead with Project Rover—someone had to decide in the very near future whether or not Rover could ever be used, and if used, under what circumstances. The commission people want to use it as a first-stage rocket vehicle. Just where one would launch such a beast with its ever-present possibility of a catastrophic explosion resulting in the spreading of radioactive materials over the landscape is not clear. The same is true to a lesser extent in connection with the use of isotopic generating units for the long-lifed satellites we expect to fly. We badly need this sort of equipment but there's no use in spending millions of dollars—and I mean millions—to develop the units if we are not going to be permitted to use them.

We had a real hassle over the Plum Brook reactor. We are trying to gain an agreement with the AEC relative to the use of this reactor for research purposes relating to aeronautics and space flight. Unfortunately, the Atomic Energy Act gives the AEC complete control over the use of reactors for research. They maintain this through their licensing powers. I say unfortunately—perhaps it is a good thing that they have these powers in an area where costs are so very great. In any event,

John McCone accused us of simply putting together another nuclear research facility. I offered to give him the reactor—facetiously, of course. We tried to point out the conditions under which the reactor came into being—unfortunately again, this was proposed and accepted at a time when the aircraft nuclear propulsion work was at white heat.[11] The matter was left that McCone and the commission were going to re-examine the entire problem and that we would get together on it at a later date.

Dryden and I met with the public information officers for an hour. This was a mock press conference and went rather well. Expressions of approval at the fact that the two top men in the agency would take an hour to visit with these people and were so frank in their comments made the whole thing seem worthwhile.

We returned quickly to the office to meet with three of the Japanese scientists and one of the Japanese embassy people to sign a memorandum of agreement on possible space research cooperation in the future.[12] They were very obviously looking for funds but our people have become quite adept at pointing out the fact that we do not propose to underwrite foreign research in this field. We are quite happy to be cooperative and to provide substantial support in the way of launching facilities and rockets but we believe that the foreign countries must provide the support for their own technical work.

Thursday, February 25: The staff meeting went rather well with a number of interesting matters under discussion. I don't recall them at the moment but I can say it was one of the better meetings. I had planned to lunch with General Persons this noon. A few minutes before noon he called to say that the vice president wanted Persons to lunch with him. A suggestion was made that I join them if my items for discussion would fit in with such a threesome. This was a ten-strike since I wanted to talk with Jerry about the problems of continuity in our top management. Dick Horner proposes to leave the agency in midsummer and both Dryden and myself, being presidential appointees, must resign at the end of the current administration. Unfortunately, the vice president had in tow Leonard Hall, the chairman of the Republican national committee. I bowed out at this point and arranged to have lunch on Monday with Jerry Persons.

I had set up a date with Secretary of Defense, Tom Gates, his deputy Jim Douglas and Dr. York for 4:00. We wanted to talk about the Kety report on the establishment of an office of life sciences at NASA and on two or three other matters. Most of the work in bio-astronautics has been carried out in service laboratories over the past many years. Their interest has arisen, of course, from the

---

[11] On Plum Brook, see Dawson, *Engines and Innovation*, pp. 155, 156, 184-185, 201, 206, 207.

[12] This was the last of several such meetings with the Japanese scientists, laying the groundwork for future cooperation in space, which has since become quite extensive. On the meetings, see A.W. Frutkin, Memorandum to Files, March 7, 1960, in Glennan subsection, J-Official-Misc., NASA Historical Reference Collection.

fact that pilots are subjected to ever-increasing stresses as planes are designed to fly higher and faster. Some fundamental work has been done, but not very much. In addition, the service laboratories have not been well supported nor have the programs been integrated. We are anxious to avoid the appearance of stepping off in a new field with the intention of taking over and absorbing some of the service laboratories.

Clark Randt had been notified of his appointment, which he had accepted with pleasure. We were just in time—he had been appointed a full professor at Western Reserve University two weeks earlier and offered a pleasant association in research there that he most certainly would have accepted had this offer of ours not been confirmed in a timely fashion. We found a pleasant reception at the Pentagon with only a few comments relating to the necessity for making the best possible use of the service laboratories. We assured the group that this was our intention—we really wanted to have their advice as to the best method of reassuring the working-level people concerning our intentions. This was arranged in a satisfactory manner.

We followed this with a discussion of the plight of Rocketdyne, which is a division of North American and has been responsible for the largest percentage of the development of rocket engines in this country. Presently, the Atlas engine having been completed, the division faces the necessity for reducing its staff to one-quarter of its maximum size. Its principal effort at the present time is on our F-1 engine.[13] York had heard of this and is attempting to find a means of supporting a sizeable staff there just to keep this team together. He made the point that a force perhaps half the size of its maximum would be sufficient to retain the skills. This is one of the interesting problems that face a governmental agency charged with maintaining a state of readiness and a reservoir of strength in a variety of fields that may be extremely important to the national welfare. The easy answer is to subsidize everyone; the tough answer is to find a rational solution for a problem of this kind.

We also talked of the Cisler report. Neil McElroy had asked Walker Cisler to look into the problems of scheduling and managing our national rocket launching complexes and the associated tracking stations throughout the world. This is a very costly business and we have had the usual pulling and hauling between the armed services. NASA has its own rather small and economically-operated launching facility at Wallops Island, Virginia, which is available to anyone and apparently represents very little in the way of a problem. Walker, a devoted public servant and a determined man, had made up his mind pretty well before he started the study that

---

[13] In the 1950s Rocketdyne had done a feasibility study for the Air Force on a single engine with a thrust of 4.45 million newtons (1 million pounds). Ultimately, the Saturn V—the largest of the launch vehicles used in the Apollo program—employed five of Rocketdyne's F-1 engines clustered together as a first stage, with a combined 33.4 million newtons (7.5 million pounds) of thrust. (Bilstein, *Stages to Saturn*, pp. 26, 58-59, 192-193, 198-199; Brooks, Grimwood, & Swenson, *Chariots for Apollo*, p. 47.) The second and third stages of the Saturn V, incidentally, also employed Rocketdyne powerplants, in this case five and one each (respectively) of the J-2, which used liquid hydrogen and liquid oxygen as propellants. (*NASA Historical Data Book*, Vol. III, p. 27.)

the real precedent for a solution lay in the War Production Board.[14]   He proposes to set up an office reporting directly to the secretary of defense that will control and manage—he avoids using some of these words—the country's launch and tracking facilities. There is much to recommend his solution but it fails completely to take into account the problems of the users of these ranges. Principally, these are the research and development agencies, and it seems clear that the centralized control and scheduling should reside in the R&D director's office. In NASA, we are happy to work with the Defense Department on this matter so long as it will provide a single point of contact, and we believe that this should be in Dr. York's office. Tom Gates is saddled with a situation not of his own making and doesn't quite know how to get off the hook. Cisler has gone to the vice president and seems to be talking to a great many people, thus making a sensible solution impossible. All of us pressed for the solution of assigning it to York, and I hope this will be done in the very near future.

During the course of the day, Colonel "Red" Blaik, former famed coach of the Army football team and now a member of the staff of Avco, came in to see me. Blaik is concerned that Avco is not getting a fair deal in its attempts to get business with NASA. Part of this, he says, stems from the fact that [Arthur] Kantrowitz, a former employee of NACA, is not viewed with favor by our staff. Undoubtedly, there is something to this statement, but I have looked into the matter sufficiently to make certain that a good portion of the difficulty lies with Kantrowitz himself. He is going to do things his way or he isn't going to play. We cannot accept this kind of situation, of course. Nevertheless, Avco is a good organization and should be doing work for us. I hope we can settle this one without too much difficulty.[15]

[Recounts Ruth's flight that day to Los Angeles.]

Friday, February 26: Ruth's real fiftieth birthday! I did not call her or make telegrams fly that way—I thought it might be considered an anticlimax after last week's festivities. At 9:00 we took off for Langley Research Center. Chairman Roger Jones of the Civil Service Commission, one of the persons who has been most helpful to me in Washington, was to make the address at a graduation of apprentices at our Center.[16] It was really a fine occasion with an Air Force band, flowers and

---

[14] The War Production Board, created in January 1942, was the World War II counterpart to the War Industries Board of World War I. Both entities converted peacetime manufacturing facilities to wartime production and worked to ensure that scarce resources were allocated in a rational manner.

[15] In fact, on 18 April 1960 NASA selected Avco Manufacturing and General Electric to carry out engineering and development studies on an electric rocket engine. Later Avco Corp. became one of NASA's top 100 contractors. (Emme, *Aeronautics and Astronautics . . . 1915-1960*, p. 122; *NASA Historical Data Book*, vol. I: *NASA Resources 1958-1968*, ed. Jane Van Nimmen and Leonard C. Bruno with Robert L. Rosholt [Washington, D.C.: NASA SP-4012, 1988], pp. 210, 213, 216, 219, 222, 225.)

[16] NASA developed skilled craftsmen through an apprentice training program, in which 367 employees were enrolled at about this time. After a minimum of 4 years of classwork and on-the-job training, these personnel received journeymen's certificates. (*Third Semiannual Report of the National Aeronautics and Space Administration . . . October 1, 1959, to April 1, 1960 . . .* [Washington, D.C.: GPO, 1960], p. 127.)

a full house. The apprentices—some thirty-five or forty of them—were fine looking young Americans. Those who spoke did so very well indeed. It is a thrill to take part in these things even though I find myself unable to comprehend the importance of an occasion of this sort before it actually happens.

I met at 1:30 with members of the city council of Hampton, Virginia, who are anxious to develop an educational center—somewhat in the nature of a museum but concerned more particularly with education— depicting the entire gamut of aeronautical and space activities. The NACA was started with its first laboratory at Langley, and the Project Mercury activities are carried on there today. There is much reason for this sort of thing and if we can find a way to do it sensibly, I hope we can support their efforts. Naturally, they want money. This is not possible in my opinion, but we can endorse what they are doing and probably provide exhibit materials. More about this at a later date.

We got away from Langley at 3:30 and I was at my desk at a little after 4:30. Again, Dick Horner returned from the Hill with much concern over the amount of time being consumed in these appropriation or authorization hearings. While sympathetic, I don't know what the answer is. We had a settlement of our office question during the day and Hugh had taken to the Budget Bureau our request for an additional 50 "excepted" positions and 10 additional positions paying between $19,000 and $21,000.[17] Just before I returned, word had come back from Stans that he would agree to 20 and 3 respectively. A brief meeting between Dick [Horner], Hugh [Dryden] and myself resulted in the determination to ask for 30 excepted positions as a compromise and to accept the 3 positions although we might try to get 5. A telephone call to Stans and he agreed to the 30 and 3. Thus ends another day with some progress made.

At 8:00, I went to Admiral Arleigh Burke's home for a formal dinner in honor of Admiral Boone, who is about to retire. He has been our representative at NATO on important matters and seemed an exceedingly pleasant fellow who did not want to retire. It was a pleasant evening—one that I enjoyed more than I usually do evenings with people I do not know very well.

Saturday, February 27: At the office at 8:30 for a meeting with about 15 people on the budget. Thomas has required that our books supporting the various elements of the budget be much more completely descriptive of the items we want to develop or purchase. A real job has been done during the past week and it appears that we will about double the volume for Thomas, although I doubt that we will improve the accuracy or quality very much. It is a real task to try to outguess

---

[17] The Space Act of 1958 gave the NASA administrator authority to appoint 260 scientific, engineering, and administrative personnel who were excepted from Civil Service laws governing appointment and compensation. Of these, 250 had a $19,000 ceiling and the other 10 had a $21,000 ceiling. This compared with a single rate of $17,500 for GS-18s under the Classification Act. With the transfer of what became the Marshall Spaceflight Center to NASA, as Glennan reveals below in part, the agency ultimately requested and received authority from BOB and the Civil Service Commission for only 30 additional excepted positions—3 with $21,000 ceilings—in addition to 18 high-level positions transferred from the Army. (Rosholt, *Administrative History*, pp. 56, 140-141.)

Congress on what it is going to want. Most of our people feel that Thomas is put out because we are able to answer his questions and he does not have the answers ahead of time with which to trap us. This may be a normal reaction on the part of the supplicants. We went on after two hours of budgeting to a session on the preparation of statements for the balance of the congressional hearings we now know about. There are four or five yet to come, and I hope we can do a better job of planning for them than we have in the past. All of our people are really overly tired, and it is beginning to show in their barbed statements.

I stayed at the office until almost 2:30, having canceled a luncheon meeting with Bonney in favor of talking with him in the office. I have been waiting for his reaction on his organizational problems. Instead of bringing this to me, he came in to tell me that he was having a great deal of trouble with Modarelli whose feelings had been hurt when we did not ask him to do all of the exhibits for the proposed U.N. Space Symposium to be held abroad later this year. Bonney also felt that perhaps we ought to dramatize a little bit more our sense of urgency in this space business. His suggestion was that I cancel all the rest of my speeches (I would love to do this) and make something of this in the newspapers. He had one other suggestion, equally absurd, which I have already forgotten.

I then turned the discussion to his problems and found that very little progress was being made. I pointed to the need for a long-range planning in motion pictures, television, etc. He finally reacted a bit more strongly than he has by saying that other people made his job difficult. I pointed out that if this were the case he either had to fight back or come to me with the problems. I don't know just where we will get off in this situation, but I told him that I was going to speak with Herb Rosen to try and improve the morale of that gentleman.

Dick [Horner] and some of his staff came in to discuss the organization of the NASA office at Cape Canaveral. It is really going to be quite an operation and will be centered under von Braun.[18] We had had a representative there in the person of Mel Gough, who has very limited capabilities but was chosen at a time when we felt that only two or three people were going to be involved in the operation. It was a bad choice, and I think the only answer today is to remove him from the scene in as thoughtful and considerate a manner as possible. These are never easy decisions, but an entire organization should not be made to suffer when one man is found to be out of place. Rather, the one man should be thoughtfully cared for and the concerns of the total organization given adequate attention.

---

[18] This organization became the John F. Kennedy Space Center, so-named by President Johnson a week after President Kennedy's death in 1963. It had been the Missile Firing Laboratory of the Army, where that branch of the service had launched its Redstone intermediate-range ballistic missiles. In 1960, it became part of the Marshall Space Flight Center. In 1962, it became a separate NASA installation under the name, Launch Operations Center. Expanding to adjacent Merritt Island, the center was responsible for overall NASA launch activities at the Eastern Test Range, Western Test Range, and its own facilities. The Apollo launch vehicles and the Space Shuttles have been launched from there. (Helen T. Wells, Susan H. Whiteley, and Carrie E. Karegeannes, *Origins of NASA Names* [Washington, DC: NASA SP-4402, 1976], pp. 149-150.)

Back home about 4 o'clock. I spent a good portion of the next two hours shining my shoes and doing a bit of thinking. At 6:20, I went to the roof and witnessed the successful launching of another 100-foot balloon.[19] This is part of our communication satellite program. This seemed to be a completely satisfactory launch and was clearly visible in the sky to the southeast. I don't know what height was reached, but I would guess that it was a 250 mile shot that had gone down range about 450 miles before it returned to the surface of the ocean. At 7 o'clock, I went to dinner with Commander George Hoover (retired), who now lives in Los Angeles and is very much concerned over the proper display of information for our astronauts. Accompanying him was the top Navy bioastronautics doctor, Captain Phoebus. I was very blunt in my comments relating to the lack of enthusiasm I had for persons who, at this late date, cast doubts as to the reliability of the Mercury system. I pointed out that we had had the advice of medical people and all of the services, and the bioastronauts themselves had agreed on the configuration to be used in their capsule.

Hoover backed away from this sort of thing and discussed more particularly the future. I agreed to have a meeting with him and others in late March to give him a chance to explain what was on his mind. Obviously, this was a gambit to get a contract. This is a natural process but it needs to be watched carefully since we can easily find ourselves buying peace from persons of this sort who otherwise might make disparaging remarks, in a completely biased manner, about programs over which they have no control or responsibility. This statement should not be construed as one resenting constructive criticism; rather, it is my strong reaction against those who would cast stones so they may gain.

Sunday, February 28: A day of work at home—outlines for speeches, outlines for congressional statements, catching up on all of the work that had to be pushed aside over the past several days.

Monday, February 29: The staff meeting was rather ordinary. We discussed, too thoroughly, the initiation of a "length of service" recognition program. After we had spent half an hour on the discussion, I discovered that our people had gone ahead and bought the pins anyway so most of the discussion was academic. This happens all too often. At 10:30, Paul Martin of the Gannett newspapers came in and took 40 minutes asking the same old questions we usually get. These include most prominently the familiar, "How far are we behind the Russians and when are we going to catch up?"

At 11 o'clock, Don Harvey and one of his associates from the Civil Service Commission came in to give me some assistance on the search for a man to replace Dick Horner as associate administrator. Dick wants to retire or resign in mid-summer, and it is going to be very difficult to get a man to fill his job in the first place,

---

19 This was the third suborbital test of the 100-foot-diameter inflatable sphere later known as Echo. Launched to an altitude of 225 miles from Wallops Station, Virginia, the sphere reflected radio transmissions from Holmdel, New Jersey, to Round Hill, Massachusetts. (Emme, *Aeronautics and Astronautics . . . 1915-1960*, p. 120.)

and particularly so since there will be a change in administration at the turn of the year. The Civil Service roster has turned up a good many names, but most of them are not really suitable for this kind of a job. I doubt that the federal service trains many people who can step in and take over a billion dollar job involving principally research and development with the operations program that goes along with such an enterprise. At ll:30, the life sciences people came in. Dryden, Horner, Randt, and I discussed some of the matters that seemed to be bothering Clark. He wants to get going without delay. Horner, in situations like this, does not seem to grab the ball and run with it. He operates in a very much lower key than I do, and I oftentimes have trouble recognizing whether or not he has the ball.

Lunch with General Persons was put off because of his having developed an infection in his hand that has required him to be hospitalized. I used the time to advantage since I had to go to the District of Columbia building to make my peace with the automobile license people. They had given me a warning because my Ohio tags seem to them to be old. Actually, they do not expire until 31 March. I hope I can avoid buying District tags, and I think my position as a presidential appointee may help me in this regard.

At 2:15, a Project Mercury briefing on the reliability program gave me some needed reassurance. It is quite apparent that the project people have been thoughtful and diligent in making as certain as one can that all elements of the system are being subjected to reliability tests. At 4 o'clock, the dentist—at last we are through. Back to the office for an emergency meeting on testimony required on the Hill tomorrow. This has to do with a number of excepted positions that we need, and it is terribly important that we avoid having the committee tell us that they must all go to von Braun and the Huntsville group. My, how they like to get into the management phases of our business!

A little late, I went to Walter Reed Hospital to talk with General Persons. My concern with him was to lay on a talk with the vice president about the necessity for continuity of management in NASA as this administration draws to a close. I recognize that Nixon may not be the next president but, at the present writing, he is as good a choice as any. Also, I asked Jerry to call a meeting of the Defense Department people, John McCone and myself to set our sights properly on the kind of testimony to be given on the NASA legislative proposals calling for the abolition of the Space Council, the CMLC, etc. Jerry was most happy to agree and I think this is on the rails again. Back home for a brief few moments before going downtown for dinner. I must return early and get myself prepared for the trip to California tomorrow afternoon.

CHAPTER FOUR

MARCH 1960

Tuesday, March 1: Up at 6:15 to get the last preparations out of the way for the trip west. Packing had been done last night but the laundry had to be made ready, the icebox cleared out, the plants watered, etc.

Things seemed to go well at the office this morning. I accomplished more than I usually accomplish in a full day. In the first place, I had organized my work rather well. Secondly, I knew I had to complete it and I went from one thing to another, getting full cooperation from all concerned at the office. Decision-making seemed to be easy and I must say that I wish all days could go as well. I managed to get in touch with Lyndon Johnson's office, Styles Bridges' office, Senator Stennis, Congressman McCormack and one or two others to tell them about the impending appointment of Clark Randt as director of our new office of life sciences. The response in each instance was good. Once again, I was amazed at the apparent spirit of goodwill expressed by these men and their obvious pleasure at the fact that I would take the trouble to call them to tell them about an action of this sort. Of such little things is success made. A visit to George Kistiakowsky covered several items of business that will be heard from later in this chronicle. A quick lunch and I was off to the airport to catch the jet to Los Angeles and San Francisco.[1]

A first-class compartment of the jet was only half-filled. We took off on time and landed about 10 minutes late at each of the West Coast cities. I managed to write about a third of a speech but found the air so bumpy for a fair portion of the flight that I couldn't write legibly. This was at 31,000 feet. Ruth, Tom and Martha met me at San Francisco.[2] I had carried out a rum pie, which seemed to survive the trip in a satisfactory manner. We had a most pleasant evening together, having dinner at the Rickey's Motel. Ruth stayed the night with me while the kids went back to their apartment.

Wednesday, March 2: Up at 7:00 and over to Ames at 8:00. It is always a great relief to get out of Washington and visit with the people who are doing the work in these great laboratories. They have plenty of problems but their spirit is good and they can see the results of their efforts. I had lunch with 16 of the division

---

[1] As appears from what follows in the diary, Dr. Glennan visited Ames Research Center and then attended a semiannual conference of senior NASA managers at Monterey. From there, he went on to Vandenberg Air Force Base, returned to Monterey, and then stopped off in Cleveland on the way back to Washington, D.C.

[2] Tom and Martha were the Glennans' son and daughter-in-law.

supervisors and spent an hour and a half in an excellent give-and-take discussion. We started off for Monterey at 2:30 and arrived at the Mark Thomas Inn at 5:00. We had stopped at the San Martin Winery to taste the product and to talk with the proprietor. It was an interesting experience but somewhat disillusioning. Manufacturing processes often are.

The Mark Thomas Inn is a delightful place and we had a scrumptious buffet supper. I was delighted to see Admiral Yeomans with whom I had many contacts during World War II. Buddy, as he is called, seems the same fine person and I am hoping that Ruth will have a chance to meet him. Arrangements were made during the day for Tommy to go down to Vandenberg [Air Force Base] with us on Friday. This should be somewhat of a thrill for him. Dick Horner and I spent a few minutes after dinner trying to grasp the sense of two or three problems that are bothering us. What a business this is! This morning, I called Harlow Curtice, former president of General Motors, asking him to be a member of the organization study committee. He did not say no and I have some slight hope that he will join us. It would be a wonderful thing if he can find the time to work on this task.

Thursday, March 3 [Dictated from notes on March 7]: The quarters provided by the U.S. Navy Postgraduate School were excellent. The meeting started out with Horner giving a review of "the management highlights" since our last conference. Dick is not a polished speaker, although sincerity stands out in everything he says. He lacks warmth and at times seems aloof and quite unwilling to come to grips with a problem in the hopes of making an early decision. I think this results largely from his desire to hear all sides of a question before making a decision. I would not call him a natural leader. Neither did I think his selection of the management highlights as pertinent and interesting as they might have been.

Abe Silverstein did a fine job in putting together the elements of the space flight program. It is a broad one and contains enough excitement for the most avid space cadet. Abe speaks well and knows his subject thoroughly. He is an old hand in government research and development work, is stubborn and likes to keep things "fluid." This characteristic was all right in NACA days but hardly fits our needs today when we have so many relationships with industry and other elements of the government. Silverstein does accept instructions and carries them out. He is really a fine leader in our program.

I had asked Admiral Yeomans to stop by for coffee and to listen to the balance of the program scheduled for the morning session. General Ostrander, on loan to us from the Air Force, is in charge of our launch vehicle program and he summarized our activities in this area before turning the meeting over to Wernher von Braun. This was one of the most straightforward jobs I have heard von Braun do. He speaks clearly, with just the right emphasis and with an abundance of good humor. He turns a phrase every once in a while and certainly has control of his audience.

Lunch at the Copper Kettle and right back to the session. Harry Goett, director of Goddard Space Flight Center, talked about his program. Harry is lacking in humor and has some of the characteristics of an old woman. One gets the feeling that he is a bit of a crybaby, although I have high respect for his organizing ability and his objectivity. He has done many good jobs for the agency. Bill Pickering then gave the group a picture of the lunar program, which is the responsibility of the Jet Propulsion Laboratory. His was a good story and served to round out the picture of our space flight activities. It is a program full of excitement but of sober responsibility, as well.

The heads of the old NACA laboratories, Ames, Lewis and Langley, stepped aside and allowed their associate directors to participate in the panel discussion on the activities of the laboratories. They are limited largely to research and development as against the operating activities of the developmental centers. Unfortunately, most of their time was taken up with a discussion of their problems in obtaining and retaining good personnel, their salary schedules, and a variety of other matters of this kind. To be sure, they do have these problems and something should be done about them. On the other hand, the assembled group learned very little about the quality of the work being done at the research centers in this session.

After the afternoon coffee break Abbot gave a report on the research advisory committee. He is a steady, sober-sided person with great abilities. He does ride his hobbies and is often unaware of the fact that he is taking a great deal of time to talk about a trivial matter. He is an old NACA hand and knows his way about. He was followed by the director of the Flight Research Center at Edwards Air Force Base—Paul Bikle. Paul gave a report on the X-15 program, which was well received. [Robert L.] Krieger then gave a short review of the capabilities of the Wallops Station and the program being carried out there.

Dinner on Thursday evening was at the Copper Kettle and was followed by a talk given by Finley Carter, President of Stanford Research Institute. Carter is quite unimpressive and I thought his discussion dull. I pricked him a bit about the non-profit status of Stanford Research and managed to get a little life into the evening's discussion. Back to the hotel and a short meeting to consider the draft of the statement I have to make before the House Committee on Science and Astronautics next Tuesday in support of our legislative proposals. We did not quite finish.

Friday, March 4: Up at 6:30 for a seven o'clock breakfast with Horner, Dryden and several other members of the top staff. We were considering the problems relating to the manpower situation at the Huntsville Center. Von Braun wants 6,500 or 7,000 men; I think he should live with his agreed complement of 5,500 men. I expect I will lose this battle. In any event, we did not settle the complement at this breakfast session. We had advanced the hour for the morning session to eight o'clock and everybody was on hand. The program for the morning included principally a discussion of the FY 1962 program. Horner served as moderator and the division directors described their programs for 1962. Discussion was very healthy and vigorous. Most of it centered around manpower shortages. I am a bit adamant on this count, believing that we want to avoid loading up the federal payroll if it is at all possible. To the extent that we avoid taking people on, we will employ the services of industry in a sensible manner. Of course, we need a fair-sized group to spend hundreds of millions of dollars, but the easy answer to this question is not the answer to be accepted. Most of our center directors believe that they can do the job faster, better, and more cheaply than industry. In some instances, I think they are right. In the long run, it is my clear conviction that they are wrong.

A trip over to the hotel at noon disclosed the fact that Ruth, Tom and Martha had arrived and were well-established in my suite. I picked up Tom and we repaired to the Copper Cup for lunch with the rest of the groups. Immediately following lunch, we boarded our bus to go to the airport. I found that I had to take a car, which had been provided for me in spite of my distaste for this protocol. We were in the airplane of the secretary of defense—a beautifully appointed DC6B. We flew to Vandenberg Air Force Base in about 45 minutes and were greeted there by Major General Dave Wade, the commander. He was tied up with another group but had arranged for our trip, which was an interesting one. We boarded our bus immediately and went to the Titan site where we walked underground through the so-called "hardened site."[3] Here, everything is underground: the auxiliary diesel power station, the storage tanks for the fuel and the liquid oxygen, the missile itself, and all of the control rooms. Were there to be an alert, the missile is fueled in just a few minutes and is lifted on an elevator above the surface of the ground ready to be launched. In the "secure" position, all of these facilities are covered by very heavy concrete doors, designed to withstand the blast effects from an atomic explosion. Obviously, this is an expensive way to live and one would hope that better methods could be found.

We traveled, then, to the Atlas sites where we found three Atlas missiles on the pads.[4] We were told that they are the only operational ICBMs in the country, although our real operational base should be ready in July. Tom asked the colonel conducting our party whether or not the Atlases carried the nuclear warhead. The answer was "affirmative." I was a little surprised although, clearly, if the 15-minute alert means anything, the warheads would have to be in place. We were taken through the various underground elements of the support facility and were properly

---

[3] The Titan was an early, liquid-fueled, two-stage intercontinental ballistic missile. The Titan I used radio guidance and employed kerosene and liquid oxygen for propellants. It was first launched from Cape Canaveral with a dummy second stage on 6 February 1959. The first successful separation and ignition of the second stage occurred almost a year later, and on 24 October 1960, a Titan fired 6,100 miles—100 miles further than any previous missile—while equipped with a simulated tactical nose cone. Later versions of the Titan were used as boosters in the space program. (Robert L. Perry, "The Atlas, Thor, Titan, and Minuteman," *The History of Rocket Technology: Essays on Research, Development, and Utility*, Eugene M. Emme, ed. [Detroit: Wayne State University Press, 1964], pp. 142-161, esp. p. 156; Emme, *Aeronautics and Astronautics... 1915-1960*, pp. 106, 119, 129.)

[4] The Atas was the first U.S. intercontinental ballistic missile and, as such, was less sophisticated than the Titan. Unlike the Titan, it was not a true, two-stage missile. Slightly shorter than the Titan I, the Atlas D was 81 feet long to Titan I's 98 feet. The Atlas, too, burned kerosene and liquid oxygen. Its first test occurred on June 11, 1957, and on December 17, 1957 a successful test firing placed the Atlas missile in the target area after a flight of about 500 miles; the rocket became in its various models an important booster for the space program. (Perry, "Atlas, Thor . . .," Emme, *History of Rocket Technology*, pp. 142-161; Emme, *Aeronautics and Astronautics. . . 1915-1960*, p. 93.)

impressed. As an aside, we learned on Sunday following our visit to Vandenberg that one of the Atlases had blown up. This apparently occurred during a fueling operation. Now I really wonder whether or not the nuclear warhead was in place. We boarded the buses and moved over to one of the launching sites for the Thor intermediate-range ballistic missile.[5] The variety of installations is really interesting to behold. All of them cost a great deal of money.

Having finished on the Vandenberg Base, we were driven to the Pacific Missile Range, which lies directly to the south of Vandenberg. This is supposed to be a range for research and development efforts comparable to the Atlantic Missile Range at Cape Canaveral. Unfortunately, the Air Force was there first and believe me, they never lose their advantage if they can. The result is that there has been a great deal of argument between the Air Force and the Navy—an argument that has bordered on the ridiculous. Presently there seems to be some sort of a truce in effect, and I hope that the situation can be improved substantially as time goes on. In any event, we were driven through the Pacific Missile Range facilities and then listened to a briefing by Admiral Monroe and his staff. He was most complimentary in his remarks about NASA—most insincerely, in my judgment. After an hour of this sort of "guff," we boarded the buses and were driven back to the Vandenberg officers club. There I was host to a cocktail party for the Air Force and Navy, and we managed to mend all of the fences that had been broken during the day. This was followed by a pleasant dinner, during which I toasted our service associates. Admiral Monroe sprang to his feet, ahead of his Air Force counterpart, to respond. The situation reminded me of two little boys arguing over which of their fathers could lick the other. About 8 o'clock, we boarded our aircraft in the hope that we could get into the field at Monterey. We were disappointed and finally wound up about 10 o'clock at Moffett Field near Palo Alto. Buses took us back to Monterey where we arrived in the fog at 12:15 a.m. Thus ended another long day.

Saturday, March 5: Up early again to have breakfast with the directors of our research centers and the directors of the headquarters divisions. I wanted to talk with them about public statements and the necessity for avoiding criticism of the military and public discussion of our budgets. It went fairly well, although I am sure that some of those present resented being "dictated" to in a matter of this sort. The morning session was one of information and questions in which Siepert, Frutkin, and Kamm participated. It seemed a little anti-climactic as a session, although the subjects covered were important ones. These included our security program, the international program, and the method of operation being followed in Los Angeles by the people in our Western Operations Office. After a coffee break, Dick Horner and Nick Golovin talked a bit about program control. I then stood up

---

[5] The Thor missile was initially developed by the Air Force as an intermediate-range ballistic missile. Tested on a launch pad in January 1957, it completed a successful, full-range flight in September of that year and a launch from its operational training site at Vandenberg on 16 December 1958. The basic rocket used liquid oxygen and kerosene as propellants. It was 65 feet (19.8 meters) long and had a body diameter of 8 feet. In a modified form, the rocket became the first stage of launch vehicles for several spacecraft.

to give a general summary and "remarks." Actually, I did not summarize anything; I simply spoke about the problems that faced us in the future and the need for real teamwork. My remarks seemed to be well-received. In any event, I closed the conference out at 12 o'clock sharp.

Admiral Yeomans had asked Ruth, Martha, Tom and myself to lunch. He also invited Dick Horner and Hugh Dryden. Dick was off to the snows for a few days of skiing, so he could not attend. It was a pleasant luncheon and a fine way to end the conference. Certainly, Buddy Yeomans has one of the best billets in the Navy. Ruth and I took off at 3 o'clock in the Convair for Albuquerque and Cleveland. We were sorry to leave Tom and Martha but these situations do have to occur, obviously. The flight to Albuquerque was a pleasant one except for the fact that, as we approached the field, we were told that we had lost one of our generators. After the landing, the mechanics at Kirtland Air Force Base set to work immediately to repair the damage. We drove in to old Albuquerque for a Mexican dinner, which I enjoyed very much. Ruth stuck to avocado stuffed with shrimp. We returned to the base to find that a replacement generator had been located. Ruth and I went to bed in our stateroom while the rest of the group—the Lewis group—sat around the officers club.

Sunday, March 6: We were delayed two or three hours in the change of the generator and finally got off for Cleveland about 3 o'clock in the morning. Ruth and I had managed a reasonable amount of sleep. We arrived in Cleveland around 8:15 in the morning with plenty of people to greet us and a car ready for me to drive home. On the way to 2965 Fairmount, I called Kent Smith to arrange an 11 o'clock appointment and found that he was waiting. Sunday was spent with Kent Smith and John Hrones going over a good many of the problems at the college. I was particularly interested in reviewing the long-range planning about which I have been concerned. This is a matter in which the Ford Foundation seems also to be interested. Later in the afternoon a group came together to greet the Russian educators at our home. This was a group of eight men who were visiting America as a counterpart group to the college presidents who went abroad in 1958. It was a really pleasant party and I was delighted to see the two or three of them that I knew. At the conclusion of this, we drove immediately to the airport and arrived at our apartment in Washington at 10:45 p.m. Thus ended Ruth's 50th birthday celebration and the third semi-annual NASA staff conference.

Monday, March 7: This was another of those days! There was a good deal of mail to be reviewed. There were not too many problems awaiting me, however. During the morning, Odd Dahl of Norway came in to see me. He told me about the death of his wife, Vesse. He is over here representing his government in an effort to develop a cooperative program with us in upper atmospheric physics. Lunch at the White House with Bob Merriam was a dull and disappointing affair. I had hoped to talk with him sensibly about the problems of presenting a unified front to Congress on the legislative program. I could not get through his head that this was really a White House problem, not one for NASA to deal with. We did make some progress before the meal was over.

At the office in the afternoon, nothing of any interest developed. I was feeling a little bit low and so decided to go to see a doctor about my head, which has been bothering me a great deal. Dick Huffman was a little disturbed at seeing the condition of my scalp and arranged for me to see another doctor tomorrow. I repaired to the apartment a little ahead of schedule to find that Ruth had gone out to have her hair done. I cancelled our attendance at a black tie dinner tonight for the science award winners put on by Westinghouse. I hope to go to bed early.

Tuesday, March 8: This was not one of my better days. I had attempted to prepare reasonably well for the hearings before the House Committee on Science and Astronautics, which start today. Unfortunately, my indisposition as a result of concern over my scalp seemed to pervade the rest of my activities. I did get through the day but without very much credit to myself. At 9 o'clock, Dr. A. W. Lines, director of the Royal Aircraft Establishment of Great Britain, came in with three of his colleagues. They are interested in visiting many of our installations, to become acquainted with satellite launching problems and with the design problems associated with the development of satellite vehicles. They seemed exceedingly pleasant. I was sorry that I had not planned to take lunch with them.

At 9:30 Golovin came in with his prepared statement on the Huntsville manpower situation. This is one of those things—von Braun has the bit in his teeth and is asking for a substantial increase in the number of employees at the Huntsville center. Apparently, he has sold Golovin, Siepert and Ostrander on the need for these additional positions. Some time ago I had set a ceiling of 5,500, whereas today the requirement is for 6,689. Obviously, I am going to give in on this because of the manner in which von Braun can bring pressure to bear in Congress and elsewhere. A year from now, he may not have the same standing. On the other hand, there is some validity in his demands as a result of the speed-up we have ordered on the Saturn project.[6] In any event, I was dissatisfied with Golovin's statement and asked him to have a new one prepared for discussion this afternoon.

At 10 o'clock, we were on the scene in the hearing room of the House Committee on Science and Astronautics. Today was a short session due to the fact that Dryden is to receive a presidential award at noon. This is a hearing on the proposed changes in our basic legislation and should prove to be an interesting one. Unfortunately, this committee is about as poorly disciplined and poorly staffed as any on the Hill. The discussion went rather well and I doubt that we lost any ground. At 12 o'clock, at the White House, President Eisenhower presented to Dr. Dryden, and four others, the Presidential Citation for Distinguished Federal Service—the highest civilian federal service award. It was a pleasant ceremony conducted in the cabinet room with a large number in attendance, mostly families of the recipients and members of the White House staff.

---

[6] Personnel (both permanent and temporary) at the Marshall Space Flight Center climbed to 6,843 in FY 1962 and reached a high point of 7,740 in FY 1966 (figures as of the end of the FY) before declining to 6,325 in FY 1970. (*NASA Pocket Statistics: History*, Jan. 1971, p. E-9.)

After a lunch at the White House mess with Secretary of Labor Jim Mitchell, I moved over to Dr. Teichmann for an examination of my scalp. Apparently I have a skin disease of the healthy person aggravated by lack of exercise, too much animal fat, too little sleep and all of the other things in which I seem to indulge myself. Dr. Teichmann was very pleasant and accommodating and gave me prescriptions with the admonition that I was to call him back in a week. He is a man who inspires confidence.

Back at the office at 3:30 to meet Harvey Pierce of the Maurice H. Connell Associates in Florida. These people seem to have done some architect-engineering work for the ABMA in connection with the development of Project Saturn. This man must have been 6 feet, 6 inches in height, a pleasant and positive talker. He had come in to determine whether or not we proposed to change the manner of contracting for construction projects in connection with Saturn. I found myself somewhat irked at the fact that he had come in through the request of Senator Holland of Florida. I told him so; there is no reason for anyone's depending on his congressman or senator to gain an audience with us at NASA. He was somewhat obtuse and we listened patiently for 15 minutes before I could get rid of him.

At 4 o'clock, Golovin came back in to discuss the Huntsville situation with several of his associates, Dryden and myself. The use of the English language to state clearly an objective seems to have escaped some of our top scientific and management people. In any event, we came to a decision on the problem before us and sent the paper back for another rewrite. On home for a pleasant dinner, a short evening, applications to my head—and bed!

Wednesday, March 9: Up at 6:30 so that Ruth could give me a half hour scalp treatment. I hope this does not continue overlong. The usual session with Frank Phillips, my assistant, with discussion of the program for the day. At 10 o'clock, we were on the stand again before the House Science and Astronautics Committee. I carried the bulk of the testimony today and managed to get through without too much difficulty. Dryden provided just the right amount of support in his own inimitable fashion. It is clear that most of the members of the committee have little or no understanding of the problems of organization and management. They seem to feel that the elimination of the Space Council must be coupled with the establishment of some other kind of a committee that would sit between the Defense Department and ourselves and report to the president on the progress or lack of progress in the two programs. It is a little difficult to deal with a matter of this kind without seeming to be egotistical or over-positive. Dryden pointed out that space seems to be considered as an area in which there must be a czar. He suggested that we didn't have such a person determining the activities to be taken on by all agencies having to do with the oceans or with aircraft activities. I thought he made a very good point—in fact, he left one of our antagonists speechless. As I turned to leave the hearing room at noon, I found that Ruth had been there for most of the session. Unfortunately, I was unable to think beyond my nose and thus let her get away when we might have visited the House gallery together. Our bill authorizing a $915 million FY 1961 program was the first item on the agenda of the House of Representatives. I thought it was going to pass without a record vote when Congressman Williams, a one-armed person of origin unknown to me, suggested the absence of a quorum. We sat through the calling of the roll twice and there were still stragglers entering the well of the House. The final score: 398 to 10. Thus we have one more hurdle behind us.

Lunch at the Statler with Mr. H. J. Heneman of Cresap, McCormick and Paget. We were discussing the Case problem of developing a long-range plan and Kent Smith's interest in having a market survey undertaken. Cresap, McCormick and Paget did a study on faculty organization and conditions of employment for us back in 1954-55 and it is my judgment that the firm could help us again. Interest having been expressed, we walked back to the office where I got in touch with John Hrones in Kent Smith's absence and set up a meeting for the morning of 16 March in Cleveland. The ball is now in their hands.

At 1:45, Dr. Furnas, Chancellor of the University of Buffalo, came in to see me. I was interested in asking his advice on the number of organizational and program users. He is a fine person and one very much interested in helping. In the middle of the afternoon, Bob Bell of our security office came in to tell me of the search for the possible security violation that resulted in a story in the *New York Times* concerning the launching by Russia of one of her long-range missiles in the Pacific two or three weeks ago. I recall that I was very startled to see the story and I am not at all surprised at this investigation.

Golovin was back in to look at the problem of staff limitations at Huntsville. Dryden and I reworked his memorandum on policy and approved it. At 4 o'clock, Robert E. Gross, chairman of the board of Lockheed Aircraft, came in to make certain that I understood the capabilities of the Lockheed Marietta, Georgia organization, which is bidding on one of our projects. It was a pleasant discussion and I hope a useful one. I often wonder what would have happened had I accepted Lockheed's offer of employment in 1956 when I was asked to take on the vice presidency and the general managership of the Missile Systems Division at a salary approximating $100,000. At 4:30, Dr. Homer Joe Stewart came in to discuss a number of matters and we set up a meeting for later in the week. Packing up the bag I left the office at 6:30 for what promises to be a long evening of work at home.

Thursday, March 10: Up at 6:30 to have a half-hour scalp treatment by Ruth, then a good breakfast and off to the office. The staff meeting this morning was a pretty good one with a review of the planning for the Monterey conference. All seemed agreed that it was the best conference held thus far. There does seem to be a need for more frequent meetings of smaller and more specialized groups. We have decided to reduce the number of staff meetings to one each week. The staff surprised me by raising a question about this, saying that even though it took an hour or two out of the day, it seemed to be the best possible way of keeping abreast of the fast-moving program. I used the opportunity to admonish them once again that these meetings were only as good as the effort put into them. It was rewarding to have this kind of discussion.

Dr. Stewart came in at 11:00, and we discussed the planning activities for the rest of this year. He is an academic person with all of the difficulties normally possessed by such a person in getting an orderly job done. On the other hand, he is thoughtful and knowledgeable. I appreciate the vigor with which he does attack a situation. He is now concerned over the degree to which we are planning far enough ahead on our spacecraft or payload projects. Certainly, if we are to do very much in this field we are going to have to have a good deal of support from industry. One is not going to build such support very solidly if one does not provide a continuing program for each of the industrial organizations involved. This, in itself, requires a great deal of forward-looking planning.

A lunch with Tom Watson, president of IBM. Having read a speech of his last December, I sent him a copy of one that I had made in the same vein. He seemed flattered that I would have read his speech and that I would take the trouble to write to him. In his talk he had raised a question about the extent to which we seem to require publicity on our shots. I spent most of the luncheon telling him the facts of life about the press. Watson has been an outspoken advocate of higher taxes so we may have more adequate defense and greater support of activities that will go to strengthen this nation in its competition with Russia. At least, he is willing to pay for it. Very few people have been willing to couple additional expenditures with additional taxes. I think this luncheon was useful. I think I was able to clear up some of his misconceptions and to give him a feeling of confidence in our organization.

Back to the office for an afternoon of dealing with a variety of situations, most of them having to do with Congress. At 4:30, several of us went over to the White House to debate office space. The Inter-American Bank has been promised the same space as was offered to us. We were able, with Bobby Cutler, to compromise our differences and, as a matter of fact, NASA came away with 8,000 additional square feet. Finally, at 6:30, I locked the desk and headed for home. After a pleasant dinner I sat down to work at a speech and some other papers. About 10 o'clock, I felt dizzy and went to bed thoroughly exhausted. I had very little sleep during the night. I am certain that some of this exhaustion is a result of inadequate attention to exercise. What does one do for time?

Friday, March 11: I should have said that yesterday was the scheduled date for the Thor-Able IV shot. They counted down to within less than a minute of ignition and had to postpone the shot because of difficulty in the fueling mechanism. When I arrived at the office a bit late—it was 8:30—the shot had gone and was well on its way. The three stages had fired properly and the payload had separated from the third stage. There was a good deal of gloom around, however, because initial returns indicated that the probe had not reached escape velocity.[7]

---

[7] This was the launching of Pioneer V, successful as it turned out. The Pioneer V probe sent back excellent data on radiation, magnetic fields, cosmic dust distribution, and solar phenomena in interplanetary space between Earth and Venus. The Able upper stage for the Thor rocket was derived from the Vanguard launch vehicle. The IV designation referred to one of four variations of the Able, having to do with weight, thrust, and engine numbers. (*NASA Historical Data Book*, vol. II, pp. 67, 72-73.)

I was feeling very rocky after the bad night and was in no mood for foolishness. I had forgotten that at 9:15, I was to appear at Station WTTG for a 15-minute tape recording on TV with Ted Granik. He does the "American Forum of the Air," which will be played on Sunday night. I was to record this with Overton Brooks, congressman from Louisiana and chairman of the House Committee on Science and Astronautics. Fortunately, just before I left the office, a rerun of the data began to give promise of more satisfactory results.[8] Nevertheless, when I reached the studio, I was in no mood for humor. Granik indicated that controversy and argument would make the program worthwhile. I immediately said that I had all that I could take of that without doing it over the air. This was one hell of a way to start a TV program. Actually, it went rather well although I am sure that the people of the nation will not feel that they have learned very much. It is too bad that shows such as this cannot be planned a little bit ahead of time. I would have organized my thoughts much better

Thor-Able IV launch vehicle with Pioneer V satellite.

if I had known the general import of the program. During the course of the recording, I received word that the probe was really on its way.

Back to the office to work over the statement for the House subcommittee on appropriations, which we meet on Monday. In drafting my statement, I had indicated six areas of policy I wanted to set forth clearly as bits of philosophy guiding the conduct of our program. I asked staff members to enlarge upon these rather brief statements. I found that this had not been done and thus will have to

---

8 This was a reference to the launch of Pioneer V.

undertake it myself. Over to the Pentagon for lunch with Deputy Secretary of Defense James Douglas, Under Secretary of the Air Force Joe Charyk, and Director of Defense, Research and Engineering Herb York. We were rehearsing material for the appearance of York and Douglas before the House Committee on Science and Astronautics on Monday. They are called to testify on our legislative proposals. Yesterday, Bill Holaday, former chairman of the Civilian-Military Liaison Committee, had testified for an hour and a half and had opened up quite a keg of nails. Some of his assertions were sensible—others quite incomprehensible. In any event, I believe we developed a bit of thinking that may be useful as we move forward. The problem of providing coordination between the Defense Department and ourselves is a continuing one. In this area, it must not only work, it must give the appearance of working.

Back to the office at 2:30 to find that a meeting called for discussion of our budget presentation had been going on for half an hour. Another hour on this and we managed to get ourselves fairly well set for the session on Monday. I then discussed with Jim Gleason, Hugh Dryden and Frank Phillips the matters about which I had been talking with Douglas. Our people do not think we ought to suggest the kind of a mechanism for coordination on which Douglas and I had tentatively agreed. I am on the fence.

A call to Douglas to find that he had been talking to Holaday and that he felt he should go ahead to make a suggestion to the committee. I am inclined, as I said above, to agree that an aggressive position is very much better than one that is defensive in this situation. Jim is to call me tomorrow to finally lay this on. This took up the rest of the afternoon. The dinner was quite wonderful, the company and the conversation excellent—a fitting end to the day, which began with a successful Thor-Able IV (Pioneer V) shot to place a 90-pound payload into orbit about the sun. I should say that the real purpose of that shot is to test our ability to transmit electronic intelligence over a stretch of fifty million miles. Think of it!

Saturday, March 12: Pioneer V is still going strong. All circuits and channels seem to be working very well. It looks as though we have a winner! After a leisurely breakfast, I met John Corson and Jack Young of the McKinsey organization at the office. Al Hodgson was with us. We spent the better part of two hours discussing the organizational study that we plan to get under way in mid-April. Jack Young showed himself to be an incisive thinker. At least, he believes in the old adage of "when in doubt, draw sabers and charge." We did come up with a schedule and plan for operations that seems to me to make sense. This could be a most interesting study.

At 12:30, I had a call from Jim Douglas at the Pentagon. He had been working with his staff people on his statement for Monday and asked me if I would come over for lunch. The statement is an excellent one although it will not please all of my people. Jim feels that he wants to make a statement to the effect that he recognizes the need for some formal method of coordination at the "management level." I couldn't agree more with him on this, and while I understand the hesitancy on the part of our staff, I am going along with Jim. Sooner or later, this sort of thing has to be discussed and I see no reason for delay. One can outsmart oneself in these political jungles. As a matter of precaution, I read back to Hugh Dryden later in the day the statement Jim was to make, and he agreed with me that it could do no harm and probably would do some good.

I got back to the apartment at 4 o'clock and just had time to relax a little bit before taking a shower and starting out for the Gridiron Dinner. This turned out to be the 75th Annual Dinner and it was a pleasant and amusing affair. The Gridiron Club, consisting of an active membership list of fifty of Washington's best known newsmen, uses this annual dinner as a background against which to poke fun at official Washington.

Sunday, March 13: A leisurely breakfast with a feeling of satisfaction that Pioneer V is still on its way. I worked most of the morning on the speech for next Wednesday and then, at 1:15, Ruth and I drove to the home of Richard Harkness who had been my host the night before. Dick is a top-rated commentator on NBC television and has proven to be a very good friend here in Washington. He was having a buffet lunch for quite a crowd of people. We stayed for about an hour and a half and I had a chance to talk with Stu Symington and with Milo Perkins—the latter a very interesting economist and former member of the Roosevelt administration.

My discussion with Symington was anything but cordial—he accused me of taking the stump against him. We wound up betting each other $20 but, not being sure what we were betting about, the $20 bills were returned to us with no blows landed. Ruth and I enjoyed talking to Milo Perkins about some of the important problems of the day. Among these, the one that interests me most is the necessity for arousing the interest of our youth in something besides the purchase of motor cars and that quiet little home in the country.

Back to the apartment for more work on the speech and to look at some of the Sunday TV programs. Tom Lanphier, late of Convair Astronautics, was on "Meet the Press." What arrogance and stupidity! I think Senator Case of North Dakota was exactly right when he said in remarking about Lanphier's "resignation," it was about time. At 10 o'clock, Ruth and I set out for a walk in what turned out to be a brisk and beautiful evening. Back to the apartment and at the books again until midnight. I seem to have much more energy than usual today. Off to bed about 12:30—quite a day!

Monday, March 14: Having done pretty well with my homework yesterday, I was in very good shape for the office this morning. Clearing up several matters with Frank Phillips, we moved off to the Hill at 10 o'clock to open the defense of our budget before Albert Thomas and his House subcommittee on independent office appropriations. He was in his usual good form—insulting a lot of good people but doing it with a smile. I think we got through the day fairly well. I broke off at lunch time to have a session with Ernest K. Lindley of *Newsweek* magazine. Lindley is a reporter on the foreign scene, of very good reputation. I wanted to talk with him about the impact made by the Russian space achievements on the peoples of other nations. I think I made some progress in convincing him that there were lots of other areas in which we were competing with the Soviet Union. At least, he expressed himself as having gained an increased respect for the kind of planning and program development we are now doing.

Back to the Hill for more of the hearings and then to the office for an hour of concentrated attention to the work at hand. At 5:15, I went up to the Hill again and spent 15 minutes with Vice President Nixon. For the first time, he greeted me as "Keith" as I walked in. He was his usual suave, well-groomed self and exhibited the usual high degree of interest in our program. I was talking to him about the necessity for planning ahead on the replacement of the top management people in our agency. I leave at the end of the year, Dick Horner leaves in mid-summer and it isn't going to be easy to get replacements. I pointed out that I had a conviction that Dick Nixon was going to be the next president and that he ought to be interested in continuity of management in this operation. He agreed and said it seemed to him to make sense that I should attempt to find a person to take on the associate administrator's position with an idea that same person would replace me when I leave the scene next January. This was my solution to the problem and Dick bought it without difficulty. He asked that I respond to any particular questions about his interest in the program with a statement to the effect he strongly supports a vigorous, imaginative program. He even went so far as to say that he would support additional money for it if the need could be shown. That wound up the day and I went home about quarter of seven to a pleasant dinner and a lot more work. It is now 10:30 and I have just finished for the day.

Tuesday, March 15: Up early for a 7:30 breakfast with Jim Killian at the Metropolitan Club. I had asked to see him when he was next in Washington to determine his willingness to testify before the House Space Committee supporting the proposed amendments to the law. I'm certain that he will be willing. I brought him up to date on all of the activities that have been under way recently, particularly the new office of reliability we have just begun to staff. Back to the office where Hugh and I spent an hour with Dr. Ed Purcell of Harvard and Dr. Donald Hornig of Princeton. Both are members of the President's Science Advisory Committee and were visiting us at my suggestion following an expression of concern by the PSAC about our method of dealing with scientific groups in general and astronomers in particular.[9] There has been some concern expressed that we were not supporting the astronomers adequately or sufficiently generously. Actually, there are five groups of astronomers attempting to put together their ideas with respect to the right kinds of experiments to be carried aloft in a satellite. As is normally the case, each of them would like relatively unlimited funding and freedom to do as they please. We

---

[9] President Truman established the PSAC on 20 April 1951 within the Office of Defense Mobilization. The White House announced an enlarged membership on 22 November 1957 and transferred the committee to the White House effective 1 December of that year. The purpose of the PSAC was to advise the president on scientific and technical issues. (*United States Government Organization Manual 1960-61*, p. 557.)

promised to look into the matter although it is my conviction that such a complaint was inevitable. Actually, I have been somewhat concerned over the extent to which we may be keeping as in-house tasks the interesting phases of design and development of satellites to be used for scientific purposes.

We had to excuse ourselves and move up the Hill to the continuation of the hearings of the House subcommittee on appropriations. Thomas was in his usual form and seemed determined to make it appear that we really don't know how to run our business. It is too bad that the work of the Appropriations Committee and of the Budget Bureau cannot be combined in some more effective way. I excused myself at 11:30 to dash to the White House where the president was going to sign the executive order naming the Huntsville center as the George C. Marshall Space Flight Center. I was particularly interested because I had suggested this name some months ago. Mrs. Marshall, the widow of the general, was escorted to the White House by General Persons. We were delayed because Chancellor Adenauer took an overly long time to hold his discussion with the president.[10] At 12:15—thirty minutes late—General Persons, Mrs. Marshall and I entered the presidential office. The president was most cordial to Mrs. Marshall. When it was suggested that an earlier meeting of the senators was overly extended because the senators had little in the way of terminal facilities, the president agreed but said that the chancellor suffered from the same disease. The president signed the order with three pens and handed one to Mrs. Marshall, one to General Persons and one to me. He took the trouble to take Mrs. Marshall around the office and discuss the paintings hanging there; once again, he showed himself to be a really fine human being.

A quick lunch at the White House mess and back up the Hill to try to finish out the day with the House Appropriations Committee. There was nothing unusual about the afternoon's performance. I left again at 3:30 to be available for a discussion at 4:30 before the Advertising Council [a public-service organization founded in 1942 by business, media and advertising firms]. Dr. Kistiakowsky and I spent an hour answering questions that had been sent in by the people attending the meetings. This is an annual session extending over two days. Cabinet members and other prominent governmental people speak frankly and off the record in an attempt to give the audience a real and factual picture of activities taking place in several areas of public interest.

Back to the office for a discussion with Horner, Phillips and Herb Rosen about appearances by the seven astronauts. It seems clear that we have "goofed" in apparently promising that one of the astronauts would appear at a meeting of the Aviation Medicine Association. I made the decision to take the rap and withdraw what seems to have been an agreement. A call to Dr. Lederer was anything but pleasant, but I shall see him next week in an attempt to mollify him and his organization. If we let the astronauts appear at one of these meetings, we will never be able to control their movements again.

---

10 Adenauer was on an informal visit to the U.S. on the way to Japan.

Wednesday, March 16: Up early to take an 8:10 plane to New York. Everything seemed fine weather-wise and I arrived at the Yale Club about 10 o'clock. Called Carter Burgess to ask him to join up as a member of the committee on NASA organization. He promised to give the matter immediate consideration. I then moved over to the Union League Club where I had a standing-room-only audience for my luncheon discussion. This was a fine audience—they were very much interested and I enjoyed the session. Dan Hickson was in attendance and we spent an hour together discussing some of his concerns about a study he was about to start with Tim Shea. A car picked me up and took me down to the Calvin Bullock Forum where I was to speak at 4 o'clock. Hugh Bullock, apparently a stockbroker, continues a practice established by his father of having important personages speak at their convenience to an audience of about 200 professional people in the Wall Street district. He has a very important collection of memorabilia from the Napoleonic and Nelson eras. This, too, was most interesting and I felt well repaid for the effort.

Back to the airport to catch the 7:25 ship back to Washington. Since my departure, snow had appeared and we had a really rough time getting back. I had called Ruth from the airport and found that she, too, had had a rough day since it had snowed throughout her trip to Cleveland. I managed to get to bed about 11 o'clock.

Thursday, March 17: St. Patrick's Day and it was interesting to see that all of the non-Irish on the staff wore the green neckties, etc. The staff meeting was taken up principally with the discussion of an executive development program. There is a great need for this sort of activity and we plan to support about five persons on full year appointments to executive development courses and about 100 each year on short course assignments. At 10:30, the source selection board gave us a presentation on the Snap 8 competition.[11] It was interesting to see the way the earlier argument with Congress on "executive privilege" in this area was already beginning to effect the sharpness of statement made by participants in this kind of an exercise.

The president of Hoffman Electronics came in to ask about the probable market for solar cells. His company has supplied about 80 or 85 percent of the requirements of NASA to date. I sent him to Dr. Silverstein. Jack Whitney brought in one of the vice presidents of North American Aviation who wanted to assure me that North American Aviation would put its entire resources behind the development of the 200 K if they won the competition.[12] This is a seemingly normal practice

[11] SNAP (System for Nuclear Auxiliary Power)-8 was a joint NASA-AEC project to develop a nuclear-powered electrical generating system to produce 35 kilowatts of electrical power and be capable of continuous reliable operation in space flight for periods of about 10,000 hours. On 28 March 1960, NASA announced selection of Aerojet-General to build the power conversion equipment for SNAP-8 and to integrate the reactor into an operational system. NASA ultimately decided in 1970 to phase out the SNAP-8 program during FY 1971, but in 1983 NASA, the Defense Department, and the Department of Energy began a new SP-100 nuclear space reactor program. (Emme, *Aeronautics and Astronautics . . . 1915-1960*, p. 121; "SNAP-8" folder, NASA Historical Reference Collection; *Aeronautics and Space Report of the President, FY 1991 Activities*, p. 100.)

[12] The "200 K" was evidently what later became the J-2 engine used in the second stage of the Saturn IB and Saturn V rockets. On 31 May 1960 NASA selected Rocketdyne Division of North American Aviation to develop a 200,000-pound-thrust engine using hydrogen and oxygen propellants; it was later designated the J-2. (Emme, *Aeronautics and Astronautics . . . 1915-1960*, p. 123.)

although it is beyond me that any significant contribution to the selection of a contractor can be made in this way. At 3:30, Clark Randt came in to discuss some

of his problems. I put him off until the next morning. At 5 o'clock, I was picked up at the apartment and driven to Balti- more by one of the vice presidents of the Glenn L. Mar- tin Company. I guess it is called The Martin Com- pany now.[13] This is the second an- niversary of the launching of Van- guard I.[14]

It was a long evening and I suppose it must have done some- body some good. I got back to the office at midnight. We were prepar- ing for a signifi-

Pioneer V satellite.

cant event which was to take place about 2 o'clock Friday morning. Pioneer V will be about a million miles from the earth at that time. I was to ask the tracking station in Hawaii to interrogate the satellite. If everything went well, we should have information back from a million miles out in space within 10 minutes telling us something about the micrometeorite count, the cosmic ray radiation intensities and

---

13 Glenn L. Martin (1886-1955), an American airplane inventor, established the company that bore his name in Cleveland in 1917, moving it to Maryland near Baltimore in 1929. Meanwhile, in 1928, it incorporated as the Martin Company. By 1960 its major concerns were the design, development, and manufacture of missiles and electronic systems for the government and nucleonics for the AEC. In 1961 it merged with the American-Marietta Company. The new firm produced chemicals, aluminum, cement, and aerospace products such as missiles, rockets, space-launch vehicles, and electronic systems.

14 Vanguard was a National Academy of Sciences-Navy project to launch the first U.S. satellite. On 6 December 1957, however, the Vanguard test vehicle failed in its attempted test launch, and an ABMA-JPL satellite, Explorer I, became the first U.S. satellite to enter orbit. Then, on 17 March 1958 Vanguard I joined it in orbit. For details of the Vanguard story see Constance McLaughlin Green and Milton Lomask, *Vanguard: A History* (Washington, DC: NASA SP-4202, 1970).

temperatures both inside and outside of the satellite. Chairman Overton Brooks of the House Committee on Science and Astronautics was to be with me. We had quite a group of reporters and television cameramen there. Precisely on the dot at 2:00 a.m., I talked with Hawaii, and within 10 minutes we were getting back all of the information we had hoped to acquire. It was a really good exercise and I think appropriately marked a significant milestone in our business. We had hoped to have the president do this but maybe there will be a chance to work him into the exercise later on. I think his advisers have been a little bit concerned that a failure would occur and thus create a negative impression. Actually, everything went well and we were very happy to see that some six hours later, Princess Margaret indulged in exactly the same exercise at Jodrell Bank in England. This is the first real attempt to get international coverage of a significant step in our space program. I got home at 4 o'clock.

Friday, March 18: After three hours of sleep, I got to the office about an hour late. This was one of the better days—maybe one doesn't need sleep. At 10 o'clock, we had a presentation on the new program control center. The boys have done an exceptionally fine job on this development. When it is fully operational, we will have in one place a complete and up-to-date picture of all of our programs with quite adequate information to permit detailed and effective management of all phases of the program.[15]

At 11 o'clock, Horner, Dryden, Phillips, Abbott, and Randt sat with me to discuss some of Randt's problems. We made the decision to change the person directing our research contract programs office and will allow Randt to manage his own research grant funds. At 1:30 I checked over a film made by Walt Disney on weather modification. It wasn't too bad but had a little of the space cadet flavor. At 2:30 Dryden, Horner, Phillips and I met with George Low of the Mercury project. We were going over the same ground we had covered two days before on the astronaut appearances. I think everything is under control now.

At 3:30, Gleason, Horner, Golovin and Dryden sat with me to plan the presentation of our appropriation legislation before the Senate committee. At 4:30, Hodgson, Phillips and I sat down to complete work on the paper to go out to our management review committee. That pretty well ended the day. I was pleased with everything that had been done during the day but found that I was beginning to feel the effects of loss of sleep.

Saturday, March 19: I got up late and went to the office about 11 o'clock to clear the desk. I took a 2:45 plane to Philadelphia. [Describes meeting with Ruth, Sally, and friends, and an evening of choral music put on by three college choruses and the Philadelphia Orchestra.]

---

[15] This is an apparent reference to the program management system and the office of program analysis and control that was a part of it. Both were established during the spring of 1960. (See Rosholt, *Administrative History of NASA*, pp. 150-151.)

Sunday, March 20: Up for a leisurely breakfast with Polly and her friend Beth. Ruth and I started out about 11:30 and reached the apartment about a quarter of three. The rest of the day was spent in leisurely living.

Monday, March 21: At 10 o'clock, Morehead Patterson came in at my request. I was asking him to become a member of the advisory committee on organization. He stayed with me for half an hour and said he would call back in the afternoon. This, he did and with a favorable reply. Later in the day, I asked Jim Perkins if he wouldn't join in this committee work and he replied affirmatively without delay. Still later, Nathan W. Pearson of T. Mellon & Sons in Pittsburgh called to say that he would join up. Thus, in one day, the advisory committee membership has been completed. A triumph!

At 11 o'clock, Abbott Washburn and Hal Goodwin of the U.S. Information Agency, Karl Harr of the White House staff met with Bonney, Rosen, Phillips and Dryden of our staff to discuss the possibilities of gaining international goodwill through the exploitation of the present success of Pioneer V and of the possible success of shots that are coming up in the very near future. There was the usual suggestion to refer to committee—this I resisted strongly. Pointing out that our public information office was charged with the responsibility of avoiding undue publicity and of controlling information relating to our activities, I insisted that some agency such as the USIA must take the creative responsibility for developing propagandistic programs for exploiting whatever successes we might have. Finally, I put the question very bluntly to Washburn: will you take this responsibility or won't you? This brought the meeting to an end with the agreement that Washburn would present a program to us without further delay. One has to get a little angry at times in this business.

I took Dryden, Horner, Abbott, Bonney and Phillips to the White House mess for lunch. We were discussing the charges made by Clay Blair of the *Saturday Evening Post* about our handling of the X-15 program.[16] A letter came to me some three weeks ago in which Blair stated that we were jeopardizing the X-15 program by replacing Scotty Crossfield with NASA, Air Force and Navy pilots. He made several other accusations at complete variance with the facts. We were getting our ducks in a row in anticipation of his meeting with us at 1:30. Blair came in and was pretty much set back on his heels by the statements relative to the letter he had written and the articles he had published recently in the *Saturday Evening Post*.

---

16 Designed in the 1950s by the NACA, in conjunction with the Navy and the Air Force, as a test vehicle for research leading to manned space flight, the X-15 was capable of speeds up to 6.7 times the speed of sound and altitudes in excess of 350,000 feet. Between 1959 and 1968, three X-15 aircraft performed 199 test flights, providing data on hypersonic aerodynamics, engine capabilities, structures, reentry techniques, and reaction controls over attitude in space. As such, it was an essential research vehicle for subsequent Apollo and Space Shuttle missions. (X-15 file, NASA Historical Reference collection; for further information see Hallion, *On the Frontier*, esp. pp. 104-129, and M. O. Thompson, *At the Edge of Space: The X-15 Flight Program* [Washington, DC: Smithsonian Institution Press, 1992].)

There is no question that public interest in the X-15 program is high. Blair seemed to be carrying the torch for Crossfield, although he was professing great interest in the research achievements to be made with the X-15. Unless the information we gave him is very erroneous, Blair's case came apart like a straw stack in a tornado. I am sure he will not let it rest at this, but I believe he went away with a certain amount of respect for the integrity with which we are attempting to run this program. I was

The X-15 research airplane.

not overly considerate in my attempts to call to his attention the kind of responsible journalism I felt must support a program such as ours.[17]

At 3 o'clock, Newell, Dryden and Frutkin came in to discuss the proposed cooperative meteorological satellite program with the U.S.S.R. Silverstein joined in and we came quickly to a conclusion on the rewriting of the proposal. It is hoped the president may take this proposal with him on his visit to the U.S.S.R. At 4 o'clock, I had a briefing by Dick Rhode and others relating to structural studies of aircraft in flight and particularly of those involving the Lockheed Electra.[18] It is quite evident that the NASA research centers do a thorough and workmanlike job in the areas in which they have great competence. Very fortunately, my concern

---

[17] Interestingly, about this time Clay Blair was helping Crossfield write *Always Another Dawn: The Story of a Rocket Test Pilot* (New York: World Publishing, 1960).

[18] The Lockheed Electra was the only large turboprop airliner developed in the U.S.A. Delivery began in 1958, but although the aircraft was efficient and offered high performance, Lockheed never produced it in large numbers because the Boeing 707 jet airliner appeared about the same time. The Electra could not compete with it commercially. The U.S. naval version of the aircraft, the P-3 Orion, performed antisubmarine patrol duties. (Laurence K. Loftin, Jr., *Quest for Performance: The Evolution of Modern Aircraft* [Washington, DC: NASA SP-468, 1985], p. 140; *Jane's All the World's Aircraft, 1955-56* [London: Sampson Low, Marston & Co. Ltd., n.d.], pp. 275-276.)

with this having been expressed in the setting up of this meeting, we were quite ready to be called by the FAA to a meeting tomorrow morning at which Lockheed, the airlines and the engine manufacturers are to come together with General Quesada of the FAA for a meeting on the whole question of the airworthiness of the Electra aircraft.[19]

Another hour at the end of the day cleared up my desk fairly well and I was able to get away at 6 o'clock to pick up Sally and Helen Hamilton at the bus depot.[20] They were in good shape and we returned home without delay. A pleasant dinner and an evening with the books finds me now at 10:10 ready to have a game of backgammon before going to bed.

Tuesday, March 22: We started the day off with a look at the film report on experiments made in connection with Project Shotput.[21] Dr. Jaffe described how the inflation of the 100-foot balloon had been made more reliable. There was a question as to whether or not we should undertake an additional shot from Wallops Island before letting the orbital shot for Project Echo go from Canaveral. After thoughtful consideration, I decided that we should make the experimental shot and run the risk that a random failure would not occur. Most of the afternoon was taken up with the meeting of the Federal Council for Science and Technology. This may be a useful exercise but I rather doubt it. There is most certainly a need for an overall control or management of science and technology in the federal government. Surely, as one looks ahead 20 years, the very large amounts to be spent in this area will begin to overshadow almost all other expenditures by federal departments except the Defense Department. I think George Kistiakowsky is working sensibly on this problem and I hope to be able to give him a little help. Following the conclusion of the council meeting, Gleason, Dryden, Horner and I discussed the preparations for the Senate hearing to be held next week.

---

19 Electra aircraft had been involved in crashes on 3 February 1959 into East River in New York (due to pilot error) and four subsequent dates by 4 October 1960, with the last crash due to bird ingestion into the engines. In two of the crashes, the aircraft had disintegrated in midair due, according to the FAA, to a structural defect that could be corrected. Pending modifications to the aircraft, amidst controversy, Quesada allowed the Electras to continue to fly at reduced speeds. (Stuart I. Rochester, *Takeoff at Mid-Century: Federal Civil Aviation Policy in the Eisenhower Years, 1953-1961* [Washington, DC: FAA, 1976], pp. 234, 280.)

20 Sally, it will be recalled, was his daughter. Helen Hamilton was apparently a family friend.

21 Project Shotput employed Sergeant-Delta launch vehicles to test payloads for the Echo project, which successfully launched a passive communications satellite on 12 August 1960. On 28 October 1959 NASA had launched a 30-meter (100-foot) inflatable sphere into a suborbital trajectory from Wallops Island (see above in diary) as part of Project Shotput. A second suborbital shot occurred on 16 January 1960. The suborbital vehicles provided useful scientific information, and further suborbital tests followed, including a test of the Italian San Marco satellite from Wallops in 1962. (William R. Corliss, *NASA Sounding Rockets, 1958-1968: A Historical Summary* [Washington, DC: NASA SP-4401, 1971], pp. 42-43.)

NASA ECHO COMMUNICATIONS SATELLITE
LAUNCHING CONDITION

ROCKET
MOTOR

SATELLITE
CONTAINER

TELEMETER
SEPARATION MECHANISM

CONTAINER OPENING
MECHANISM

EJECTION CONDITION

SEPARATION SPRING
CONTAINER HALF

UNFOLDING
SATELLITE

CONTAINER
HALF

L-1007

Schematic drawing showing the launch and ejection sequence for the 100-foot inflatable sphere employed in Project Echo. The sphere was folded inside a container of 26 inches in diameter until injection into orbit, when the container separated from the burnt-out third stage of the launch vehicle and divided to free the satellite, which then inflated.

Wednesday, March 23: Up early and off to the Capitol Hill Club where I had breakfast with fifteen or twenty young Republican congressmen. Apparently, some eight years ago, as freshmen congressmen, these people started a club to which they invited the responsible leadership of the executive branch each day for breakfast. As I understand it, they take a new name each year. This year it is the SOS Club. Congressman Perkins Bass was my host and I was able to get about an hour's worth of solid discussion with the Republican congressmen. At 10:00, Dr. Lederer of the Aerospace Medical Association came in to protest my ruling that none of our astronauts could appear at the association's meeting in Miami. Unfortunately, there had been a mix-up on this matter and I was, in effect, withdrawing a permission previously given and on which the association had based a certain amount of publicity. I tried to keep my temper but finally was unable to stand it any longer and told Dr. Lederer that I was tired of the protestations of his organization about the quality of their professional operations. It was so obvious that they were attempting to make a sideshow out of the astronauts that there was no reason for speaking further with him about it. We parted in assumed friendliness, but I have now passed the word that, under no circumstances, will there be a breach of the instruction I thought I had given many months ago.

At 11:00, several of us went over to the motion picture projection room to view the results of efforts made by Hugh Odishaw of the National Academy and Arnold Frutkin, formerly of the Academy and now one of our own people, in preparing short motion picture subjects intended to arouse the interest of high school students in careers in science. Generally speaking, the results were quite excellent. It is obvious, however, that professionals in the field need to be brigaded with the top scientists if a really good job is to be done. I was interested in seeing this exhibition because of our own need for working in this area.

At 2:00, Arch Colwell of Thompson-Ramo-Wooldridge came in to gain information about how we were evaluating bids. Not having heard anything from the firm's bid on the Snap-8 program, he and his colleagues were worried. Unfortunately, I could not tell him that his company had lost this particular bid and that we would announce it in the next couple of days. Later in the afternoon, Dryden, Horner, Abbott, Ostrander and Finger came together to present a proposition relating to the management of the Rover program. I bought it and then tried to get John McCone to set up a meeting to discuss this matter with him. No luck.

Dinner at the Metropolitan Club with Jim Killian and the trustees of the Institute for Defense Analyses. A pleasant evening but not one that will be long remembered.

Thursday, March 24: The staff meeting this morning was an excellent one. Two papers were dealt with rather quickly and then a brisk discussion around the table brought out a good many bits of information to which all had not had access. Sometimes, I believe these staff meetings are really worthwhile. At 10:30, Admiral Rawson Bennett and several colleagues and people from industry gave us a presentation on the ANIP program.[22] Actually, I was confused when I started and I was more confused when Bennett and his people finished. This is presumably a new method of presentation of an operating situation to a pilot or an astronaut. None of this came through and I will have to follow up with Admiral Bennett.

At lunch with Dryden and Horner, I discussed at some length the problems that bothered me about the public information office. We are in agreement that exploitation in educational activities is to be removed from the PIO responsibilities but how this is to be accomplished is still a mystery.[23] At 2:00, Clarke Newlon, editor of *Missiles and Rockets Magazine*, came in to ask me to attend a dinner at his home in honor of the editorial advisory board for his magazine. This turned into a

---

[22] The acronym stood for Army-Navy instrumentation program.

[23] On 31 May 1960, NASA created an Office of Technical Information and Educational Programs, responsible for acquisition and dissemination of technical information such as scientific reports and for educational programs explaining NASA's activities. ("Shelby Thompson," biographical file, NASA Historical Reference Collection.)

discussion of the propriety of his having Wernher von Braun as a member of that board. I promised that I would look into this matter.

At 3:00, Arnold Frutkin and I visited with Secretary Herter and several members of his staff. We were discussing the desirability of undertaking a cooperative program in meteorology with the Russians. Herter is a very pleasant person who makes one feel very much at ease. It was obvious that there was an intent to be cooperative—a real interest on Herter's part in developing such a program. However, he expressed concern over the possibility of a misunderstanding of such a bilateral U.S.-U.S.S.R. operation since we are actively participating in U.N. discussions of worldwide cooperation in space activities. I left it that Secretary Herter would look into the matter further and then would talk with me about the desirability of going to the president with this proposal. I don't want to take it to the National Security Council before having had a positive reaction, one way or the other, from the president.

A little after 4:00, Hugh Dryden and I went up to the Hill to attend a reception arranged by Congressman Brooks and the House committee in honor of their advisory panel of eminent scientists. It was a pleasant affair. A little bit late— almost 6:00—I saw Jason Nassau at the Roger Smith Hotel. We had 15 minutes of conversation; then I went directly on home. I avoided a reception at the Shoreham— just too tired.

Friday, March 25: This was a red letter day for Sally, Helen Hamilton, Ruth and myself. We took off in the Convair for Huntsville at 9:00. Several members of the staff were with me. We were going to visit the community for the first time and I was to make a speech that evening. We were met at the airport by the president of the Huntsville Industrial Expansion Committee, which was to be our host. The committee had set aside the governor's suite at the hotel. [Following lunch,] Ruth and the children were taken through the Redstone Arsenal where they were given a complete tour of the Saturn project. In the meantime, I had gone with General Ostrander and Wernher von Braun to the Arsenal for a conference that lasted about two hours and wound up with a short visit with General Schomburg.

Back to the hotel about 4:45 and a coffee session with the members of the Huntsville Industrial Expansion Committee. There was a good bit of sparring but, I think, a great deal of useful exchange of views on the developmental problems of the region. This is really an outstanding community in the sense of its desire to be helpful to the industries of the area. Bedecked with orchids and gardenias, the girls were ready for the reception and dinner at 6:30. I spoke briefly to the assembled multitude—there must have been 250 of them—and my remarks seemed to be well received. They gave to Ruth a beautiful handmade quilt and to me a piece of one of the original wooden water pipes from the area, suitably inscribed. Immediately following the dinner, the girls changed clothes and we were taken to the airport in a cavalcade properly escorted by motorcycles. What a day! We arrived home at 2:30 in the morning and I managed to get off to sleep about 3 o'clock.

Saturday, March 26: The day was spent, that is from 10:30 on, with John Hrones and Kent Smith in discussions of the studies that are necessary to be made on the [Case] campus. Helen had to go on home to Cleveland at noon and Polly arrived from Swarthmore at 1:15.

Sunday, March 27: This was Sally's sixteenth birthday. She got up when I arose to go downtown for a further session with John and Kent. She opened her presents. It is really a great privilege to have these young ladies with us for a few days. Maybe I'm getting a little bit older and more mellow, maybe they are getting a little older and more responsible—in any event, I am proud and happy when they are here with us. Actually, this goes for all of the children—I wouldn't play favorites at all. I finished up with the boys at the office about noon and took them all to the airport for their return trip to Cleveland. Coming home, we had some soup and then I took the family out for a drive in the hope that Sally could have a chance to try her hand at driving. She is well coordinated and did excellently in the few minutes devoted to that part of the activities of the day. Dinner was fun with a birthday cake topping off the ceremonies. I have done very little work today—actually, I am proud of that fact.

Monday, March 28: This was another one of those Mondays. I doubt that I got very much of any real use done even though we had a full afternoon of Senate hearings. After a hurried luncheon, I repaired to the Senate Office Building to be the principal witness at the hearings on our authorization request for the fiscal year 1961 budget. Senator John Stennis of Mississippi is the chairman of this subcommittee. He is a fine gentleman. We went up a few minutes early in order to pose for pictures along with a model of Pioneer V. I was on the stand for most of the afternoon and I think I managed to do a reasonable job. A reading of the transcript later on indicated that my grammar has not improved very much but I think I got most of the thoughts across satisfactorily. On turning around during the course of the hearing, I saw Ruth, Polly and Sally in the audience. Later information indicated that Polly and Sally had not been able to get into the Library of Congress because Sally was not a college age gal. When thus frustrated they had called mother and gone shopping. Back home for dinner and a rather restless evening.

Tuesday, March 29: This was a rather full day. I was not in the best of moods when I arrived at the office at 8:30. Going up to the Hill at 9:00 for a TV recording with Congressman Bruce Alger of Texas, I was somewhat less than civil to Herb Rosen. I did manage to get through the recording satisfactorily, however. Back to the office for a variety of meetings and then over to the Hamilton Hotel for a luncheon with the Sloan Fellows.[24] This was a good hour of give-and-take—I was very frank and I hope that some of my words do not come back to bite me.

At 2:45, Jack Parsons of Ames Research Center came in to present to me a memento of the first experiment in bringing back from a satellite orbit a simulated capsule. Actually, it was a very small button of nylon-type material. The experiment had been done through use of a light gas gun and only 2-1/2 percent of

---

[24] The Alfred P. Sloan Foundation started a program for Sloan Fellows in 1955. It consisted of young faculty members at institutions of higher learning who received grants for basic research.

the specimen had been lost during the heating of reentry.[25] At 3 o'clock, another meeting with Bonney about his problems. I agreed to approve the manufacture of 12 models of the Pioneer V space probe. We talked over some of his organizational problems—I think he is slowly coming around to a realization that it just isn't going to work, that there will have to be a revision in organization and responsibility. Up the Hill at 4 o'clock again for a TV taping with Senator Keating of New York. Out to the home of Consul General Fiorio of San Marino for cocktails with [Luigi] Broglio of the Italian space committee. A crowded and noisy throng from which I escaped rather promptly. A hurried dinner with the children and off we went to the Yale Drama School production of *John Brown's Body*.[26]

Wednesday, March 30: An 8:30 meeting this morning with, Randt, Horner and Dryden to discuss the desirability of our providing support for and using the services of the Armed Forces-National Research Council Bioastronautics Committee resulted in a negative decision. We should be appointing our own advisory committees. I managed to get hold of Dr. Bronk to set up a meeting to discuss this matter with him. He is president of the National Academy of Sciences, the parent organization to these committees. A brief meeting with Bill Pickering corrected some misapprehension he had about the testimony he was supposed to give on the Hill this morning. I think Bill wants to do what is right. At 9:45, a meeting with Kistiakowsky to discuss possibilities for replacement of Horner and myself. This did not turn out very well.

At noon, Dick, Hugh, Frank and I ate a sandwich luncheon while we tried to prepare a bit for the meeting to be held later in the afternoon with Jim Douglas and the other people over at the Defense Department.[27] I think we made sufficient progress, but it is clear that we need more time to debate some of these issues. The meeting with Jim Douglas and the others at the Pentagon was clouded a bit by the news that Roy Johnson had testified on the Hill in a manner quite injurious to Herb York. Some of these people who have no responsibility ought to learn to keep their mouths shut or at least to talk in a constructive manner. The meeting with the Defense Department people went rather well, and all of us came away satisfied. I got away from the office a little bit early and came home to find that Sally had made the dinner—sauerkraut, wieners and a pudding cake that was quite delicious. More work brought the day to an end about 10:30.

Thursday, March 31: The staff meeting at 8:30 went off fairly well. We were discussing grievance procedures and then had about an hour of discussion around the table on matters of interest to all concerned. I had to break it up at 10 o'clock because of a date to speak to the Foreign Service Institute of the State Department. About 20 senior foreign service officers plus a few military types make

---

[25] Apparently, this had occurred on 27 February 1960 in a simulator. (See Emme, *Aeronautics and Astronautics . . . 1915-1960*, p. 120; on the gas gun, see Edwin P. Hartman, *Adventures in Research: A History of Ames Research Center, 1940-1965* [Washington, DC: NASA SP-4302, 1970], pp. 236-239.)

[26] A long narrative poem by Stephen Vincent Benét, dramatized by Charles Laughton in 1953.

[27] The people referenced by first name were Dick Horner, Hugh Dryden, and Frank Phillips.

up this class of some 25 people. I was going to speak for only 10 minutes and then introduce Frutkin. Instead of this, I spoke for 25 minutes and then spent another 20 minutes answering interesting questions. It's always a pleasure to do this when there are real, solid people in the audience. Back at the office to speak briefly with the chairman of our advisory committee on loads. Ed Gray of Boeing spoke frankly and well about his concern and that of his committee on this matter. All of this has to do with the kinds of problems that may well have been involved in the recent crashes of the Lockheed Electras.

I dashed off to lunch at the Washington Hotel where a meeting of the Washington section of the American Rocket Society was to be addressed by Senator Wiley of Wisconsin. He was late but gave a good speech. It didn't say too much, but the manner of delivery and the atmosphere created was worth watching.

Back to the office in a hurry to talk to Max Lehrer of the Senate Space Committee and Jim Gleason. Max wants the authorization bill language to provide for the authorization of $50 million over and above the amount we have thus far requested. I doubt that it will get us any place but it is a gesture in the right direction anyway. It will amount to a slush fund or emergency fund if it is passed.

At 2 o'clock, Senator Javits of New York with four congressmen and five labor union people headed by Victor Reuther, came in to talk about the Buffalo area as a labor distress area. Actually, they were trying to get us to pay particular attention to the proposal of the Bell Aircraft people on our S-IV Saturn stage and the 200 K engine.[28] I thought Senator Javits was very decent about the whole thing. This is one of those situations where overly much dependence upon the federal government has led to a catastrophe with the only remedy available apparently in the federal government. These are captive organizations, not private industry. This is the case with most of the aircraft industry. I am reasonably certain that we are involved in a slowly changing relationship between industry and the federal government. The entire question of large government expenditures on health, education, welfare, defense, space, atomic energy, etc., involves a reexamination of the way we do business. Perhaps Tom will have some part in the solution to this problem.[29]

Detlev Bronk came in at 2:30 to advise with us on problems we are having in attempting to deal with the Armed Forces-NRC Bioastronautics Committee. Det is a wise person full of experience in government matters. He agreed with our concept that we were now sufficiently mature to have advisory committees of our

---

[28] On the 200 K engine, see note 12 in this chapter. The S-IV Saturn stage was the second stage in the Saturn I launch vehicle used in the early Apollo launches. It consisted of four Pratt & Whitney engines using liquid oxygen and liquid hydrogen as propellants and had a total thrust of 80,000 pounds (355,800 newtons). The last 4 of 10 launches for the Saturn I were considered operational, but none carried humans. They did, however, prove the viability of a clustered-engine concept and paved the way for the later Saturn IB and Saturn V boosters. For the most succinct and comprehensible discussions, see the *NASA Historical Data Book*, vol. II, pp. 54-55, and vol. III, pp. 26-27; Brooks, Grimwood, and Swenson, *Chariots for Apollo*, p. 47; and Roger E. Bilstein, *Orders of Magnitude: A History of the NACA and NASA, 1915-1990* [Washington, DC: NASA SP-4406, 1989], pp. 78-79. For the full details, see Bilstein, *Stages to Saturn*.)

[29] This is a reference to Glennan's son.

own—that we should look to the National Academy of Sciences for ad hoc advisory committees or special purpose committees that can free-wheel without assuming undue responsibility for program matters. Starting at 3:30, Dryden, Horner, Phillips, Silverstein and Ostrander continued until about 6 o'clock on a variety of problems. It is interesting to note the changing attitude toward von Braun—all are concerned about some of his shortcomings but all seem determined to help him be successful in his new responsibilities.[30]

About 6:30, over to the Statler to attend for a while the reception given by Senator Young of Ohio. Actually, I think Senator Young was acting for Thompson-Ramo-Wooldridge in hosting a party in honor of Pioneer V. Space Technology Laboratories, a subsidiary of TRW, was the prime contractor on this particular space probe.[31] It was a fine job and deserves lots of plaudits. What was behind all of this display I am not quite sure. Back home in a hurry to get into tails to attend the White House musicale at 9 o'clock with Ruth. She had made herself a new dress that was quite pretty—a sort of deep blue brocade. I daresay there was not another homemade dress in the entire house of some 400 "little cabinet" members. We were there only an hour and a half with the president, the vice president and their wives in attendance. Carmen Cavallero played the piano accompanied by a couple of other Latins. I was not overwhelmed by the music, but it was a pleasant experience.

On the way home, we stopped at the Carlton Hotel on the off chance that we might find Dave Wright of Thompson. As I opened the door, there he was along with several of his colleagues from the TRW organization. We had a liqueur with him and then on home to find Sally ready for bed although we didn't make it until some time after midnight. Enough for tonight—tomorrow morning at 5:44 a.m. the first real meteorological satellite is scheduled to be launched.[32] I'll keep my fingers crossed all night.

---

[30] The reference to von Braun's shortcomings concerned the implications of his belief in meticulous attention to small details and in proceeding in rocket development incrementally. As a corollary to these beliefs, the Germans sought to control all aspects of their work by doing as much of it themselves as possible rather than handing it over to contractors not only to develop and build but also to coordinate. Dr. Glennan and his successor, James Webb, on the other hand, worked successfully to contract out the bulk of NASA's work and to build up the capabilities of the U.S. aerospace industry. The Air Force had also used the latter approach in developing its ballistic missiles. The competing viewpoints later came to a head in the famous controversy between the von Braun team and George Mueller, who had worked with the Air Force missile program, over all-up versus stage-by-stage testing of the Saturn V. Here, von Braun also represented the overall NASA approach, not just that of the Marshall Space Flight Center, but the Mueller approach won out because of the need for speed to achieve President Kennedy's goal to reach the moon by the end of the decade. (For excellent treatment of these competing cultures within NASA, see Howard E. McCurdy, *Inside NASA: High Technology and Organizational Change in the U.S. Space Program* [Baltimore: Johns Hopkins University Press, 1993], pp. 16-17, 37-38, 92-97, 167.)

[31] The headquarters of TRW, Inc. was Cleveland, Ohio, the state from which Young was a senator.

[32] On 1 April 1960, a Thor-Able rocket did, indeed, launch the first known weather observation satellite, Television Infrared Observation Satellite (TIROS) I. It took some 22,500 cloud-cover photographs on a global scale from roughly 450 miles above the Earth until 29 June 1960 and was regarded as beginning "a new era of meteorological observing." (Emme, *Aeronautics and Astronautics . . . 1915-1960*, pp. 121, 146.)

CHAPTER FIVE

APRIL 1960

Friday, April 1: It wasn't April Fool's Day, after all! An hour late— but very positively—Tiros I took off from Cape Canaveral and went into orbit. It was about the most perfect launch we have had. I had not bothered to go to the office at 5 o'clock but called in at about 7:00 just in time to hear that the third stage had ignited properly. I was to have breakfast with Dave Wright and Dean Wooldridge at the Carlton at quarter of 8:00. Just before I left the house, the word came that the satellite seemed to be in orbit. Because Jodrell Bank in Manchester, England, was talking with Pioneer V, it could not track the Tiros satellite. While at breakfast, I got the word that it was definitely in orbit and that it should be over New Jersey and that it would be interrogated within 10 minutes. The satellite was supposed to go into a circular orbit of 380 miles above the earth. It actually had an [initial] apogee of 407 miles and a perigee of 378 miles. Its velocity at injection was within 21 feet per second of the desired velocity some 26,600 feet per second. On the first pass, it was apparent that a picture was being transmitted back to the earth. This was sent to us on facsimile and was so good we were a little startled. It was a picture taken obliquely looking westward from New Jersey toward the center of the country and revealed a cyclonic disturbance that was actually in being at that time. Naturally everybody was excited.

I got to the office to share in some of the excitement and to help arrange the press conference set up for 11 o'clock. In the meantime, Dr. Lines of Great Britain came in to tell us of the results of his trip of some three or four weeks around various NASA installations and some of the industrial organizations in the United States. He was really quite amazing in the enthusiasm he displayed. He said he was unable to believe thoroughly the broad scientific base that had been developed to support our satellite space program. He was astounded by the variety of technological approaches being made to various problems we have been encountering. He was absolutely certain that great benefits would come from the satellite program to other technologies outside of the space field. It was a heart-warming session that erased some of the cares from the day, which was already fairly bright.

Just before 11 o'clock, the Italians came in to sign an agreement calling for cooperation in sounding rocket launchings. They are buying six Nike-Asps and will launch certain payloads prepared and supplied by the U.S.[1] At 11 o'clock, we had the first pictures we were willing to show to the president, and I called to find that he was in the National Security Council meeting. He asked us to come right over and five of us did. We broke into the meeting and were very well received. I arranged with the president for pictures to be taken later in the day. We got back to the press conference, which I started off with a very brief statement, and everything seemed to be going well. A slight altercation with the United Press International and the Associated Press people occurred when they tried to out-guess us as to whether or not we had satisfactory pictures. I asserted that everything was going well but that we would not state what the pictures were or whether we had them really until we knew. This caused some hard feelings but it was all smoothed over in a satisfactory manner.

I called Albert Thomas to urge him to be liberal with our construction and equipment budget. Word had reached me that he was going to cut out the office building at Huntsville, which would cause great hardship there. In the meantime, Lyndon Johnson and Overton Brooks had received word of the success of Tiros I and were clamoring—and I mean clamoring—for some of the pictures. By this time, we had received a sequence of 4 pictures taken 30 seconds apart that clearly showed the Gulf of St. Lawrence, with the satellite passing over it at some 5 miles a second. These were prepared for a montage negative and arrangements were made with the president for a picture with him at 5 o'clock and a general release to the press at the same time. The pictures were to go to the Hill to Johnson and Brooks at 5:10 p.m. All of this worked out well and I think we managed to get through the day without further damage.

Just to keep our minds on our business, we sat through a source selection board report on the development of the plasma jet engine, which took an hour and a half.[2] When that was finished, Dick Horner talked with me seriously about the fact that he has put his house on the market and plans to leave as of 1 July. This puts the problem directly up to me and I must find a man to replace him. I have made note of this before, but the task is always a little more difficult than one wants to admit.

[1] The Nikes were early U.S. guided missiles developed primarily for defense. With other rockets added to the solid-propellant version of the Nike, various sounding rockets became available. The Asp (atmospheric sounding projectile) was a sounding rocket in its own right, first tested in 1956. Combining it with the Nike increased the weight it could lift from 13.6 kilograms (30 lbs.) to 27 kilograms (60 lbs.) and the height it could reach from 40 kilometers (25 miles) to 260 kilometers (160 miles). (William R. Corliss, *NASA Sounding Rockets, 1958-1968: A Historical Summary* [Washington, D.C.: NASA SP-4401, 1971], pp. 24, 32, 80.) Using the Nike-Asp and also the Nike-Cajun, in 1961-1962 the Italian Space Commission measured the upper atmospheric winds through ground photography of illuminated sodium vapor released from the rocket. (Frutkin, *International Cooperation in Space*, p. 56.)

[2] A plasma engine used a heavy gas such as nitrogen, compressed and accelerated in a cylindrical magnetic field. There, molecules were broken into electrons and positive ions, with the flow constricted to a thin cylinder at the axis of the electrodes to emerge as a jet of extreme temperature passing through the engine's nozzle. This appears to have been the concept involved in two contracts NASA awarded on 18 April 1960 to Avco Manufacturing and General Electric for a one-year competitive project to develop a laboratory model of a 30-kilowatt arc-jet (electric-propulsion) engine. (Emme, *Aeronautics and Astronautics . . . 1915-1960*, p. 122; NASA, *Third Annual Report in the Fields of Aeronautics and Space*, 18 January 1961, p. 19.) Throughout the 1960s and into the 1970s, Lewis Research Center investigated several types of electric rockets, including an ion (electrostatic) thrustor, an electro-thermal thrustor, and electromagnetic or plasma accelerators, but the projects seem to have ended in the 1970s, although related efforts revived again in the late 1980s or early 1990s. ("Electric Propulsion" folders, NASA Historical Reference Collection.)

On home to a dinner of pheasant cooked in white wine and a bottle of lovely white wine. Just after reaching the apartment, we went up on the roof and witnessed the successful launching of a 100-foot balloon from Wallops Island. It was a beautiful sight as it rose into the air some 200 miles and then disappeared into a cloud bank as it moved out to sea some 500 miles.[3]

Saturday, April 2: I stayed home all day working away at a variety of problems. First, I wrote or rewrote a letter to Congressman Quigley that I hope will answer the questions he has been bothering me about. The business of commenting on the statements of General Medaris and Dr. Pickering is not too pleasant but I think I made sense of it.[4] In the mid-afternoon, Walter Bonney called to say that he had pictures from the high resolution camera in Tiros I. Naturally, pressures had been applied to the public information office by the newspapers wanting copies of pictures from this camera. The problem is that these pictures cannot be released until it is determined that there is no possibility of [the cameras that took them] being used for a reconnaissance satellite. The boys brought the pictures out and it was so obvious that they were not in the class of reconnaissance [photos] that I felt we could release them immediately. I called Allen Dulles and General Cabell to tell them of my decision and arrangements were made for the release. The remarkable thing about this is that none of the pictures were printed. We had satisfied the requirements of the press and it found no significant "sex appeal" in these particular prints.

Sunday, April 3: After a hearty breakfast, I started in on the paper work again. This was broken up in the middle of the afternoon when I took Sally over to the Pentagon parking lot to give her an opportunity to practice driving. When we returned, we found that Polly had called—her plane was very much delayed so that she was not about to get back to Philadelphia from Cleveland. Throughout the rest of the afternoon and evening, I attempted to purchase most of the Bell Telephone Company as we tried to get her cared for. Finally, she put up for the night with the Borcherts and will take a plane to Philadelphia tomorrow.[5] We went to the Richmonds' for dinner. It was a pleasant evening but we were glad to get back home. Somehow or other, there isn't much resilience in this old frame these days.

Monday, April 4: Once again, Monday seemed to be a bad day for me. I just don't understand it, but I go in each Monday morning with a scowl on my face and meanness in my heart, apparently. We were greeted by the boys from the Tiros group who had made up a weather map from the pictures taken over Russia, Africa,

---

[3] This was the fourth suborbital test under Project Shotput of the 100-foot-diameter balloon later called Echo, which in this test rose to an altitude of 235 miles and successfully inflated. (Emme, *Aeronautics and Astronautics . . . 1915-1960*, p. 121.)

[4] As Glennan understood it, "each [of the two men] would employ but a single agency to carry out both military and non-military activities in space," with Medaris choosing the DoD and Pickering, NASA. Glennan showed the drawbacks of both proposals and argued that "we are well on the way to achieving a satisfactory management-level coordination that will work." (T. Keith Glennan to James M. Quigley, House of Representatives, 4 Apr. 1960, Glennan subsection, NASA Historical Reference Collection.)

[5] On the Borcherts see note 9, Chapter 3. Below in the diary, the Richmonds were friends in Washington.

Europe and the United States. It was really quite a map and correlated well with the
standard weather maps presently being distributed. Nevertheless, I felt we should
not release it until we had talked with the State Department and with the CIA. We
were able to take a copy of the weather map to the Hill; it matched the four pictures
printed last Friday afternoon, and these were explained to the House Committee on
Science and Astronautics before I began my testimony. I think they did well in the
important department of impressing the congressmen. [See photo, next
chapter.]

We were asked to come up to the Hill a little bit early this morning so the
congressmen could be photographed with a Tiros payload. What a crush! Each
wanted his picture taken with the payload and most wanted me in the picture with
them. Fame is fleeting—each of us should keep this in mind. My testimony went
rather well and the questions were mild and rather easily handled. All in all, I think
it was a good exercise and perhaps we will now get on with the business of having
the amendments to the [1958] law [creating NASA] accepted.

During the afternoon there were the usual visits from industrialists such as
those from Kaiser Steel who wanted a little bit of an edge on the Saturn launching
pad at Canaveral. At 2:30, we went over to the Atomic Energy Commission to meet
with John McCone and General Luedecke to discuss the management arrangements
for Project Rover. Much to my surprise, they bought it without too much difficulty
and indicated that they would immediately present our proposal to the Commission.
John was impressed by the results of the Tiros flight. He is really a strange person
with whom to deal. He takes strong positions but is willing to change them. I think
he takes the positions largely in order to force decisions out of other people. I like
him and hope one day to be able to understand him better.

At 5:15, we gathered together the people from State, CIA and other
elements of the intelligence community. We told them what we planned to do with
the map that had been developed by our people, and they all agreed that it was
satisfactory for us to proceed. Thus ended another day, although I managed to spend
a good hour and a half at home putting the final touches on my income tax return
for 1959. Then to bed after packing for the trip to Houston, Texas, tomorrow.

Tuesday, April 5: To the office for a few minutes before taking off at
10:15 for Houston, Texas. I traveled via Eastern Airlines Electra—not without a
little trepidation. We have been involved, to an extent, in the studies of the disasters
that have plagued two of these Lockheed Electras in recent months. In this instance,
the flight was pleasant though long. We ran into 70 knot headwinds and took about
six hours to make a trip which normally would take very little more than four. I was
met in Houston by Mike McGuire, who is chairman of the great issues committee
at Texas A&M College this year. He had with him two or three cadets who were
members of the great issues committee. At the airport in Houston I was met by
Mr. O'Leary of the *Houston Post* who asked me the usual questions. He got the
usual answers, which he did not want to believe and thus was admonished to accept
what I had to tell him or write his own story. He wrote the story all right—he quoted
just those words.

A 35-minute flight via Beachcraft took us to the Texas A&M campus, where I immediately went to a seminar and talked to graduate students for 45 minutes. Then a brief tour to the country club for a drink, back to change my shirt and off to dinner with the president of the college and members of the faculty and student groups. At 8:00, I spoke to a crowd of students and faculty that must have numbered better than 1,500. I was at it for better than 1 hour and 20 minutes and seemed to have satisfied the crowd pretty well. After the speech the students wanted to take me around the campus a bit, so I did walk with them for perhaps 45 minutes before going to bed. It was an interesting day and I was glad to have the opportunity to speak to this particular audience.

Wednesday, April 6: Up at 6:00 to be flown to Houston where I was to take an Electra at 8:00 with a scheduled return to Washington at 1:25 p.m. We had a stop at New Orleans where it was discovered that one of the indicator lights on the propeller feathering mechanism was not working properly. A 4-hour delay ensued and it was 5:30 before I was able to get back to Washington. I went directly home because we were due at Anderson House for dinner with the president of Colombia as guests of Secretary Herter. We donned white tie and tails and evening dress and enjoyed a rather formal evening. A brief statement by Secretary Herter was responded to by President Lleras in a most friendly fashion. Much comment was made about Tiros I and Pioneer V. Success is a wonderful experience in these circumstances.

Thursday, April 7: The staff meeting went off pretty well and we moved from that immediately to the presentation by the source selection board of its recommendations on a contractor for the nuclear rocket project engine nozzle. We have not yet found the right way to handle this situation but I think we are making progress. It appears that it would be best if the evaluation groups and the source selection board acted really as an evaluation body and left recommendation out of their discussion. This would leave the administrator and his immediate assistants more maneuvering room in which to make a judgment.

A number of the boys came in to talk about some of the security aspects of Project Tiros. There is really no problem, but other people are attempting to make one for us. It looks as though we will have to meet with the Central Intelligence Agency and other members of the intelligence community to iron out the problem. After a quick lunch I had a visit from several people from Ford Aeronutronics.[6] They were making certain that we understood the urgency with which they looked at some of our problems. I went through my standard discussion of the organizational study we have underway and what I thought it could mean for future contract operations involving industry. They went away, seemingly satisfied.

At 2:15, Harold Goodwin of the USIA came in to talk with me about some of our educational and technical information problems. He seems a very forthright

---

[6] In the summer of 1959, NASA had contracted with the Aeronutronic Division of Ford Motor Co. to perform a computational study of radar and trajectories for tracking and data acquisition purposes. Later, Ford Aeronutronics provided the lunar capsule subsystems for Ranger 3 through Ranger 5, all failed attempts to collect data about gamma rays and the moon itself, although later Ranger missions, as discussed below, were successful and important. (*NASA Historical Data Book*, Vol. II, pp. 314-316, 560.)

and capable person with plenty of assurance. He does not himself seem to be interested in a job with us. At 3:00, we set up the first length-of-service award ceremony in the USO Auditorium next door. It went very well and I think everyone had a good time. I was not able to stay very long because of the problems with the release of information from Tiros and so returned to the office to meet with six or seven of the top staff. Karl Harr of the White House staff came over to plead that we do nothing that would compromise the Department of State's position in this matter. I suggested he come along with us to General Cabell's office where we had a discussion until about 6:30, ending in an agreement that the pictures from Hawaii would be reviewed and if found to contain no compromising information, the whole matter would be dropped and we could proceed at will. We were so certain this would be the case that I finally agreed to this procedure. Off to the apartment then to get packed for the trip to Cleveland tomorrow.

Friday, April 8: In at the usual time, after packing a dinner jacket and other clothes for Cleveland, to prepare for a visit with the president at 10:45. John Corson and Al Hodgson came in to talk about the program being put together for the first meeting of the organizational advisory committee next week. It seems we are in pretty good shape for the meeting. Jim Perkins had called to say that he needs a bed on Friday night and this has been cared for in our apartment. Over to the White House at 10:45 to visit with the president for a few minutes. This was in the nature of a progress report but I did want to bring him up to date on two or three problems. I gave him a statement on the status of our appropriations and the legislative amendments that are yet to be acted upon by Congress. He was interested in the progress on Pioneer V and Tiros. I told him about the State Department concerns and he approved the actions we had been taking.

I told Mr. Eisenhower about Project Echo and what I hoped for it. In all of these activities, he expressed a great deal of interest. I did talk with him about the unfortunate connotations placed upon Tiros by some of our newspaper friends, in particular John Finney of the *New York Times.* The president expressed himself strongly on this matter and when I told him that I was to be on "Meet the Press" on Sunday, he said he would think the less of me if I didn't speak out strongly against this sort of thing. I told him about the problems I was having with John McCone and Senator Anderson on Project Rover. Once again, the president said he would take a hand if it became necessary—he might have to call in John McCone with me. I told him I would do my best to get this settled otherwise. It is obvious that he recalls Clinton Anderson's operation on Lewis Strauss.[7] I bade him farewell, hoping that he would have a good weekend.

---

[7] Senator Anderson had led a fight against President Eisenhower's appointment of Admiral Lewis Strauss to be his secretary of commerce. Admiral Strauss (U.S. Navy, Reserve) served as chairman of the AEC from 1953-1958 but declined reappointment because of the hostility of Senator Anderson, who as a member of the congressional watchdog committee for the AEC had frequently clashed with Strauss. Then in 1959, Anderson and other senators succeeded in defeating Strauss' appointment to be secretary of commerce, much to the president's chagrin. Long before the confirmation fight began on 17 March 1959, Strauss had taken his oath as secretary of commerce on 13 November 1958. He served in that capacity until the final vote against him, which did not occur until 19 June 1959. (Dwight D. Eisenhower, *The White House Years: Waging Peace, 1956-1961* [Garden City, N.Y.: Doubleday, 1965], pp. 392-396; Ambrose, *Eisenhower the President,* p. 530.)

He told me he hoped to get away for a week in Augusta starting next Monday. All in all, it was a good session.

Back to the office for a telephone call from Station KING in Seattle, Washington. I did a 15-minute radio question-and-answer session with a man by the name of Al Wallace. It went reasonably well and I had a good opportunity to express my thoughts strongly about the manner in which the press was misleading the public on Tiros by calling it a reconnaissance satellite. At lunch, I talked with Senator Francis Case of South Dakota and then went off to the television room in the Capitol to make a recording for him. Actually it was a TV and radio simulcast. Immediately following that one, Senator Clifford Case of New Jersey came in to have his picture taken with a model of Tiros and with me.

Finishing that operation, I got back to the office just in time to pick up Ruth and go over to the airport to make the trip to Cleveland. We arrived there in good shape and picked up Sally. After a pleasant dinner, we took off for Sally's school and enjoyed—really enjoyed—a joint concert involving the girls from Hathaway-Brown and the boys from University School. Back to the house for a good night's sleep.

Saturday, April 9: [Discusses some Case business.] Last night, Tom Morrow had called from Detroit via Washington to see if I would see him in Cleveland on Saturday. I had arranged a meeting at 12 noon at the Union Club and drove down to have lunch with him. He simply wanted to make certain that I understood how serious Chrysler Corporation is about the contract for upper stages of Saturn. It was the usual discussion without anything new coming from it. With Ruth who had come downtown with me, I went over to Halle's and bought a topcoat. Back to the house to wait for Mr. Bacome to discuss the furnishing of the seminar room we are building in the basement. Then, a shower and preparations for dinner occupied my time. The meal was a pleasant affair with a great many opportunities to speak with good friends. I did manage to escape as soon as the dinner was disbanded and dashed home to have a bit of rest.

Sunday, April 10: We took Sally back to school and dashed out to the airport where we joined a good group of 12 or 14 NASA people who were on their way to Washington and Langley. The trip back was a pleasant one and we arrived at the apartment shortly before 3:00. I forgot to say that almost as soon as I had reached Cleveland I had a call from Walter Bonney saying that the pictures taken over Russia had turned out much the same as those taken over other parts of the world. This means that we no longer need worry about the screening of these pictures by other agencies.

Walter came out about 4:30 in preparation for my appearance on "Meet the Press." We checked over a few facts and statistics and then started over to the studio. Ruth came along and I guess she enjoyed the experience. I was being interviewed by John Finney of the *New York Times*, Peter Hackes of NBC and Bill Kines of the *Washington Evening Star* along with Lawrence Spivak [producer of "Meet the Press"]. This is a strange program in that anybody that takes part in it seems to be quite worried about the questioning. I must say that I was and nothing that happened at the studio before the program gave me any feeling of confidence. It turned out, however, that after muffing one or two questions, I seemed to deal with the rest of them rather easily. I did have a chance to get back at John Finney by giving him a rather cryptic answer to a question he asked. He was once again trying to make something of Tiros as a reconnaissance satellite and started to ask questions about the desirability of agreements being reached between nations about the use of reconnaissance satellites. He asked me if I wouldn't be indignant at having a Soviet reconnaissance satellite orbiting over United States territory. I said that there was no real reason for the Russians doing this, all they had to do was read John Finney's newspaper and they could get all the information they desired. This stopped that part of the questioning.

Monday, April 11: We started out the day with a brief discussion of the Centaur Project.[8] I was anxious to understand what was being done and was given a discussion by Ostrander and Milt Rosen.

At 11:00, General Luedecke of the AEC came in with General Ostrander. We discussed the problems of attempting to arrange our management affairs so that the Joint Congressional Committee on Atomic Energy would be satisfied and we would be able to get on with the job. John McCone is out of the country and I have just found that Senator Anderson is leaving tomorrow for the west coast. I decided I would have a chat with Anderson and try to correct the misimpression he has about the qualifications of Harold Finger, whom we have selected to head the office of space reactors.

Later in the day, we had quite a discussion of television programs but came to no decisions. It is becoming increasingly apparent that I cannot continue with Bonney as the responsible person in the entire field of public information and other areas of education and technical information. Late in the afternoon, Wexler and Reichelderfer of the Weather Bureau plus several of our people came in to discuss the proper handling of public information on Project Tiros. The Weather Bureau people are apt to be a little enthusiastic—more than enthusiastic—about the prospects that are available to us with Tiros. They all agreed, finally, to play down these stories and to be as factual as possible in their discussion of Tiros.

Tuesday, April 12: At 10:00, Dryden, Horner and Bonney came in to discuss a proposed exhibits program. This was again a very unsatisfactory discussion. I cannot accept the recommendations that Bonney brings me—part of it is just an antipathy towards him because of my lack of confidence in his ability to plan.

---

[8] The Centaur rocket was an upper stage, used with the Atlas booster to launch the Surveyor lunar probes (1966-1968) and other scientific satellites. Originally intended for earlier Mars and Venus spacecraft (1962-1965), the Centaur experienced launch development problems and was not available until 1966. It was the first American launch vehicle using liquid hydrogen as a propellant, which explains some of its early problems, the result of heat transfer between liquid oxygen and liquid hydrogen fuel tanks, causing evaporation of the liquid hydrogen. (*NASA Historical Data Book*, vol. II, pp. 42-44.)

The rest of the day was not too exciting although we did have a good discussion with the Columbia Broadcasting System people at 3:30. Sig Mickelson, president of one of CBS's divisions, came in to talk about doing a series of programs with us. I finally agreed that we would wait a month for their development of story lines, etc. At 4:30, I spoke for an hour and a half to a group of Holyoke and Amherst seniors and juniors. They are studying political science and were spending their Easter vacation in Washington. It was an interesting question and answer session— I hope the kids learned something of value.

Wednesday, April 13: This was another rather full day. Starting off at 8:30 with General Ostrander coming in to discuss with me the qualifications of Harold Finger for the space reactors branch job, I had a few minutes with a Dr. David Abshire who seems to be a member of the staff of the House Republican Policy Committee. It wants from us two or three statements relating to our accomplishments in space and the basic policies we are attempting to follow. Presumably, these will become part of its store of speech materials. I think we can provide some help.

At 9:30, I visited Senator Clinton Anderson in the new Senate Office Building to discuss with him the problem of setting up an effective organization for nuclear rocket development. He spent 15 minutes going over his arguments with Lewis Strauss, almost in an apologetic fashion. He told me how he happened to become embroiled in the argument over the confirmation of Strauss. He now states that his reason for running for the Senate again in the State of New Mexico is that Strauss had threatened to put a quarter of a million dollars behind Anderson's possible opponent. I had taken with me a set of charts outlining the division of responsibility between the AEC and NASA for the development of a nuclear rocket system. I then told him what we were planning—jointly with the Atomic Energy Commission—in the way of a management structure. He seemed to accept all of this but was concerned about the eagerness with which Finger would tackle the job. Apparently, someone has been talking negatively about Finger, and Anderson, while unwilling to identify any such activity, said that he had gained a very bad impression of Finger's enthusiasm in the course of one brief statement made by Finger at a Joint Committee on Atomic Energy hearing. I finally said, "Clinton, I am asking you to take my word in this matter. Finger is a good man and will do a good job. He will be in no position to do the program any harm and is the best man we can select for the post." Anderson turned to me immediately and said, "I will take your word," and we shook hands on it. I moved immediately to the Capitol and did a 10-minute telecast recording with Senator Barry Goldwater of Arizona. Goldwater seemed almost shy although he is one of the most outspoken of the Republic senators.

Back at the office I called together Ostrander, Dryden and Phillips to tell them of my discussion with Senator Anderson and set about drafting a letter to confirm our discussion. At 11:45, I met Morse Salisbury at the Statler for lunch.[9]

---

[9] Morse Salisbury was then assistant to the general manager of the Atomic Energy Commission.

I was seeking his advice on the organization of the public information and technical information services. Our activities in these areas have never been adequately handled and a much more significant effort is to be required. I have spoken earlier in these pages of my dissatisfaction with and lack of confidence in Walter Bonney for the overall job. Morse, a very quiet and unimpressive man of long experience in a variety of governmental agencies, suggested Shelby Thompson, his deputy, as a possible top man for us. Much to my surprise, Morse suggested that he himself might be interested. I am going to set up meetings for both of them with others of our top staff.

A brief meeting with Gleason, Horner and Dryden to review the legislative program resulted in my decision to fight to the best of our ability the cuts proposed by Albert Thomas. We have just learned that he has cut some $38 million from our request for $915 million. It seems a little silly to be arguing about that sum within such a large sum, but we are already over the proverbial barrel through our lack of experience in estimating sufficiently conservatively to cover the constantly-rising costs in this business. A meeting at 2:00, with a variety of people involved, considered the matter of source selection board criteria. This is a subject that is taking up a great deal of our time these days. The business of trying to be completely fair and objective in the selection of companies to do business with our organization, and the problems of keeping a complete and open record of all deliberations, is not an easy one with which to deal. I suppose all of us are a little scarred by the memories of the "executive privilege" hearings.

At 4:00, I called together the top supervisory staff to read a mild bit of the "riot act" because of the obvious lack of enthusiasm in planning for the joint Federal Crusade for International Agencies and the National Health Fund campaign. None of these are activities about which one can wax very enthusiastic, but they are part of the business of living in a democracy. Each of us has responsibility in these matters and leadership must be provided. I felt the response from the staff was very favorable even though I was a bit brutal in my discussion.

This was the evening on which James "Scotty" Reston and Milo Perkins and their wives were to come to dinner. Ruth and I had been looking forward to the evening with a great deal of pleasure. Scotty Reston is the head of the Washington Bureau of the *New York Times* and one of the most highly regarded of the Washington correspondents. Milo Perkins is a "hold-over" from the New Deal days who, with his wife, now indulges in making economic studies for other nations for a handsome fee. They seem to keep quite a staff of people operating at their ranch in Arizona and spend a fair amount of their time in South America. Milo and his wife came early—or Scotty and his wife were late. In any event, we had an opportunity to talk with Milo and Karen for about forty minutes before Scotty came in. This was pleasant enough although not terribly exciting. When Reston came in we began to get down to business and discuss politics, leadership or its lack, foreign policy, and a particular concern with Latin and South America. As usual when one anticipates the pleasure he is going to take from an evening, we were somewhat disappointed. Perkins talks a great deal and seems genuinely interested in the work he is doing. He is an uplifter of better than average quality although I wonder a little how dedicated he would be if he could not afford it. On the other hand, Reston is a serious person who seems to labor over every thought he has and every attempt to express a thought. Ruth suggests that he writes so much that he finds it difficult to express himself orally with ease. I lost none of my respect for him in the course of the evening, and we were both happy to know Mrs. Reston. All in all, it was a good evening.

Thursday, April 14: The staff meeting this morning was nothing out of the ordinary. I do believe these meetings are serving an excellent purpose, however. At 10:00, Mr. Paul Mellon and Mr. Hughes of the Mellon Institute in Pittsburgh came in to see me. I had been alerted by Warren Johnson and by Bill Baker of the Bell Telephone Laboratories that they wanted to talk to me about the possibility of my accepting the presidency of the Mellon Institute in Pittsburgh. This is a very well-paying job—somewhere in the neighborhood of $60,000 base salary—with excellent pension and fringe benefits. It is a job that carries with it high respectability and a substantial position in the community. It was obvious that they were serious about this matter and that they would want an answer shortly. I told them that there was little chance that I could be interested—that I had already said to my board at Case that I would return there. There was just a slight chance that the full board would find difficulty in accepting the conditions under which I have said I would return. These conditions were accepted enthusiastically by the executive committee but I have asked Kent [Smith] to place them before the entire board. There is nothing unusual in them. I simply want the board to know it is in for some tough times during the next five years; that it will be involved in raising large sums of money; that I believe strongly in [Vice President for Academic Affairs] John Hrones and will support his desires to build a strong graduate school; that the alumni situation must be straightened out. I will let the Mellon people know next week about my decision. I talked with them about the possibility of their being interested in Dick Horner. I think Dick would do a good job, although his experience in dealing with community affairs is almost wholly lacking. However, I think my first responsibility is to Case.

At 2:00, we had a source selection board presentation on Project Sunflower.[10] Thompson-Ramo-Wooldridge won the competition. These are interesting exercises that take a good bit of time but are very necessary. Following that session, Dryden, Horner, Siepert, Ulmer and Keyser came in to talk about the effect of the budget cuts. I asked that material be prepared for me so that I could move into this situation immediately after the weekend. At 4:00, Tim Shea came in to spend an hour discussing some of my problems and telling me of his own plans. Much to my surprise, Tim is planning or at least would like to plan to leave the Western Electric Company a little ahead of his retirement date. He is now 62 and thinks that

---

[10] Sunflower was a project to develop a spacecraft solar powerplant. NASA announced on 19 April 1960 that it had awarded the contract for the project to Thompson-Ramo-Wooldridge. (Emme, *Aeronautics and Astronautics . . . 1915-1960*, p. 122.)

he could make a greater contribution by having a part-time association with some organization and then spending a good bit of his effort in community and national affairs. I agree that this would be a good thing for him to do—perhaps I can help.

Friday, April 15: This was the day our advisory committee on organization started its deliberations. The committee was prompt in attendance and all members were present. It is chaired by Chancellor Lawrence Kimpton of the University of Chicago and its members are Elmer Lindseth [president of the Cleveland Electric Illuminating Co.], Charles Stauffacher of Continental Can, Fletcher Waller of Bell & Howell, Nathan Pearson of T. Mellon and Sons, Morehead Patterson of American Machine & Foundry, and Jim Perkins of the Carnegie Corporation. We had set up a full day of briefings for them so that they might get a look at the total program and at the people who had principal responsibility for its administration. I thought all of our people did very well and the committee seemed to gain in interest as the day wore on.

I have asked them to look at our organizational setup for the following reasons:

1. I do not expect to be here after January 20, 1961, and I want to make as certain as possible that the organization I turn over is structured sensibly and that no roadblocks have been erected that would prevent effective administration by my successor.

2. I know of some problem areas, which I will describe to these people and ask their help in providing solutions.

3. I am certain that there must be problem areas I have not identified that will come to their attention. I have asked that they be alert to discover such situations and make recommendations.

Among the issues and problem areas that have bothered me are the following:

1. The proper balance of work to be done in our own laboratories and through contract with industry. By this, I mean the proper balance in terms of total amounts of work involved and the character of the work involved. It is natural that our scientists and engineers want to keep to themselves all of the interesting and creative problems, while farming out to industry only the repetitive and straight production items. This does not make sense to me.

2. Our program planning and program management and control activities are not properly organized or integrated. The first was organized by me when I came here and has worked only reasonably well. The second is a feature of Dick Horner's and looks to me to be an excellent operation. They need to be brought together—probably as a single unit through the associate administrator.

3. The current split between headquarters responsibilities and field responsibilities.

4. The effectiveness of our present relationships with other agencies of government, and with Congress as well as the public.

5. An identification of the trouble spots in terms of top personnel replacement.

It was a full day but nobody wanted to quit. We had dinner together at the Carleton Hotel and then Jim Perkins came out to the apartment with me. He spent the night but before going to bed we had a long discussion of some testimony he is planning to give to Scoop Jackson's Government Operations Committee.

Saturday, April 16: We were up early to have breakfast at the apartment with John Corson. Jim Perkins had asked me to sharpen up my problems, and I thought we'd best have a talk about it before revealing to the committee exactly what was in my mind. I think Perkins does this sort of thing rather well, although at times his type of questioning is bothersome to a person like myself. He is an excellent member of any committee. Ruth gave us a good breakfast and I was able to organize my thinking in a sensible manner.

Back to the office at 9:00 for another full day of discussions—at least, the staff and I stayed through the noon hour while the committee carried on until about 3:00 in the afternoon. It was obvious that they were getting increasingly interested and recognized more and more the seriousness of the problems I was posing for them. Elmer Lindseth stopped by after finishing and he came out to the apartment with me for a visit before going to Waynesboro, Virginia, for a weekend with his married son. During the course of our conversation, it became apparent that I should have stayed a little longer with the group. They had become very much concerned over the program planning and program management aspects of the problem. Apparently, John Corson had not indicated to them the manner in which we were building our long-range program. I was able to convince Elmer that we were very much better off than he thought, but then I had to turn around and say that I was concerned about the organizational mechanics we have in this area. I'll have to get at this on Monday. I am very much satisfied with the beginning of this exercise.

Sunday, April 17: Up at 6 o'clock to go to a Sunrise Service at the Arlington National Cemetery. In many ways, this was a real travesty. I had not realized that the service at Arlington was sponsored and conducted by the Knights Templars of the United States. This is a fine organization of the Masonic faith, I believe, but an Easter Sunrise service ought not to be conducted as this one was. There was no opportunity for the audience to participate other than in standing and sitting. The Army choir and the Marine band were equally bad over the loudspeakers, which were poorly maintained and operated. The sermon seemed to be addressed to the Knights Templars and not concerned with the religious significance of Easter. All in all it was less than a satisfactory beginning to the Easter Day.

Back home for a good breakfast of the things that had been ordered by Polly from her mother. She enjoyed finding her basket in which Ruth had placed a black dress purse. We thought a bit about the other children but there isn't much one can do about it when they're so far away from us. The rest of the day was spent in reading and writing. Polly left for Swarthmore at 5 o'clock. Then the apartment became quiet once again.

I should like to put down here another small bit of information about the events of the past week—particularly those involving Senator Anderson. On the afternoon of the day I had talked with him, Jim Ramey, executive director of the Joint Congressional Committee on Atomic Energy, called to say that he had taken the senator to the airport and that the senator had given him a brief review of our discussion. Would I fill him in? I read him the letter I had written to Senator Anderson confirming our discussion and explained as best I could over the telephone, offering to have him come down next week to get a full understanding. I thought his questions were reasonable, although he seemed to be inordinately concerned about Finger. Perhaps he is part of the "plot" to get Finger.

On Thursday morning, bright and early, I had a call from New Mexico. It was Senator Anderson, who had just heard from Jim Ramey. Jim was very much disturbed over the arrangements he thought I had concluded with Senator Anderson. I replied by reading my letter to the senator. Anderson said, "Well, now this doesn't sound so bad; this is exactly what we had agreed to, except that I wish you would change one or two words." I patiently explained again that Finger was a good man— that it was absolutely necessary that he take this job with the support of Anderson, and that unless this could be accomplished, we would have to seek some other solution—in my opinion, one not so satisfactory. Anderson again said he would rely completely on my assurances and that he would call Ramey to tell him so. This is an example of the manner in which staff members of congressional committees cause trouble and misunderstanding. This sort of thing happens all too often in Washington. The executive director of one of the important committees of the Congress must have an inordinate sense of power. He has no responsibility, is neither elected nor appointed, but is a hired hand. Yet, he wields an uncommon amount of power. Some are excellent, most are mediocre, and some are downright sinister.

During the week, I notified the staff that I would be away for 10 days in May and 10 days at the end of July. The first holiday will be in Florida at the home of Don and Eleanor Adams, where Ruth and I will be alone and completely quiet, I hope. The second, a less restful but equally rewarding experience, will be a stay of 10 days at the Bohemian Grove in California.

Monday, April 18: This morning wasn't as bad as the Mondays usually are. I felt rather relaxed going to the office, but this didn't last long. Apparently, Tiros I and Pioneer V continue to operate satisfactorily. Six thousand pictures have been received from Tiros I, and it is becoming almost impossible to cope with the tremendous amounts of data being received. I had spent a good bit of time yesterday attempting to write a letter to Lyndon Johnson to give him the information he wants in order to make some important news next Thursday. He wants to announce some coming event in the space business in such a manner as to indicate his great interest in the program. I am happy to do this with him because he has been most helpful. This morning, then, was spent with a variety of people checking over the various words I was using and getting the whole operation lined up. This included coming into agreement with the staff of the Senate Space Committee, including Lyndon Johnson's executive assistant. It went well, everything considered.

I have been having a good bit of difficulty bringing into focus the kinds of arguments that may be used on the floor to counter the cut recommended by the House subcommittee on appropriations. It is said to be a losing game to attempt to beat Thomas on the floor and that it is very much better to argue the point with the Senate and then bring the two bills back together for conference treatment. I am inclined to go along with this—especially since the Senate Legislative Committee seems to be intent on giving us authorization for $50 million more than we have asked.

The morning went rapidly and I took lunch quickly at the White House with Hugh Dryden. I took off at 1:15 in a small Air Force plane to fly to Harrisburg, where I was being met by a car that would take me to the Hershey Hotel in Hershey, Pennsylvania. It was a really rough trip both ways but not too long. I spoke in Hershey to 40 or 50 editors and editorial writers of the Scripps-Howard newspapers. Louis Seltzer introduced me in very glowing terms. It was an exceedingly useful trip and I think I made some good friends among the editors of the Scripps-Howard papers. I came back home about 7:00 p.m.

Tuesday, April 19: At 9 o'clock, John Corson of McKinsey and Company, his associate Jack Young, Horner, Dryden and Hodgson came together to discuss the results of the first meeting of the advisory committee on organization and of the planning for the meeting to be held in Huntsville on 6 and 7 May. All agreed that a good start had been made but it seemed apparent that some confusion had arisen out of the fact that we had not given the group a proper briefing on the planning function, which has taken up so much of our time this past year. I had left the meeting to allow them to carry on in an executive session. I think this might have been a mistake—at least, questions were asked by two or three members of the committee that could have been answered in such a way as to avoid their proceeding under misapprehensions. One does not want to stifle suggestions, comment, criticism or expressions of concern. Nevertheless, it is desirable to avoid misconceptions on which may be based later investigations that are apt to be of very little value.

I asked Corson to come back with a plan to provide a chronology dealing with our planning activities and to lay on a session that would detail the activities we have been undertaking in this area. Associated with this presentation would be one on the program management plan just now under development. With this presentation, I think we can get the study back on the track. Indeed, it seems to be necessary if they are to listen intelligently to the statements that will be made by Pickering, von Braun and Goett. I want them to get the full import of the concerns that will be expressed by these men who are in charge of our field development centers. However, I want them to have this against an understanding of our program planning and management.

At 10:00, I listened to the presentation of the Saturn source selection board. This is apt to be one of the most controversial and difficult of the source selections that I have to make. It actually covers the contracting for the so-called S-IV stage of Saturn. This is a four-engined, 80,000 lb. thrust, liquid hydrogen-liquid oxygen stage that will find use in all of the various models of Saturn. The company winning this competition is assured of some reasonable edge in subsequent competitions for other stages for this launch vehicle system. Presently, Convair in San Diego has an edge since it is our contractor on the Centaur stage. This unit, being developed for use with an Atlas, is also to be the top stage for the Saturn.[11] It is the first unit in this country to make use of liquid hydrogen as a fuel.

A lunch with Dr. York for the purpose of discussing his viewpoint with respect to Joe Charyk and others who might be candidates for Dick Horner's job or ultimately for my job. He brought up the name of Rube Mettler of Space Technologies Laboratories, of whom we had not thought. He would be a really fine man if we could get him.[12] York seems to think Charyk a good man. I told him of Tommy White's interest in the job.[13] His response was less than entirely enthusiastic but indicated the desirability of some consideration. Herb York is desirous of going back to a university on some basis such that he can undertake a substantial amount of consulting so he may build up a backlog of funds over a period of five or six years. He then wants to come back into public service. He is a genuinely dedicated person with great ability. He has done a tremendous job—one that seemed almost impossible and perhaps still is— without doing more than acquiring the usual amount of enemies. I am glad he is coming to Case for the commencement. Ruth deserves the credit for this suggestion.

Back to the office for a meeting with Gleason, Johnson and Nunn on the legislative amendments we have before the House committee. The staff has asked for our judgment and drafting assistance on several provisions they think the committee is going to insist upon. The real issue is whether or not my intention to appoint a general advisory committee to me and the actions we already have underway to establish an activities coordinating board with the Department of Defense should be cast in concrete in the law. The committee is probably going to insist upon it. Actually, in the very early drafts of our proposals last November and December, I had included sections in the law to do just this. The White House staff had insisted they be taken out on the basis that authority to appoint advisory committees and coordinating boards already existed— there was no need to write it into the law. I agreed on our providing to the committee staff language that would be satisfactory to me. I insisted that the DOD have a chance at this language before the House committee put it into print. We will, however, bypass the budget bureau and the White House since we are acting in an unofficial manner on this particular problem.

---

[11] As things turned out, the Centaur (developed by Convair Astronautics Division of General Dynamics Corporation) was not the upper stage for the Saturn I but did provide a basis for the development of the RL-10 engines (by Pratt & Whitney) that were used in the S-IV stage of that launch system. (For this complex evolution and its further details, see Bilstein, *Stages to Saturn*, pp. 131-140, 188-190.)

[12] Ruben F. Mettler (1924-) had been a division director for Hughes Aircraft Co., a special assistant to the secretary of defense, and an assistant general manager for Ramo-Wooldridge Corp. before becoming president and director of Space Technologies Laboratories, Inc. in 1958.

[13] Thomas D. White (1901-1965) was an Air Force general who served as chief of staff of the Air Force from 1957-1961.

We had a meeting at 3 o'clock on Pioneer V, for the purpose of attempting to determine at what point in time and space we would turn on the 150-watt transmitter. I have been worried that the scientists would push the 5-watt transmitter much too far and would, one day, find themselves without the ability to turn on the 150-watt transmitter. There seems to be no real probability of this but, in order to establish a proper time for reviewing the situation again, we agreed that the higher-powered transmitter would be turned on at the 10-million-mile mark or when the signal had reached a minus-155-decibel level.[14] I hope the scientists were reasonably well satisfied, but I doubt it.

By this time, I had missed a date I had made at the National Science Foundation to look at some more films prepared by the Educational Testing Service for use in high schools. This evening we had dinner with Admiral Bennett and his wife and Admiral James Russell [vice chief of Naval Operations] and his wife. It was a pleasant evening without too much excitement.

Wednesday, April 20: For some reason this day was a busy one without too much to show for it. Will Mickle, editor of the *Huntsville Times*, had a date with me at 10:00, which he did not keep. He had had trouble finding a room the night before through some mix-up involving hundreds of students visiting the great city of Washington. I was able to say hello to him but not to give him any time when he did come in.

At 11:00, Arthur Hermann, education columnist for the Gannett newspapers, came in to talk with me about the changing scene in education. I hope I gave him an encouraging story. I had a chance to speak of Case and its forward-looking program and I am anxious to see how the column will appear in the papers of this particular chain. At noon, I went over to the luncheon of the Advertising Club, which was held in honor of Leland I. Doan, president of Dow Chemical Company. It was pleasant to see him and to listen to the tributes paid to him. Apparently, this is the beginning of a week of promotion involving the use of Dow products by the Hecht Company.

A hurry-up trip to the Pentagon found me meeting with Dick Horner and Phil Taylor, assistant secretary of the Air Force. We were there to try to get a bit of background on the engineering and manufacturing load presently facing Convair and Douglas. This has to do with the decision we must make about the Saturn S-IV stage. The balance of the afternoon was spent in attempting to study the Saturn problem properly. I went home for a few minutes and returned to the Hecht Company with Herb Rosen to be on time for the television program on which I was to appear along with Lee Doan and others. This was about the most mixed-up operation I have ever seen. I had written a short piece—about five minutes—to use on the program, and I must say it was the only part of the program that had been prepared. There were many nice statements made about it; I hope it had some value.

---

[14] Pioneer V had two transmitters for sending data back to Earth, with the 5-watt one acting as the driver. On ground command, the power could be increased to 150 watts. (*A Record of NASA Space Missions since 1958*, Alfred Rosenthal, comp. [Greenbelt, MD: Goddard Space Flight Center, 1982], pp. 26-27.)

Thursday, April 21: Both Horner and Dryden were away and the staff meeting went off with a good bit of useful discussion. We finished up the staff paper on research advisory committees. The exchange of information was useful and pertinent. Immediately following the staff meeting, Johnson, Gleason and Nunn came in with the final revisions of the text for the House committee staff. Once again, I insisted that the Defense Department have a chance to look at the text. At 11:00, Jim McDonnell of St. Louis came in to report on Project Mercury. At least, that's what he told me he wanted to do. About three or four minutes was taken up with that aspect of the visit and the balance was a low pressure sales talk on the capabilities of the McDonnell Aircraft Corporation and the competition for the Saturn S-IV stage. Almost all of the competitors have now been in to see me under some pretext or other.

Immediately after lunch, Golovin, Rosenthal and Lilly came in to discuss the administrative personnel ceilings at Huntsville for the balance of this fiscal year. We have been able to set aside 100 positions but the people at Huntsville say it is a wholly impossible task—that they will need 500 plus positions before 1 July. This is the first inkling we have had of any difficulty of this sort. I suggest that there has been some bad planning at some point in the past.

At 6:00, I went over to the Washington Hotel where I was the recipient of the "Man of the Month" Award of the National Aviation Club. A great many people came in—only a few of them known to me. Most of them seemed to know me, but that is not unusual in Washington. A very short ceremony accompanied the presenting of the plaque and I responded in kind. It was a pleasant affair and I imagine it must do some good. I feel a little silly accepting awards of this kind, which supposedly single out a person who has done something unusual. Perhaps the fact that I have stayed in Washington this long on this trip is unusual. Back home for a visit with Ruth and a pleasant and light supper.

Friday, April 22: Will Mickle came in on time this time, and we had a few minutes together. It was obvious he had nothing particular on his mind—at least he was disarmingly clear in his avoidance of any questions that might make headlines in his newspaper. He simply wanted to assure us of his assistance at any stage of the game. At 11:00 Jim Gleason came in to talk over the legislative schedule for the next two or three weeks. It is obvious that I must get up on the Hill and do a little bit of "politicking." I'll get started on this Monday.

I called Dick Harkness and was able to get him to have lunch with me at Jack Hunt's Raw Bar. Dick had to do a radio commentary on the arrival of President de Gaulle but this was to take place 10 minutes before 1:00 within two blocks of the restaurant. I saw the very large crowd—avoided it—and arrived at the restaurant in time to get a good table. Dick soon appeared and we had a pleasant lunch. Ruth and I have eaten here several times but have never seen Jack Hunt. Shortly after Dick had come in, a heavy-set man in what appeared to be work clothes stepped up to the table and said, "You are Dr. Glennan and Richard Harkness, aren't you?" He went on to say that he was Jack Hunt and "wouldn't we be his guests?" He was really very pleasant and seemed pleased that we had come in. Dick Harkness has some problems with his youngster who wants to go on for graduate work in physics but is having a little difficulty gaining admittance to the two or three schools he has selected on the basis of their reputations. Apparently, his academic record is good but not exceptional.

At 2:30 John Corson and Don Stone came in to interview me on the organizational advisory committee's program. I don't think I helped them much, but we did clarify a few problems. It seemed to me that they were striking at some rather unimportant targets and I spared no words in telling them so.

At 4:00 Shelby Thompson came in and we had a talk about the possibility of his joining up with us as head of our public and technical information activities. This is the job about which I have been worrying for several weeks. He is going to think the matter over—he is under a severe mental strain because of the serious illness of his wife.

Saturday, April 23: This was a really busy day. We got up rather leisurely and had a good breakfast; then I attempted to do a bit of work. Most of it had to do with trying to get started on a speech for the Foreign Policy Association in New York on 11 May, but I managed to get a good letter written to Vannevar Bush. I hope he will respond in kind. [After a visit to the farm of friends near the new Chantilly airport,] we dashed back to the apartment and prepared for a call at the French embassy where we were invited—along with 2,000 other people— to a reception for President de Gaulle. This turned out to be a pleasant affair and was exceptionally well handled. I did not know too many of the people there, but we enjoyed a couple of glasses of champagne and some of the very excellent French pastries.

Leaving there, we dashed back to the apartment and changed again—this time into black tie—for a dinner with Deputy Secretary of Defense Jim Douglas and Administrator Pete Quesada of the Federal Aviation Administration. This was a dinner given for several of their friends—perhaps 50 in all—preceding a concert of the Air Force music organizations. This was a benefit performance for the Air Force Wives' Association fund. It was a hot night—the thermometer had stood at 91 during the day. The auditorium was not cooled and everyone suffered accordingly. What a day! Ruth had to change her dress four times and we managed all of our engagements without excessive strain. I did not get much work done, but I think that's not too bad an idea for this weekend.

Sunday, April 24: Again, we had a leisurely breakfast and then set out for the airport, where I was to find my dinner jacket that had been left there some two weeks before. I had needed it last night but was able to "make do" with a white dinner jacket and the trousers of my full dress suit. We drove on to Huntlands where we spent the day with George and Alice Brown and his brother Herman and his wife. Being with the Browns is a real joy. They are very relaxing and make one feel completely at home. I am sure that we will enjoy using the house during the summer and for once, I have no compunctions about it. They are so genuine in their desire to have us make use of the facility that I think it would be an affront if we had the opportunity and did not accept.

Back at the apartment finishing up this recording and trying to clean up the desk. I did forget to tell the tape about a really interesting situation that happened this past week. When Congressman Albert Thomas desires to do something on the floor of the House about a bill he is presenting, he seems to go into cahoots with Congressman Gross. At least, it appears that way, for Congressman Gross is continually raising "points of order" on particular items of particular bills. When our 1961 appropriation bill was being presented, it contained $10,000 for extraordinary expenses—these being entertainment expenses for specific and proper purposes. This represented a cut of $10,000 from the amount we had asked. It so happens, the Senate has not as yet passed the authorization legislation. Thus, a point of order could have been raised to stall the entire appropriation action within the House of Representatives. Gross this time raised a point of order only about the $10,000. Thomas admitted that the point of order was valid and the chair sustained it. Actually, this action had the effect of restoring the full $20,000 due to a legalistic twist in the legislation.

It seemed wise to call Thomas and bring this to his attention—in fact, to invite his suggestion that we act in good faith and operate at the $10,000 level if at all possible. When I asked the question of him he stated that our people were completely wrong—that Gross' action had stricken the entire amount from the bill. I said, "Well, then Albert, I guess we will have to go and attempt to get it back on the Senate side. Will you be willing to assist in this?" Albert responded in the affirmative. Next episode—I ran into Thomas at the concert on Saturday night. He stopped me and said, "You know, Doctor, your budget man is legally exactly right but, morally, he is entirely wrong." I would have liked to exclaim, "the pot calling the kettle black!"

Monday, April 25: I was a little late getting in this morning and arrived to find that Johnson, Gleason, Horner, Dryden and others were waiting for me to discuss the draft bill prepared by the House Legislative Committee. We agreed that it did no violence to any of our principles. I think that Johnny Johnson has done an excellent job in containing what might otherwise have been a difficult situation. His advice is always sound—at least, it seems that way to me. A little bit late, we listened to the source evaluation committee's report on a lunar hard-landing capsule project.[15] This is one of the Jet Propulsion Laboratory's responsibilities. JPL presented a good report and it does not seem that it will be difficult to accept its recommendation—at least its evaluation.

At 11:00, Maurice Taggert came in to urge consideration of his request for a job. He is the man who has been recommended to us by John McCormack. He

---

[15] This was evidently a reference to an element of Project Ranger, in the early years a troubled program of lunar exploration but one that was highly successful in its last three flights, serving to provide information on the lunar surface that was essential for the later Apollo program and helped to further scientific study of the Moon. (R. Cargill Hall, *Lunar Impact: A History of Project Ranger* [Washington, DC: NASA SP-4210, 1977].) At this particular meeting, Dr. Glennan selected the Aeronutronic Division of Ford Motor Co. to develop and produce the first survivable capsule for landing instruments on the Moon. (Emme, *Aeronautics and Astronautics . . . 1915-1960*, p. 122; R. Cargill Hall, *Project Ranger: A Chronology* [Pasadena: JPL/HR-2, 1971], p. 168.)

is a Boston Irishman without formal education who has worked his way up in the civil service to grade 15. He is now in the Boston office of the Internal Revenue Service; he wants to change again and come back to Washington at his same grade level. He is so certain of his own capabilities that it was difficult to talk with him, but I hope I discouraged his anticipation of any early decision. This is one of those problems that I will have to see John McCormack on and hope that he is a little bit more reasonable in this one than he has been previously. Politics may be the art of compromise but it is also the art of influence.

At 11:30, von Braun, Horner and Ostrander came in with Hugh Dryden to deal finally with the Saturn S-IV stage competition. I gave my reasons for selecting the Douglas Company as the winner and found, to my satisfaction, that all present agreed with me. This is not an easy decision to reach—the competitors, Convair and Douglas, were quite close together. The overriding concern is that of providing an opportunity for some semblance of competition.

Rube Mettler of Space Technology Laboratories came in at noon for lunch with me. I was trying to interest him in taking on Dick Horner's job. Mettler has made close to a million dollars out of his association with his firm, but he has so committed himself with his superiors and with Dr. Louis Dunn that it is only reasonable to expect that he must respect those commitments. He did not have any particular suggestions that would be helpful in this situation. Hence, we must look to the next man.

Having swallowed my disappointment on Mettler, I repaired to the Pentagon to visit with Joe Charyk. He is deputy secretary of the Air Force and seems to enjoy his experiences here in Washington. He has been strongly recommended for the Horner job. Once again, he has accepted so many responsibilities in connection with his present job that he believes he should not leave it until the end of the current administration. I cannot argue with a person in this situation—he has spent the better part of a year getting to the point where he now believes he can make real progress. To change at this time seems unwise. Once again, we must look elsewhere.

I have been having difficulty in securing clearance in the matter of the AEC-NASA relationships. Attempting to avoid a clash with Clinton Anderson and some of the committee members on the Joint Congressional Committee on Atomic Energy seems almost a full-time job. I was able to get hold of Norris Bradbury [director of the Los Alamos Scientific Laboratory] today and he confirmed my impression that we are in complete agreement on the pace and course to be followed in the Rover project.[16] I will now try to get John McCone on the line and attempt to clarify and then crystallize the management picture. Norris says that he has as

---

16 In the joint NASA-AEC Project Rover to develop a nuclear rocket, the AEC was responsible for reactor development and NASA for "integration of the reactor into engines and vehicles." (Undated article by Harold B. Finger, "Nuclear Energy for Advanced Space Missions," in "Nuclear Propulsion: Project Rover" file, NASA Historical Reference Collection.) The Los Alamos Scientific Laboratory had primary responsibility for AEC's mission in the project to develop a rocket reactor. (NASA, *Third Annual Report in the Fields of Aeronautics and Space,* 18 Jan. 1961, p. 31.)

much trouble with McCone as I do. The ways of important men are often difficult to fathom.

Homer has had his center directors in today and we are going to his house for buffet supper tonight. I am looking forward to this.

Tuesday, April 26: The dinner at Horner's house was very pleasant last night. He built the house two or three years ago. It is situated in a very attractive location, is air-conditioned and well suited for modern living. I should think he might be a little unhappy about leaving it. We did not stay overly long but had good conversation while there. I had called a meeting at 8:30 this morning to deal with the award of the Saturn S-IV stage contract and the lunar hard-landing capsule contract. Von Braun and Pickering were added to the group so that they might, in effect, be a part of the decision-making process—at least they could protest if they so desired. In this instance they were both happy with the decisions I had made. Having established this fact, we got on with the business of determining who would carry out what part of the program set up for notification of the winning contractor, the organizations that had lost in the competition, the interested congressmen, the press, etc.

I had asked Bonney to prepare draft stories for our review this morning. Really, it is almost unbelievable that such inadequate releases could be prepared. I don't want to be unreasonable but one wonders what is necessary to develop a sense of responsible reporting in an office such as our public information office. We did a bit of drafting at the table and were able to build on the framework given us so that a reasonably satisfactory release was available in each instance. I had to call Donald Douglas and I asked Homer to call the Ford people about the lunar capsule. Marvin Stone, a new Washington correspondent for the *U.S. News and World Report*, came in to see me. He is a friend of Senator Hartke's and seems a responsible person. I had to cut the visit short and take him down to see Bonney because I had to be up on the Hill to do a little "politicking."

In rapid succession, I saw Congressman Anfuso, Congressman Roush, and Congressman Dave King. I was attempting to get them to understand why it seemed unwise to me and to NASA to include in the legislation sections relating to the establishment of a general advisory committee and the aeronautics and space activities coordinating board. These could be covered adequately by strong statements in a report prepared by the committee to accompany the proposed legislation. The real problem, of course, is that the committee wants to have some hand in the draftsmanship and it is improbable, therefore, that we will have our way. I called Congressman King out of the executive session of the committee to talk with him. This was a mistake—or was it? Almost immediately, word was sent out from the committee room asking me to appear. I found myself on the witness stand for almost an hour. It was a good session. The committee was in good spirits and I enjoyed the debate. As a matter of fact, I found out later that they accepted my strong recommendation on the general advisory committee but I did not win on the activities coordinating board.

Anfuso, Roush, and King all gave me strong support. The value of maintaining a good relationship with these people is demonstrated over and over again. At 11:30 I hastened over to the Senate television recording studio to do a 15-minute program with Senator Saltonstall. He is a pleasant fellow and I think doing these TV recordings with people of this sort has real value for myself and the agency.

For the first time, I really used the automobile telephone to good advantage. I was late but wanted to get Donald Douglas on the telephone in California before going to lunch. By the time I had returned to the office, he was on the phone and I was able to tell him the good news for his company regarding the Saturn S-IV stage.

Lunch at the Army and Navy Club with Admiral Chick Hayward. He has been criticizing our lack of attention to research and development in the aeronautical field. I shamed him into agreeing to make the trip to Langley with me on Thursday. At 1:30, I went to the Federal Council on Science and Technology where we heard John McCone speak about the high energy physics program of the nation and raise some questions about the materials research program. In the first instance, it is clear that the high energy physicists have an inside track but it isn't quite far enough inside. Actually, the country is facing an expenditure in excess of $200 million a year in this field, and this expenditure will continue as far as one can determine into the future. When he came to the materials program, John stubbed his toe. Only his agency has failed to pick up its responsibility in this matter. I hope he got the point. The council had had a request from the National Security Council to make a series of studies. As we looked them over, we asked that the NSC reconsider its request. There is just too much to be done between now and the end of this administration to make a bunch of studies—particularly, if the studies are not to be carried out by the people who have responsibility at the present time.

At 5:30, I was asked to be over at the White House at 5:50. Getting to the White House a few minutes early, I was able to talk over my problems with General Goodpaster. It was interesting to see that the office was kept advised by radio-telephone of the progress of President Eisenhower from Burning Tree Golf Club clear into the White House. I did have about a half an hour with him and discussed the probability of giving him a sound proposal for a cooperative project with the Russians. He asked some searching questions, expressed himself strongly about Khrushchev's recent public utterances and then said, "Go ahead and get this prepared—I think it's a good idea." I told the president about the long distance repair job done by one of the Space Technology Laboratory people on Pioneer V.[17] He was delighted. Back home for a good dinner with a feeling of satisfaction at the results of the day's work.

---

17 On 24 April 1960, NASA announced that Robert Gottfried of Space Technology Laboratories, Inc., had successfully compensated for a faulty diode by working out a new translation code for a telebit channel that collected information from sensors before transmitting it back to Earth. Dr. Glennan congratulated Gottfried "who reached 5.5 million miles into space to clear a trouble that threatened the continuing performance of Pioneer V." (News Release, Apr 24, 1960, in Pioneer V folder, NASA Historical Reference Collection.)

Wednesday, April 27: For some reason, I misplaced a piece of paper on which I had made a list of a good many things that I wanted to accomplish today. Search as I will, I cannot find it. I'll make short work of the record of this day's work because it wasn't too impressive. At 9 o'clock, I took a look at a movie made by the Air Force entitled, "A Survey of Astronautics." It was a pretty good description of the NASA program. How does one keep ahead of the military in matters of this kind? The implication, of course, was that all space activities were to be undertaken by the Air Force—therefore, join the Air Force.

Over to the White House to talk to Goodpaster about the necessity for clearing with a number of people on the cooperative proposal for the Russians. This will be done through the Operations Coordinating Board. Frank Phillips takes on that job and I am sure it will be done well. A quick lunch while listening to the recording of the president's press conference and then off in a hurry to the Senate TV recording rooms again to record a 15-minute tape with Senator Hruska of Nebraska.

From this session, I went directly to Goddard Space Flight Center where I took part in a ceremony commemorating the first anniversary of the establishment of the laboratory and awarded several service pins.

Back to the office to visit with a man Jim Gleason wants to hire. He seems like a good man and I approved this transaction. At 4:30, Horner, Dryden and others got together with me to review several problems. High on the list was the nuclear rocket program. I hope one day we get this settled. Now it is 10:30 and I have been at work practically continuously trying to get the desk cleared up and all of the paper work done so that I can leave for three or four days without a feeling of work incompleted.

Thursday, April 28: We took off from Butler at 8:30 for a day's visit to Langley. Ruth went along with me and was entertained by two or three ladies from the Langley Research Center who took her on a tour of some of that part of the country. Accompanying me were Abbott, Admiral Hayward, Capt. Keene of the Navy Bureau of Weapons and Capt. Myers of the Office of Naval Research. I had wanted to make this trip to have a look at the aeronautical research being carried on by NASA. Admiral Hayward had been critical of our activities—stating, out of ignorance mainly, that we were dropping the ball with respect to aeronautical research. The day was a successful one and the boys at Langley gave a good account of themselves. Discussion was had of our work in the VTOL/STOL type aircraft, structures, Mach 3 configurations and a variety of allied subjects.[18] I have had a note from Admiral Hayward admitting his ignorance of the true facts. It is still not clear that we are doing as much as we should in this field, however.

At 3:10, Ruth and I took off for Boston in the NASA plane. We were to spend the night with General and Mrs. [James] McCormack. The flight was a good

---

[18] VTOL and STOL are acronyms referring to "vertical takeoff and landing" and "short takeoff and landing," respectively. They refer to capabilities of aircraft other than helicopters to take off and land either vertically or on a very short runway. Mach, of course, refers to velocity in relation to the speed of sound in a given medium (e.g., about 741 m.p.h. in dry air at 32° F. and at sea level).

one although we were unable to land at the Boston airport because of fog. We did land at Hanscom Air Force Base at Bedford. This was only 40 minutes away from Boston and caused us no inconvenience. We had a pleasant evening with the McCormacks and General James Doolittle who was in Boston for the same purpose as I—namely, a meeting of the MIT visiting committee on sponsored research. The dinner was as lovely as Mac's dinners always are and the wine was superb.

Friday, April 29: The day was taken up entirely with the activities of the visiting committee. In the morning, we spent most of the time at Lincoln Laboratories. There, in contrast to the situation I had noted two years ago, there has been developed a fine program of basic research with perhaps 50 percent of the effort of the entire group given over to this activity. There is an atmosphere of competence and academic quality about the entire operation. They have transformed themselves from a burgeoning project-oriented group to one in which stability of personnel and quality of research are the principal themes. Much of the responsibility for this must go to Carl Overhage and, I expect, to Jay Stratton. In any event, there is credit enough for everyone in this situation.

Late in the morning we drove to the Instrumentation Laboratory, which is headed by Stark Draper. Draper is the most knowledgeable man in America, if not in the world, in guidance equipment. Most of the gyro-stabilized and inertial reference guidance equipments now in use or in development stem from his original work. He knows this and, while he was not in attendance on this particular occasion, most of his audiences know it before he is finished. I gained the impression that the Instrumentation Laboratory had really run its course as an academically associated laboratory. I say this in spite of the fact that a good deal of educational activity seems to be going on there. The entire operation appears to be that of a factory, however. I am sure that the income to MIT is substantial—I believe it's in the neighborhood of $2 million a year. In spite of this, however, I and my associates would recommend limiting its growth and allowing it to phase out gradually as personnel attrition begins to occur. I hope that our report to the corporation will state this rather clearly.

After a quick lunch at the Instrumentation Laboratory, we visited several other laboratories including the Computer Laboratory, a biological element of the Institute on Communication Sciences, the Physics Research Group, and an Icing Phenomena Laboratory. We were not particularly impressed by the last of these—it would fit better into an industrial organization's operations. After having completed our visits, we repaired to the Algonquin Club for a drink, dinner and some discussion. It was during the course of this discussion that we agreed that relatively harsh words should be said about the Instrumentation Laboratory. Much of this was centered about the fact that Stark Draper is an individualist—one who has never been willing to concern himself about his successor. I added my own strong convictions about the nature of the task being done as one that stretches the concept of an academically-associated research activity.

Returning to Mac's home, I found Ruth, Eleanor and Mac still out at dinner. They came back; we had a nightcap and were off to bed. Once again, I found myself thinking over a day at MIT as an experience of real significance. This is a truly great institution—it needs to watch the course of its growth if it is not to fall into bad habits with respect to its extra-educational activities. It has the same standing in its field as Harvard, Yale or Princeton in the large university field and Swarthmore, Pomona, Oberlin, etc. in the small college group. It commands respect as few institutions do. How to continue to warrant this respect without becoming supercilious on the one hand or bowing to crass materialism on the other is going to be a problem.

Saturday, April 30: We were up early, had a nice breakfast with the McCormacks, accepted a gift of a couple of bottles of wine, and were off to the airport and to Cleveland. We landed at 11:45 and, picking up a car from NASA, drove directly home. After picking up Sally, I had a brief lunch and prepared for a meeting with Kent Smith and Fred Crawford.

CHAPTER SIX

MAY 1960

Sunday, May 1: Ruth, Sally and I went to early church so that I might be back at the house for a meeting. John Hrones came over late in the afternoon and we went over a number of problems. It is a very satisfying relationship we have. I hope that there continues to be the mutual respect that now exists.

Monday, May 2: Up at an early hour and out to the NASA hanger to meet the AEC people for a trip to the Plum Brook reactor site. John McCone was unable to make the trip because of his involvement in the negotiations of the nuclear test ban but he arranged for Jack Floberg, a good friend and another of the commissioners, to be present. Accompanying Floberg were General Luedecke and [Director of the Division of Reactor Development] Frank Pittman. Abbott and Horner were in attendance as well. An Army helicopter took us over to the site. It is an excellent reactor, well-designed and getting very close to the point where it will go "critical." I think the AEC people were a little bit surprised at the quality of the design job, the planning for operational controls, the obvious attention that had been given to experimental methods, etc. There is a real question about the use to which NASA can put this reactor. The AEC holds control over all research work involved in reactor design and this means fuel elements, as well. All we are allowed to do, really, is to place specimens in the radiation field inside the reactor. And may it be said here, the AEC is very jealous of its prerogatives.

After a quick visit to the rocket systems test area—we test medium size rocket engines at the Plum Brook site—we jumped into cars and drove back to the Lewis Research Center for lunch. We did not take the helicopter back because Dick Horner felt it was unsafe and in hands that were not as well-trained as he thought necessary. After lunch, we made a quick tour of several of the wind tunnels to look at the models of missiles and space flight vehicles now being tested at Lewis. We also saw the training device used by the astronauts to test their reactions and give them experience in orienting themselves in the face of violent tumbling.

Tuesday, May 3: Up at 5:30 and out to the NASA hanger to fly to Washington in a Convair. Several others availed themselves of the opportunity for a ride, and we reached Butler Hanger in Washington at 8:30. I went directly to the Pentagon for a meeting of the Naval Research Advisory Committee of which I am a member. This took up the rest of the morning and I found the discussion rather interesting. Nothing can be said about it in these pages, however.

I had lunch with Dryden and Horner to tell them about my discussion with Henry Reid at Langley. Reid proposes to retire but would like to work for another year or two on some projects of his own. He is now the director of Langley Research

Center and will step down immediately so that Tommy Thompson, his associate director, can take over. All of us believe this to be a good move—particularly if accompanied by a determination on the part of Thompson to immediately set about finding three or four candidates who can be watched closely as possible replacements for him. Thompson is now 60 years of age.

Further discussion of the problems of finding a satisfactory solution to our public and technical information problem resulted in a hope—a pious hope—that we might be able to pull off a reorganization that would place Bonney under a new man who would head the entire information services activity. The leading candidate for that job is Shelby Thompson of the AEC. He will be in later today to visit with Dryden and Horner. Another matter under discussion was the result of discussions in the plane returning to Washington between Horner, Luedecke, et al. It seems quite apparent that there is no real disposition to accept a NASA man as head of the combined operation on Project Rover. Someone has really poisoned the ground on which Harry Finger walks. I am just about as much determined to ensure there is no solution that will not be fully acceptable to NASA since we are the ultimate user of the Rover device. It was agreed I would do my best to have a further discussion with John McCone and attempt to resolve this problem. He is not an easy one to pin down and then to keep pinned.

At 2 o'clock Eugene Zuckert, an old friend and fellow commissioner on the AEC back in the early 50s, came in to talk with me. What he really wanted was to find out whether or not it was useful for his associates in a small company in Pittsburgh, started by Gordon Dean, to make a bid on an operating contract for the Plum Brook reactor. Word had gotten around that some of the very large companies were bidding and Gene wanted to know whether the small ones would get a fair shake. He was assured that this would be the case but I did ask others on the staff what might have given rise to the rumor. They were mystified and later on, I clarified the whole issue with Zuckert. His people better bid if they want to get the job. Nothing attempted—nothing gained!

W. W. Baker of the *Kansas City Star* came in at 2:30 for a brief visit. These newspaper chaps from the Middle West are uniformly pleasant to deal with. I hope I gave him some background information that will allow him to do a better reporting job in the future. At 3:30, Shelby Thompson came in and spent an hour with Hugh, Dick and Frank Phillips. He did a low-key selling job but was obviously in command of the situation. A later check showed that someone—I think Al Hodgson—had been operating on his own to look into Thompson's reputation and background. Finding a negative reaction from some of his pals at the AEC, Al had told Dryden, Horner and Phillips about Thompson's shortcomings. In spite of this negative approach, all later reported themselves well satisfied with Thompson's discussion and appearance. The next move is up to me.

At 5 o'clock, T. S. Dixon of Rocketdyne came in to assure me of the determination of his organization to win the 200 K engine contract. Home a little after 6:00 with a real load of work to get behind me before the next day.

Wednesday, May 4: Arnold Frutkin was in at 8:30 to talk over the speech I am to give the Council on Foreign Relations in New York next Tuesday. Arnold has done a good job of writing up a story on our international cooperation activities. At 9 o'clock the source evaluation board on the 200 K engine reported. This is a stinker, in the vernacular—five companies bid and three of them are very close together at the top. In fact, they are so close in the technical evaluation that it is almost impossible to choose among them. The same is essentially true in the business evaluation except that one of them bid $138 million, a second bid $69 million, the third bid $44 million. These bids are really estimates of the total cost of the project since this research and development work is always handled on a cost plus a fixed-fee basis. The costs do give an indication of the extent of experience of a company in undertaking a difficult task of this sort. For instance, one of the companies that was not in the running bid only $24 million dollars. While the highest one is undoubtedly high, the lowest indicates a complete lack of understanding of the difficulty of the job. I took the reports and will now have to sit down with myself in an attempt to find a proper answer to this question.

At 11 o'clock, Bob Abernathy of NBC came in to discuss the show we are doing together on 14 May. This is a good operation, in my opinion. It will, for the first time, concern itself as a total show with the reasons for our being active in the space research field. Dryden, Horner, Siepert, Johnson and Ostrander sat with me at lunch to review our source evaluation and our source selection processes. These are never simple matters to deal with and we certainly are having our problems. We are not certain whether to indicate to the prospective bidders something more of the criteria against which we will evaluate their proposals. We are not certain of the extent to which we ought to "debrief" the unsuccessful bidders. It is a fact that if 10 people bid, 9 of them are going to be unhappy because only 1 can win. With the 9 having representatives in Congress, it is almost inevitable that some charges of favoritism, lack of objectivity, etc., will be tossed our way.

At 2 o'clock, Dryden, Frutkin and I attended a meeting of the principals of the Operations Coordinating Board at the New State Building. Our purpose—to review with the OCB the proposal we hope to hand to President Eisenhower for a cooperative program in meteorological satellites with the Soviet Union. We have been working on this for some time—I have already reported having a visit with the president about this matter. Most of the members of the OCB were favorably inclined once they knew that this was to be prepared as a proposal to be held in readiness for possible submission to Khrushchev. John McCone, however, rather strongly objected on the basis that the question of the resolution to be expected from the television cameras in the meteorological satellites might be subject to review by Khrushchev. We had proposed that we continue to use the cameras in Tiros I because these have given us quite adequate cloud cover pictures without any possibility of recognition of bases, cities, etc., on the ground.

Both Allen Dulles and John McCone wanted us to propose reciprocal representation at the launching sites of the two countries involved. This seems to

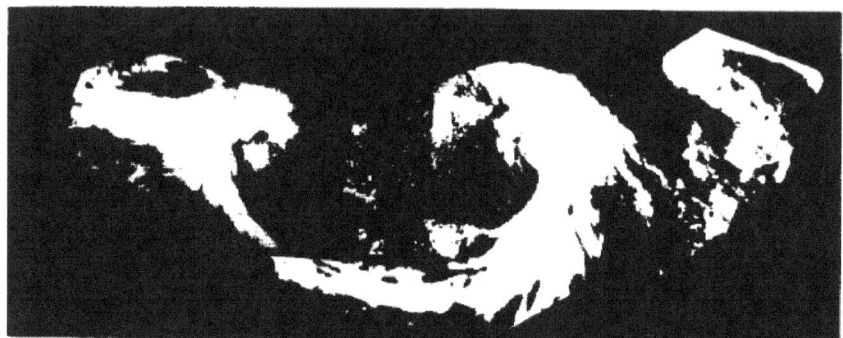

MOSAIC OF TIROS PHOTOGRAPHS

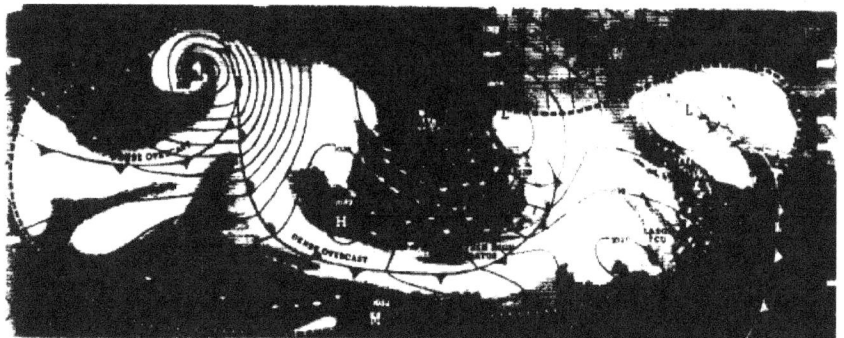

WEATHER MAP, MAY 20, 1960, WITH TIROS CLOUD DATA

THE BEGINNING OF GLOBAL WEATHER PHOTOGRAPHY
TIROS I, MAY 1960

NASA HQ61-89
REV 10-10 66

me to be completely unacceptable—the millennium has not yet arrived. If, indeed, we were to take this position with the idea that we would have a backup position in which we would avoid representation at each other's launchings, it seems silly to make the proposal in the first place. In the light of other events taking place at this very time, I think my judgment is a valid one. In any event, we came away a little crestfallen but more thoroughly aware of the difficulties of gaining agreement on matters of this kind in a government such as ours.

At 3:15, Dryden, Horner, Gleason and I met with the Michigan congressional delegation at the Capitol. As I have noted earlier today, companies have friends in Congress. In this instance, the Michigan delegation was questioning our denial of the Saturn S-IV stage to Chrysler. Chrysler apparently has made misrepresentations on the Hill that have caused these difficulties. I have called Tom Morrow [of Chrysler] to ask him just what he hoped to gain by this kind of action. He tried to beg off from any responsibility for it but refused to support the evaluation process we use as I had hoped he might. I finally broke off the conversation with him by suggesting that if he didn't believe in the system of competitive bidding, I hoped he would avoid making any proposals to us in the future. As is evident, I was mad! In any event, I think we were able to give the Michigan delegation a good bit of satisfaction—particularly when we told it that the bulk of the work Chrysler would have undertaken would have been done in Florida. This fact had not been related to the delegation, which was concerned only with the labor situation in Michigan—not Florida.

Back at the office for a discussion with Jim Gleason of the program to be followed in the Senate Appropriations Committee hearings. Dryden, Horner and I will carry the ball in this one, which is now set for 17 May; this means I must come back from Florida for that day. At 5:00, Dryden, Newell and I discussed the handling of the space science panel meeting set for tomorrow noon.

Ruth and I were expected to attend the United States Chamber of Commerce dinner meeting at 7 o'clock. I just had to call at 6:15 and say that it was impossible for us to be present even though I was to be at the head table. I was completely worn out—went home and went right to bed but managed to do quite a bit of work by catching up on my reading.

Thursday, May 5: Staff meeting at 8:30 with other meetings during the morning on a variety of problems including particularly the public and technical information problem and the budget review planning.

At noon, the members of the space science panel of the President's Scientific Advisory Committee came in for lunch. They had sent us a long memorandum of complaint about the manner in which we were dealing with the scientific community, so called. Homer Newell did a fine job of answering, very patiently, each of their complaints. Lloyd Berkner was his usual dominant self, but we managed to deflate him a little bit during the course of the discussion. Actually, the scientific community, as such, is a bunch of spoiled individuals—the higher they rise in the hierarchy the more spoiled they become. In this instance, at least two of the men present spent most of the time arguing over projects in which they were involved. Conflict of interest? Not at all—they are scientists! In spite of these statements, these are good people and their voices must be heard. We are doing our best to accommodate their interests and to handle our program so that a maximum number of them are involved. On the other hand, the responsibility for the make-up of the program and the expenditure of the money cannot be delegated to any group outside of the government.

Dashing for my plane for New York, I stopped in with Gleason and Johnson for a few minutes to check over some papers—there seems to be time to do these jobs well. I took the plane to New York with a stop-over at Philadelphia where Ruth joined me. She had driven up to Philadelphia to leave the car there, for she plans to come back tomorrow and drive on to Cleveland in order to provide home and fireside for Sally and Polly over the mock political convention weekend at Case. We arrived in New York on time and there was ample opportunity for me to get a haircut, which I had been putting off for a week. It turned out that our planned dinner had grown into quite a dinner party. I think there were 14 in all. A fine discussion was had throughout the evening. We didn't break up for bed until midnight—not much rest for the wicked.

Friday, May 6: Up at 6:15 to see that Ruth catches a bus at 6:50 so that she may get down to Philadelphia in time to pick up the car and drive to Cleveland during daylight and in time to help the girls. This was accomplished without too much difficulty and I walked back to Grand Central Station to make certain about my train arrangements for New Haven. At 10 o'clock, off to New Haven and a meeting of the university council. I had lunch with Charlie O'Hearn and Phil Pillsbury of the Minneapolis Pillsbury flour interests. The lunch at Mory's was pleasant, noisy and not quite the same as it used to be when I was a kid in college. The university council met at the building newly occupied by the Yale University Press. It is a rebuilt bakery and is rather attractive considering the use to which the building had previously been put. [Following the meeting of the council, an advisory group to the Yale president, and dinner], I boarded the sleeping car at 11:15 and immediately took a pill and went off to sleep.

Saturday, May 7: Into Washington three-quarters of an hour late, I had just time enough for a cup of coffee and a sweet roll before meeting a group at the office. We had gathered together the directors of the various divisions, Dryden, Horner and myself to discuss budgetary and program planning for the coming year. It must be understood that at any one time, we are dealing with three budgets or at least with elements of three budgets. We are attempting to spend the money in the current year sensibly, to defend the budget for the next fiscal year, and we are well into the planning for the next or third fiscal year. This meeting was intended to deal with the problem of planning our operations in the face of probable congressional action on our FY 1961 budget and developing guidelines for the FY 1962 budget, on which I am to have a preview late this month.

After three hours of it, I think we could agree that we made some progress. Bob Abernathy came in to go over the script for the television show next Tuesday and Wednesday. We are to rehearse on Tuesday morning and tape the show on Wednesday night. It will be televised on "Worldwide 60" on NBC on Saturday night, 14 May. I think the script is making good sense. Since Ruth was away, I stayed at the office working on a variety of problems until about 4 o'clock. Dashing home, I had another visit with Bob Abernathy to clarify a few points, made up a dinner of tamales, did some more work on my speech for Tuesday next and off to bed about 9:30.

Sunday, May 8: This was another of those days spent at the work table. Everything went quite well and I think I managed to get quite a bit done. It rained most of the day and when Ruth and Polly arrived at 4 o'clock, they told me they had driven through hard rain most of the way from Cleveland. We got Polly on to the 4 o'clock train and came back to the apartment for a quiet dinner. Ruth is obviously tired after her weekend, which had apparently been a very successful one. Both

Sally and Polly seemed quite happy with their part in it. At least, this is what I was told.

Monday, May 9: Once again, at 8:15 in the morning, Arnold Frutkin came in. He seems to have a penchant for keeping engagements with me in the early morning. This time, we were discussing the problems of Project Comet, the possible cooperative meteorological program with the USSR. I have come to the conclusion that it makes no sense to go forward with this—that it ought to be deferred until a later date. Frutkin was at first inclined to disagree but finally came around. Dryden was quite sure this was the proper course to take. I will get the message back to Gordon Gray of the NSC and try to button this project up for the moment.

At 9 o'clock, Dryden, Horner and I met with John McCone and General Luedecke at the AEC. Once again we were trying to settle the management problems for Project Rover. McCone stated clearly that there was a very real question about Harry Finger—that he had been characterized as a person who wanted to "kill" Project Rover. No one can quite determine who started this rumor but it is obvious that Finger would have six strikes against him in undertaking this job. John McCone suggested that we ask our staffs to nominate possible candidates for the job of managing the Rover program in the AEC and NASA. We will nominate Finger and perhaps one other. I agreed to this process and insisted that it be finished by 1 June—that is, that we have a decision by that time. I then asked John to consider the use of the Plum Brook reactor. We were willing to give it to them— I am not sure all of my people would agree with this but it is a gambit I think worthwhile taking. On the other hand, we recognize the right of the AEC to deny us the use of the reactor for reactor development work. We offered, too, to operate the reactor on request from AEC— they would have to fund their requests. It appears that the best thing to do is to go ahead with our own work and to accept work from the AEC that it will fund. I think this may be the only way we can tackle the job at the moment; perhaps we can move later on to a more desirable form of organization. We have advocated a steering committee composed of AEC and NASA people who would determine the use of the reactor for whatever purpose. Why the AEC doesn't want to do this, I will never know. John asked me to stay after the meeting and handed me an "eyes only" paper that set forth his arguments against Project Comet. I did not tell him that we had effectively put the project to bed an hour or so earlier.

At 12:15, I had lunch with John Corson at the Metropolitan Club. I was anxious to have his evaluation of Shelby Thompson and found that it was really very much better than I had expected. I had asked Dick Horner to come along to lunch so that he could hear this evaluation firsthand. We also talked about the meeting of the Kimpton advisory committee on organization, which had taken place at Huntsville on Friday and Saturday last. Obviously, the directorate in Washington must concern itself with the development of good relationships with this advisory committee. There seems to be some question as to the division of effort we have decided upon between the various development centers. We agreed that a meeting on Wednesday of this week would be helpful in the preparation of a paper to be given to the Kimpton committee in advance of its meeting at Langley late this month.

Directly following lunch, we had a briefing by the inventions and contributions board. These poor devils have had to review 2,000 proposals for awards arising out of supposed inventions or contributions. It seems to me that we will never get any place with this kind of an operation, and I have suggested that we seek remedial legislation. In the meantime, we will go ahead with the activities at a low key. At 3:30, John Johnson brought in a report of the long range planning committee. This is a group that is attempting to look into the economic, political and social implications of space research. Already, it has contracts with the Rand Corporation and with the Brookings Institution.[1] I agreed to a further contract with a Cambridge, Massachusetts, group of Harvard and MIT professors who will look into the public acceptance aspects of our activities.

Tuesday, May 10: Out to the NBC studios at 8:30 for a rehearsal of the "Worldwide 60" program. This lasted only an hour and a half and I was able to get back to the office in time to do a little bit of work. At 11:30, Dryden, Horner and I visited with Secretary Jim Douglas at the Pentagon. Dick, Hugh and I were attempting to get Jim Douglas, Herb York and Jack Stempler to agree on the paper setting up the Aeronautics and Astronautics Coordinating Board.[2] We almost made it! Some corrective language is necessary and I have left it with Dryden and Jim Douglas has left it with York to come up with the answers and put the damn thing to bed.

After a hurried lunch, I boarded a plane to go to New York to speak to the Council on Foreign Relations. I stopped at the Yale Club and had a shower, a good massage and a light sleep. At the Council on Foreign Relations I found a number of good friends and was able to talk quite forcibly to this group of some 60 people. The dinner was pleasant, the questions were pertinent and I felt well repaid for the effort. I did not read the speech; rather, I followed it as a general outline and did quite a bit of extemporaneous talking. I arrived back in Washington at midnight, satisfied but much the worse for wear.

Wednesday, May 11: At 9 o'clock, Allen Puckett of Hughes Aircraft came in to see me. I am trying to get him to take over Horner's job on 1 July with the possibility that he might replace me next January. I had not met him previously but had excellent reports on him from a number of people. I was delighted with this man—we spoke for fully two hours with Hugh Dryden coming in for the last half

---

[1] The Rand Corporation was an organization headquartered in Santa Monica, California, that engaged in non-profit research for the Air Force and other agencies and organizations. It began as Project RAND in 1946. The Brookings Institution was a non-profit research institute founded in 1927 in Washington, D.C., by philanthropist Robert S. Brookings to provide public service through research and education in the social sciences.

[2] NASA and the DOD entered into an agreement establishing the AACB in September 1960. It was responsible for facilitating the two agencies' planning to avoid undesirable duplication of effort and promote efficiency, for coordinating activities of common interest, for identifying common problems, and for exchanging information. (Rosholt, *Administrative History of NASA*, p. 172.)

hour. I think he would make an excellent man for us; he is well-heeled scientifically and has had substantial responsibilities administratively. He is young, attractive and asks the right kind of questions. Obviously, he has a great future ahead of him in industry. To come with us he must be willing to break off all his industrial connections, take on a job where he does not know who will be his boss eight months from now, risk the possibility of an adverse election situation—all in all not a very pleasant prospect. Further, he probably makes twice as much as he can possibly make with us, maybe more. I intend to call his boss, Pat Hyland, to urge him to help us. I doubt that this will be successful but at least I must try.

At 11 o'clock I taped a 15-minute program for the Voice of America. I hope these operations are useful—the people involved tell me they are. Lunch with Gleason, Dryden and Phillips served to clear up some of the problems that have arisen regarding our appearance before the Senate Appropriations Committee. Immediately following this I visited with Maury Stans at the Bureau of the Budget. We have been cut $39 million by the House. The Senate has approved or authorized $50 million more than we had requested. On the 17th, we go before the Senate Appropriations Committee to try to convince it to restore all the cuts made by the House. I think we will be successful in this. Nevertheless, in a conference between the House and the Senate we will lose something from the total amount requested. Thus it becomes a point in strategy to have the Senate appropriate more than we have requested. I cannot urge this publicly because I would be guilty of breaking the president's budget. Nevertheless, I must so conduct myself that the Senate feels it desirable to give us more money than we have requested. If this can be pulled off, the conference between the House and the Senate may then result in our getting the $915 million we had requested in the first place. Stans was very sympathetic. I can do almost anything short of committing murder or breaking the president's budget. With his understanding in this matter, I think we may accomplish our objective.

I hurried from his office to the Statler Hotel where I had to make some opening remarks at a meeting of the Catholic Press Association. As usual, they were a half an hour late but we managed to get the task over with in time so that I could retrace my steps to the office and meet with several people regarding the development of effective and proper relationships with the General Accounting Office. We have decided that we should be completely open and above board with them but we must insist on knowing what they are after. I don't know whether we will win in this argument—at least we are going to try.

At 4 o'clock, John Corson, Jack Young, and others met with the program directors to discuss the organizational matters brought up by the Kimpton Committee at Huntsville last week. This was not as fruitful a discussion as I had hoped. Everyone is so completely tired out that to even think about organizational changes or organizational methods is a real imposition. Nevertheless, it must be done and I hope that we made some progress.

At 6:30, Ruth and I appeared at the NBC studios to participate in the taping of "Report from Outer Space." Among those on the program were David Brinkley,

[William White] Howells of Harvard, [Robert] Jastrow of NASA, Dee Wyatt of NASA, Frank McGee and Peter Hackes of NBC. There is no need to recount the horrors of this evening. It was interesting but it took five hours before we had finished. Arriving home at 11:30, Ruth finished her packing for the morrow and I sat down at my desk to try and complete some of the activities I must button up before taking off for Florida. Finally, at 1:45, Ruth appeared to drag me off to bed by the hair. I had finished almost all of my work—at least, in some fashion.

Thursday, May 12: Up at 6:30 to pack my bag and to get off at 7:15 for a breakfast with Shelby Thompson. I neglected to say that last night I had a talk with Walter Bonney about the proposed reorganization of the public and technical information activities of NASA. I told him of the desirability of a single organization and found him, not wholly to my surprise, willing to give up his job rather than to report to someone else. In spite of all his protestations about the importance of serving his country, I find that he is just as much of a straw man as any of the rest of us. I was quite frank in my appraisal of his lack of ability to take on the total job. Once again, I can only characterize him as a "great big teddy bear." In some ways he does a quite acceptable job. Creativity seems wholly lacking as is the ability to manage a group of men. This discussion—not to say argument—did not add to my equanimity for the balance of the evening.

At breakfast, as I was saying, I found that Shelby Thompson was willing to look at the public information office as one that could be separated from the total information services operation at NASA. As a matter of fact, the entire breakfast conversation led me to the conclusion that we would be well-advised to attempt to hire Thompson without further delay. A discussion later in the day at which I bared my soul completely to Dryden and Horner resulted in an agreement on our part to move forward. We will make Thompson the director of the office of educational and technical information services, leaving Bonney as the public information officer. His wings will be clipped and I doubt that he is going to thoroughly enjoy the position in which he finds himself. At the end of the year, if Thompson has proved his adaptability in this new field, we may be able to put these two operations together. This will be true particularly if he has gained the confidence of the rest of the organization. To put him in now might well be to doom him to failure because of the lack of complete acceptance of his professional capabilities.

Dick, Hugh and I spent the rest of the morning talking about a variety of things that had to be covered before we left Washington for Miami. I pointed out to them I had spent most of the night trying to figure out how to make a decision on the 200 K engine. I proposed that we give to each of the three leading contractors a contract to further develop their ideas. Each of these might call for $250,000 and we would expect to have a solid report in our hands within four months. It seems to me that this might give us a better idea of the real capabilities of each of the companies—at least it might separate them a little bit more so a decision could be made that would stand up to public scrutiny. Much to my surprise, there seemed to quite a satisfactory acceptance of this proposal.

I dashed off to the airport with Ruth only to find that the plane would be delayed about an hour because of rain. Added to this delay, headwinds brought us into Miami two hours late. Nevertheless, the sun was shining and the vacation has started with prospects for a real rest.

At this juncture, I would like to be able to review the happenings of May 1 through May 11 in respect to the U-2 incident. Unfortunately, security restrictions prevent this. Strangely enough, although NASA has been "clobbered" a bit, I find myself somewhat relaxed about the whole operation. Once again, the volatility of the American press and of American and congressional opinion has been most astounding. There continues to be general agreement that the timing of this particular flight was not a wise one. In the handling of the incident, it is now apparent that we—that is the United States—spoke too soon. Outside of this, the real crime was getting caught—circumstances being what they are in this world. Or, rather, it is a fact that we have always stood for honesty in our dealings with other nations and the fact that we have denied and then confirmed these activities makes our whole posture questionable.[3]

Friday, May 13: This was pretty much a day of rest and getting oriented to relative leisure. We walked a good bit, took the sun, and spent almost half the day on the phone to Washington clearing up matters that were hangovers from earlier in the week.

Saturday, May 14: Beautiful weather. It inclined us to rise relatively early. I started on the speech for the Tulane commencement. In half a day I achieved an outline of sorts and also blocked out a farewell statement to the graduating class at Case. Throughout the day, we listened to news broadcasts about the continuing possibility of difficulties at the summit arising out of the U-2 incident.

Sunday, May 15: I was awakened at 1:00 a.m. to be told by Frank Phillips that the USSR has put a four-and-one-half ton manned capsule into a 200-mile orbit

---

[3] President Eisenhower had been strongly affected by the Japanese surprise attack on Pearl Harbor in 1941 and was determined not to be similarly surprised by the Soviet Union now that it possessed atomic weapons. Consequently, he had authorized Lockheed to develop a high-altitude, single-engine reconnaissance aircraft called the U-2 that could fly great distances at altitudes above 80,000 feet. Once the aircraft was developed, Eisenhower kept himself intimately informed about its activities and controlled them tightly. He authorized the first flight in late June 1956. To American surprise, the Soviets were able to track it and several subsequent missions, which they protested privately but vigorously. Eisenhower then kept flights to a minimum, but did authorize flights on 9 April 1960 and again before 1 May, the eve of a summit meeting with Khrushchev and the leaders of France and Britain, de Gaulle and Macmillan. He assumed that if the Soviets shot down a U-2 they would never admit it because doing so would reveal that there had been other flights they had been unable to hit. However, on 1 May a young CIA pilot named Francis Gary Powers went down in his U-2 inside the Soviet Union. Assuming that he was dead and the plane destroyed, on 5 May Eisenhower approved a statement that a NASA U-2 aircraft doing meteorological research was missing over Turkey, allowing people to infer that it had strayed into Soviet territory. As Khrushchev released more details, including that Gary Powers was still alive and parts of the aircraft captured, Eisenhower finally had to admit publicly that the overflights had been occurring for some time under his orders. As discussed below in the diary, this situation either caused or allowed Khrushchev to wreck the summit and Eisenhower's hopes for détente and disarmament. (For a good brief discussion of these events, see Ambrose, *Eisenhower, the President*, pp. 227-228, 340-341, 569, 571-579; for a more detailed analysis, see Beschloss, *Mayday: The U-2 Affair*.)

above the surface of the earth. Actually, the man in it is a dummy according to the first news broadcasts. Our own stations confirm that there is something up there. It is said by the Russians that they have no plan to recover the capsule but that they will bring it back into the atmosphere and have it burn up on the way down. This is not wholly possible with a device as large as this and built of the materials that would be involved. Why didn't the Echo shot work?[4]

Here we are again—the Russians are successful in launching something for Ike's benefit as he steps out of his plane in Paris. They really seem to have much better control of their activities in this field than we do, as yet. We worked a good bit more on the speech; finished up the Case farewell statement. Several calls from Washington kept me pretty busy, but it was a quiet day.

Monday, May 16: [After listening to news broadcasts about the U-2 affair and the summit,] I was able to finish both of the speeches and now have them ready for typing when I go back

Failed attempt to launch an Echo satellite because the second stage of the Thor-Delta launch vehicle malfunctioned.

to Washington later this week. I worked over the statements to the Senate subcommittee on appropriations and relayed my comments back to Washington in a telephone call lasting three-quarters of an hour. Believing that I might be able to convince Pat Hyland of Hughes Aircraft that the world situation required that he

---

[4] On 13 May 1960 an attempt to launch an Echo satellite failed because of the malfunction of the Thor-Delta launch vehicle. This was the first launch with the 3-stage Thor-Delta, and the second stage was unsuccessful. (*NASA Historical Data Book*, Vol. II, pp. 77, 81, 371.) As Glennan suggests below in the text, the launch date was the day before Eisenhower flew to Paris for what proved to be the abortive summit with Khrushchev.

release Allen Puckett to NASA as a replacement for Dick Horner, I called Pat. Unfortunately, he beat me to it. He simply stated that he had given up two very good men recently, that his company was in trouble and that the organizational problems were centered around the leadership of Allen Puckett. Accordingly, I had to back down. This is a real disappointment.

Tuesday, May 17: Up at 7:00 for another beautiful day. I rewrote portions of the Tulane speech and believe it is somewhat improved. The news from the Summit is all bad. There remains that gnawing doubt that we have appeared to the rest of the world as just another ordinary nation mouthing platitudes and moralities but indulging in a variety of activities of doubtful character. It is clear, on reflection, that we might have been taking much more of a chance in sending U-2s over Russia than we were willing to admit. If a Soviet plane were to come over our part of the world, I doubt not that we would have alerted our SAC force and started it on the way. Because the bombers can be recalled, this would have been a sensible thing to do. If the Russians had wanted to look at the U-2 as an invader, could they not have been justified in launching missiles toward this country? These are difficult questions to answer. Ruth bought some red snapper and baked it. It was good.

Wednesday, May 18: Another beautiful day and little activity that imposed any great strain on us. Ruth took me to the airport to catch a 5:25 flight to Washington. Actually, instead of a DC-7B, I found myself in an Electra. We arrived in Washington 1 hour and 20 minutes late. I found a great deal of mail at the apartment and a quantity of material brought out from the office for me. It was 2:00 a.m. by the time I turned off my light.

Thursday, May 19: Up at 6 o'clock to finish up some of the work I had started earlier in a review of the mail last night. I was at the Statler at 7:30 for a breakfast meeting with Hugh and Dick. We reviewed our strategy for the Senate hearing, which is to start at 10 o'clock, discussed some of the problems arising out of the U-2 incident, reviewed the employment of Shelby Thompson as head of our division of educational and technical information services, and generally got caught up on a variety of matters. An hour-long meeting of the staff produced no important new events. A review in that meeting of the launching schedules for the balance of the year shows that almost all of the flights have slipped from two to four weeks. I have cautioned strongly against allowing this to happen without every effort being made to counter the tendency to become complacent over technical difficulties.

We started the hearing with Senator Magnuson in the chair and Senator Ellender in attendance. This is the subcommittee of the Senate Appropriations Committee that handles the budgets of the independent offices. Very few of the members attend these meetings. Lyndon Johnson did come in and made a strong statement in favor of a full restoration of all the cuts made by the House. I must say that he continues to stand up for us when the need arises. I tried to read my statement but found that it was almost impossible. They kept interrupting me and asking questions that had to be answered. Actually, Hugh and Dick never did get their statements made; they were simply put in the record. I believe that our story was well covered, however.[5]

The rest of the day was spent in attempting to clear up my desk, which had become pretty well cluttered, and in planning for future activities. We determined that I should make the trip to Langley next Thursday to meet with the Kimpton Committee in a discussion over alternative methods of organizing our activities. This seems perfectly proper to me and will give me a chance to go to Wallops Island for an inspection, as well.

The plane was about an hour late in arriving in Washington, it was crowded and dirty, and I could get nothing but tourist accommodations. In spite of this, I was glad to be back in Miami and see Ruth about 11 o'clock Washington time. She had apparently had a quiet day although she was a little bit disgusted at the delay at the airport.

Friday, May 20: Another beautiful day—a day of work, for I had brought back from Washington the entire contents of my desk. I did not seem tired after having spent about six hours on it and Ruth helped me in revising one section of the Tulane speech, which seemed to be a little bit less clear than I had liked. I don't seem to be able to get started on the speech for Eau Claire.

Saturday, May 21: Up at 7:30 and right to work on more of the papers that need attention. I did finally get a bit of a start on the Eau Claire speech, took some sun, and started packing for the return to Washington tomorrow. Ruth has been quite busy cleaning house and doing the washing so that we can leave everything shipshape for the Adamses when they come down early in June. It has been a wonderful vacation for us and a great help to have "courtesy of the port," so to speak.

Sunday, May 22: We were to take off at 8:25 a.m., so I left a call for 7 o'clock. At 6:00, National Airlines called to say that the plane would be delayed an hour. I changed the call by three-quarters of an hour. Even then, we were delayed until almost 11:00 since we taxied to the end of the flightline and then had to return to have a sparkplug changed. It was a pleasant flight, however, and we did reach Washington about 4:00 in the afternoon. Thus ended a pleasant 10 days; we finished it by calling each of the 4 children and we really enjoyed having a visit with them.

Monday, May 23: I started out early in the morning with a meeting with Arnold Frutkin. I have probably said before that we seem to have our meetings the first thing in the morning, and I think this is a pretty good idea. I was anxious about the reported concern of the Mexicans over the agreement they had signed with us for a Mercury tracking station. I found that there was really nothing more nor less than a newspaper reporter's desire to have something spectacular to say as a result of the summit conference.

---

[5] See above, entries for May 11 and U.S. Congress, Senate, Subcommittee of the Committee on Appropriations, *Independent Offices Appropriations, 1961*, Hearings, 1960, 86th Congress, 2nd sess, pp. 237-287. The ultimate NASA appropriation for fiscal year 1961 turned out to be $964 million, less than the $970 million the Senate approved but well above the original request for $915 million and the $876 million the House had allowed. (*NASA Data Book*, Vol. I, p. 128, and *Hearings* just cited, pp. 237, 245.)

At 9 o'clock, Golovin came in to see me at my request. I was anxious to know what his plans were since he has been deputy to Dick Horner these past several months. Obviously, he does not belong in the long-range picture at NASA; yet, it is not too good to contemplate his leaving at the same time as Horner. I think he will delay until August 1 or perhaps until October 1; this is a decision yet to be made.

A discussion with Jim Gleason confirmed my determination to visit with Styles Bridges and Congressman Brooks. I am trying to build a real fire under the Senate in the hopes that it will appropriate more money than we have asked in order that the result of the conference between the House and the Senate be an appropriation of at least as much as we have requested. I can't do this openly but, believe me, I am going to try to do it in other ways. At 10 o'clock, Hodgson, Young and Corson came in to discuss the forthcoming meeting of the organizational advisory committee. I am going to Langley next Thursday—I had wanted to go to MIT for an educational meeting that sounded very interesting—in order to spend some time with this committee. At 11 o'clock, Charlie Robbins of the Atomic Industrial Forum came in to talk about the desirability of his organization doing something about a public meeting on the usefulness and characteristics of nuclear propulsion for rocket vehicles. I told him about our planned meeting for industry and I think that I discouraged him from doing anything strenuous since we are carrying the ball in this instance.

I had a very pleasant lunch with General Ostrander. He is really a fine person—perhaps a little bit less aggressive and forceful than one might like, but I am delighted that he is in the organization.

At 2 o'clock, Dick Mittauer and Herb Rosen came in to discuss a proposed interview with a producer and camera crew from the British Broadcasting Corporation. Apparently they are coming over here to do a show or perhaps two shows on space and missiles. We will do our best to keep them from combining the two. Apparently, I will have to do a tape interview for this particular show. At 2:45, Siepert and Bonney came in and I found that, in my absence, there had been plans made to release the news of the appointment of Shelby Thompson as director of our new office of educational assistance and technical information programs. It is really quite amazing to see the manner in which these people forget the individual in matters of this kind. Apparently, many of our people would have read about the proposed transfers for them long before anybody had talked to them. As might be expected, the signals were changed.

At 3:45, Dr. Randt, Dr. [Freeman H.] Quimby and [Alfred M.] Mayo came in to visit with me. The latter two are new appointees in the office of life sciences programs. They appear eager and are certainly very able. Randt is doing a fine job at bringing together a good top team. At 4 o'clock, General Ostrander, Elliott Mitchell and Abe Hyatt came in to discuss again the 200 K engine. As a result of my questions and stated assumptions about the differences and likenesses of the Rocketdyne and Aerojet proposals, a new analysis had been made of the actual differences between them. Whereas before this analysis, it had been assumed—

actually stated—that the Rocketdyne proposal was substantially less costly than the Aerojet simply because of the assumption that less testing time would be required, it now appears that equalizing the testing time would still leave Rocketdyne at least $11 million lower than Aerojet. Since both figures are then reasonably well within the "ballpark" of estimates provided by our own people, it appears that Rocketdyne should get the nod.[6]

At 5 o'clock, Ed Goodheart and Albert Lowe of the Austin Company came in to see me. Really, they were just like vultures. They were assuming that the failure of the summit meeting meant that we were probably going to go faster and spend more money, and they wanted a part of it. I laughed them out of the office! I told them that if anyone proposed that we should speed up what we were doing, I would fight it and would quit before I would allow hysteria to modify the program we have laid on. Thus ended the first day back from the holiday.

Tuesday, May 24: We started out the day with a source selection board presentation on the ion engine. In this instance, it is going to be relatively easy to select the winner since there is a very much greater separation between the competitors than was the case in either the 200 K engine or the Saturn S-IV stage. At 10 o'clock, I had a meeting with a variety of people over the Saturn selection papers. We have really put everything into the record and made the record available to the General Accounting Office as well as Congress. Nevertheless, I was concerned over the adequacy of the statement in the record. We agreed that nothing should be done but that we ought to look hard at improving the procedures we are using.

At noon, I had lunch with Mervin Kelly. I wanted to ask him for suggestions as to a replacement for Dick Horner. He did come up with some suggestions, and as a result of this discussion, I have called Fred Kappel, president of AT&T. I am going to be able to see him next Friday at Holmdel.[7] The cordiality and helpfulness of Kelly and the entire Bell System group is always a wonderful inspiration to me. At 1 o'clock, we went to an early meeting of the Federal Council for Science and Technology. Roger Ravelle gave a good discussion of the importance of United States participation in the international Indian Ocean oceano-graphic program, and I was sufficiently interested to move approval in principle of United States participation. The rest of the meeting went rather well with some interesting discussion.

At 4 o'clock, Young and Hudson came in to give me some papers they had prepared in anticipation of the Thursday meeting with the advisory committee on

---

[6] A memorandum in Glennan's files for 31 May 1960 (now in the NASA Historical Reference Collection) analyzes the Pratt and Whitney, Aerojet, and Rocketdyne proposals on the 200 K liquid hydrogen-liquid oxygen engine project. All three were judged capable of delivering a satisfactory engine, but the Pratt and Whitney proposal was priced at more than twice the cost of the other two. Since, as Glennan says, Rocketdyne was lower than Aerojet, he determined to negotiate with the former for a contract to develop the engine.

[7] The small town of Holmdel, New Jersey, was home to one of the Bell Laboratories. Below, the Federal Council for Science and Technology had been established by Executive Order 10807 of 13 Mar. 1959 "to promote coordination and improve planning and management of Federal programs in science and technology." (*United States Government Organization Manual 1959-60*, p. 541.)

organization.[8] At 5:30, I had a visit with Senator Styles Bridges. This was a pleasant half hour and I was really pleased with the reception. Styles agreed with everything that I was doing, gave me some good advice and was quite flattering in his comments regarding the manner in which I have been operating personally. He did call attention to the fact that the Saturn S-IV stage competition was not helping us on the Hill, but he said that there was nothing to do but to bow our heads and accept the criticism for the present time.

Wednesday, May 25: At 9 o'clock, I had an hour with Overton Brooks in his office on the Hill. He was complaining about the fact that Wernher von Braun was unwilling to come to his state, Louisiana, for a speech but that he was quite willing to make speeches in New England for $2,000 a speech.[9] An interesting world, isn't it? Actually, Jim Gleason and I had a good hour with Brooks and I think it was a useful meeting. Back at the office to meet several representatives from the North American Phillips Company who came in at the request of Congressman John McCormack. A hurried lunch was followed by a long meeting with the program directors on the 1961 program. It appears that we are at least $20 million short of funds and that we will have to work like the devil to stay within our budget even though we are given more money than we have requested. It seems so very difficult for anybody to cut a piece out of the program and yet they are all very prompt to indicate a shortage of funds. At 5:30, we had a presentation on the geodetic satellite. This has been a question for a long time and it looks as though it might turn out to be a $5 million program. We certainly must determine whether or not to go forward with it. It is apparent that the Defense Department has some activity in this field but it is classified and not something that can be discussed here.[10] Home about 7 o'clock with some attempts to do some work, but the mind will not carry on very well after a day like this.

Thursday, May 26: Up at 6:15 to have breakfast at 7:00 with Chancellor Larry Kimpton of Chicago University, who is the chairman of our advisory committee on organization. He was attempting to get his mind in tune with the

---

[8] Like Young, Hudson presumably worked for McKinsey & Co., the firm doing the study of NASA organization and contracting procedures.

[9] Von Braun was in demand as a speaker and may have commanded this much money for some of his speeches. He did speak at Framingham College in Massachusetts on 3 May 1960 and at the Waldorf-Astoria in New York before the American Newspaper Publishers Assn. on 28 April 1960. But on 30 September 1960 he also spoke before the American Ordnance Association in Shreveport, Louisiana. (Von Braun speech file, 1960, NASA Historical Reference Collection.) Whether the latter speech was coincidental or a result of Bridges' complaint may be impossible to ascertain.

[10] In November 1960, the DOD assumed responsibility for the geodetic satellite program in a direct administrative transfer from NASA. According to *Aviation Week* (21 Nov. 1960, p. 34), "From a military standpoint, the satellite has the potential to fix with great precision the location of any possible target by triangulation of simultaneous pictures taken from ground stations of the high intensity lights which will flash from the satellite." NASA had planned to launch a geodetic satellite "to establish more accurately the shape of the earth, locate and map gravity changes, analyze moon mass and establish the gravitational center of the earth-moon system," according to the same article, but the agency stated it could conduct the program only if the data obtained from the satellite could be made available to the world scientific community. Presumably, this would have compromised military secrets, although the article does not say this.

problems he seems to find in understanding relationships between our research centers and our development centers. It was a pleasant breakfast but I'm not sure that I helped him very much. We went to the airport, picked up Morehead Patterson and flew down to Langley. There we met the rest of the committee and after a visit to various areas in the laboratory, we met in executive session for a discussion of the committee's concern over the organizational arrangements linking together the headquarters and the three development centers.[11] I had gone down to handle this conversation because it seemed to me important that they get some straightforward answers and opinions from the man who had asked the questions in the first place. Actually, it was impossible to stick with the subject. However, I do believe that we cleared up some misconceptions and obtained a useful result.

I climbed into an Aero-Commander and flew back to Washington at full speed to attend a special meeting of the cabinet to which I had been invited. President Eisenhower wanted the cabinet and the heads of the independent agencies to hear directly from him, Undersecretary Merchant of the State Department, and Special Advisor Chip Bohlen, the story of the summit conference. It does seem clear, the president said, that Khrushchev had determined to torpedo the conference before he left Moscow. The president stressed several times the very strong and warm support and loyalty given to him by de Gaulle and MacMillan. He said that there was never a moment's hesitation and that this made his task the more easy. Merchant gave a chronological account of events [ending with the release of a communique by the U.S., Britain, and France] that the summit had been a failure and that Khrushchev was completely to blame in this situation.

After this discussion, Chip Bohlen took up the discussion and made these points:

1. It seems clear now that in March and April, Khrushchev concluded that he was not going to get what he wanted out of the summit conference. His Baku speech indicated this and indicated his pessimism about the possible outcome of the summit.

2. Khrushchev had become thoroughly aware of the determination of the allies to stick together on their program through speeches made by Dillon and Herter.

3. There seems to be clear evidence of a strong reaction against Khrushchev's personal handling of foreign policy in the Soviet Union. The Communist Chinese have expressed themselves strongly as being in opposition to his attempts to win over the West. The army seems clearly

---

[11] The research centers were of course the three inherited from the NACA—Langley, Ames, and Lewis, with the Flight Research Center at Edwards being subordinate to Ames. Wallops had been an independent installation since 1 May 1959, but it did not receive the designation of "center" until 26 April 1974. The three development centers were presumably Marshall Space Flight Center (although the formal mass transfer of personnel and facilities from ABMA did not occur until 1 July 1960); Goddard Space Flight Center in Greenbelt, Maryland; and the Jet Propulsion Laboratory (although it was a contract facility, not a true NASA center). The Manned Spacecraft Center in Houston, later the Johnson Space Center, did not yet exist, and the facilities at Cape Canaveral did not become a separate Launch Operations Center—later Kennedy Space Center—until 1 July 1962. (See Rosholt, *Administrative History of NASA*, esp. chart on p. 342; Helen T. Wells, Susan H. Whiteley, and Carrie E. Karegeannes, *Origin of NASA Names* [Washington, D.C.: NASA SP-4402, 1976], pp. 139-160.)

set against Khrushchev and continues to resent the dismissal of Marshal Zhukov [Soviet minister of defense] more than a year ago.

4.   All of these matters made the possibility of a Soviet success at the summit so remote that the U-2 incident became exactly the device necessary for Khrushchev to use in a strong propagandistic manner.

Bohlen indicated that it was clear Khrushchev came to Paris with a set brief and that no changes were made when it was released. It is the opinion of the State Department that had the U-2 incident not been available, the Russians would have found another way to torpedo the conference. It was stated that Khrushchev expected that he could split the Allies—that, indeed, this was the reason for his coming to Paris a day early. He miscalculated completely and the strong support of de Gaulle and MacMillan for Eisenhower upset Khrushchev's apple cart. Bohlen called attention to the fact that there was really no change in the Soviet foreign policy. He said that he had never in his life studied a document such as the transactions of Khrushchev's press conference on Thursday in Paris in which a man moved so boldly over the total field of battle without once overstepping and suggesting a change in the foreign policy posture of his country.

As a sidelight, the lights went off in the hall in which the press conference was being held during the course of the conference. Khrushchev commented that this was a fault of the capitalistic system. The president said he had found it impossible to avoid laughing out loud during the first meeting with Khrushchev over one of the statements that was so patently erroneous. Eisenhower commented, somewhat ruefully, that one does not laugh at international conferences. Another matter of note was the fact that Khrushchev appeared at no meetings without Foreign Minister Gromyko and Marshal Malinovsky [Zhukov's successor as minister of defense]. This is a complete change from earlier meetings between the heads of state where Khrushchev had insisted on meeting without others being in attendance—that is, other than the one interpreter for each principal. The suggestion was made that Khrushchev was under wraps and that these two men were there to see to it that he stuck to the line that had been decided upon in Moscow.

The president looked tired, somewhat haggard, and acted very much like a man who is determined to take his share of the blame for everything that had happened. He is really a wonderful person and I continue to gain in my admiration for him.

Friday, May 27: Most of this day was taken up with a flight to Holmdel, New Jersey, where I met with the president of the American Telephone & Telegraph Company, Fred Kappel. I was attempting to get some advice and help with respect to my problem of replacing Dick Horner. It was, indeed, a pleasant experience. Along with Kappel, I found Jim Fisk, the president of Bell Telephone Laboratories, and Eugene McNeely, executive vice president of AT&T. They were not only interested; they were really quite sure they wanted to help me, and I left after several hours with the feeling that, if human beings could solve my problem, these men would go a long way toward providing the necessary assistance.

I had an opportunity to visit the Holmdel Station of Bell Telephone Laboratories where the eastern terminus of our Project Echo satellite system is located. These men have done an exceedingly good job in the tradition of scientific and technological research for which the Bell Telephone Laboratories is justly famous. The way they continue to look ahead several years at both the technical and economic problems leads one to believe that, even with a quasi-monopoly, this nation can outstrip any other in the communications field. Certainly, the Telephone Company acts as though it were competing with everyone under the sun.

My return to Washington was not too pleasant. The plane had dropped me at Holmdel but had gone on to Monmouth County Airport because of threatening weather. When I arrived at that field, we were held on the ground for almost an hour before we could get clearance from the FAA in New York. I returned home about 6 p.m. and spent most of the rest of the time working on matters for the morrow.

Saturday, May 28: At 9:00 I had a meeting with Jim Gleason and John Johnson relating to the statement I am to make a week from now to the Senate Committee on Aeronautics and Space Sciences. This has to do with the legislative proposals we have made that have thus far passed the House committee—not the House of Representatives. At 10:00 Homer Joe Stewart and Dick Horner came in and we spent the next three hours in reviewing, project by project, a number of the tasks that have been undertaken by various elements in the NASA organization. It was not a wholly reassuring session, believe me. Actually, however, I suspect that we are doing about as well as anyone would under the circumstances. Having started out with a program fairly well established by others, we have been attempting to develop an organization, develop a program, and make sense out of all the things we are doing while growing from $100 million a year to $900 million a year in a 22-month span. I did finish up about 2:30, went home to clean up, and then took the airplane to New Orleans.

Sunday, May 29: I was quite relaxed this morning and waited with some pleasure until Joe Morris [chair of the physics department at Tulane] came to pick me up at 10:15. We drove to a restaurant known as Brennan's, which seems to be famous for its Sunday morning brunches, and we enjoyed the menu very much. Rather than run around the countryside, I went back to the hotel for a brief conference and then loafed the rest of the afternoon. At 6:45, Joe picked me up and I had dinner with the acting president of Tulane and the chairman of the various engineering departments, Maxwell Lapham. This was not a very inspiring meeting. Lapham is a really fine person, however; I could enjoy him under any circumstances.

Monday, May 30: This was commencement day at Tulane! I was picked up at the hotel at 8 o'clock and went to the Morris home for coffee. The commencement was set for 9:45, but there was the usual delay in getting the thing under way. It was held in the Tulane gymnasium, which had been air-conditioned with portable conditioning units for the occasion. Some 950 youngsters were graduating. It was not a particularly inspiring ceremony, although Lapham did an excellent job of making it as intimate as could be done under the circumstances. I felt that my talk did not go over as well as I had hoped. Part of the reason for this was that the loudspeaking system was not the best. I promised to send them some information on the system we have put into the gymnasium at Case.

We then had just time enough to go down to the center of town where I purchased a gift for Ruth—a café brulôt set. I think we will have fun with this as an after-dinner or after-dessert operation.

Back to the Morrises for a quick lunch and then to the airport where I took my plane to Washington and Joe took his to New York. My flight was a pleasant one and I arrived back at the apartment about 10:15—not too tired but not very ready for a full night of activity, either.

Tuesday, May 31: I visited with General Persons for a few minutes this morning to talk about the plans we had been developing for entertaining Sir Bernard Lovell of the Jodrell Bank Observatory in England. This discussion was a pleasant one and I secured all the approvals necessary. I went in to see Tom Stephens, the president's appointment secretary, and set the date of 7 July for the visit of Dr. Lovell. Just before 12 o'clock, I got in to see the president for 5 minutes and found that he would be delighted to receive Lovell. I suppose this sets in motion a chain of protocol about which I will now have to learn. A meeting with John Johnson clarified the 200 K engine deal and we will make the announcement later in the day. At 4 o'clock, I met with Senator Saltonstall to discuss our appropriations bill. This business of attempting to get the senators to appropriate more than we have asked for so that we may, indeed, acquire as much as we require is a task. I hope I live through it.

At 5 o'clock, Hugh Dryden and I had a visit with Allen Dulles and several of his staff about several of the matters that have been in the press these last several weeks. Hugh will testify before the Senate tomorrow and I am concerned about what happens in the future. The future is not as yet settled! Now I must get back to work in order that I may go away tomorrow—to Case and its commencement—without feeling that I am leaving the job untended.

CHAPTER SEVEN

JUNE 1960

Wednesday, June 1: The morning was given over pretty much to a discussion with John Corson and our senior staff of the results of the meeting at Langley last week of the advisory committee on organization. It is apparent that progress is being made but that the subject is an exceedingly big one for any group to really understand sufficiently to make intelligent and useful comments. Unfortunately, Abe Silverstein takes a somewhat dim view of the whole operation although he has been cooperating fairly well. He spoke out this morning in response to a question of mine and said that he felt that they couldn't tell us anything we didn't know ourselves. His inexperience in the use of committees to make us do some of the things we ought to do but will not—even though we know we should do them—shows through in remarks of this kind.

We took off at 11:45—the Yorks and myself—for Cleveland.[1] It was a pleasant flight and Ruth met us at the other end. I left Herb at the house while I went right on down to Case for meetings relating to the student housing situation and the schedule as planned for the Greenbrier conference on resources. I have great confidence in John and Kent in their ability to clarify this situation. What a pleasant evening it was at the Skating Club as guests of Kent and Thelma Smith! The honorary degree group and the sponsors made a pleasant family affair of it. I think Kent and Thelma made everyone feel at home and quite relaxed. We had a chance to spend an hour and a half with the Yorks when we returned to the house but were inordinately tired when we turned in about 11:30.

Thursday, June 2: Up early so that Sally could drive to school in the Falcon and I could get back for breakfast with Herb and Sybil. Then, Herb and I went down to the campus where Herb had a series of meetings with several of our top people. I managed to get to the bank and do a few things on my own but the time went very fast. The luncheon before commencement was very pleasant and I was delighted to see Elmer Lindseth, Alex Brown, Sid Congdon and Herb Erf in attendance. They, together with Si Ramo, Kent Smith, Fred Crawford and myself made up quite a party of trustees. Oh yes, I forgot to mention Sam Emerson, who should never be left out of anything. Commencement went off very well and we managed to escape the rain. The honorary degree candidates were all well received and I managed to get through mine without falling down or breaking down. I was very much touched by the fact that the faculty and the students honored me by rising and applauding vigorously. There is something very precious about an educational institution like Case. I am inordinately proud of my degree.

---

[1] Herbert York and Glennan himself were receiving honorary D.Sc. degrees from Case.

Time was passing and we had to make tracks back to the house, pack, and get down to the Union Club in time for a quick dinner, and then on to the plane to return to Washington. Thus ended another memorable day.

Friday, June 3: Over to the White House at 9 o'clock for a cabinet meeting where the president spoke vigorously about the responsibility of this administration to provide a budget for the next administration that is responsible, sensible, and realistic in all respects. He noted that pre-Korea, the Department of Defense had a budget of $12.9 billion and it now has a budget of $41 billion. He pointed out that a goal had been set early in his administration to "free the economy" from controls, to provide a tax cut, and to hold the national budget to a level of $16 billion dollars. Each of these goals was reached with the exception of the budgetary one. By 1955, he continued, the scientists and other groups began to point out the necessity for large expenditures, particularly in the field of missiles. In 1957 Sputnik startled the nation and occasioned a great deal of chest-thumping.

For these and other reasons, we find ourselves operating from a platform of fear, not just fear of the Soviet Union but fear of pressure groups in the fields of health, education, science, etc. This is certainly the case and we all ought to recognize it. He stated that we have begun to accept inflation as a natural way of life. He is proud of being called conservative and hopes that we will continue to be conservative when this means conserving the things that count and doing only those things that should be the problem of the central government. This goes back to the statement of Lincoln that the central government should do only those things that the people cannot do for themselves or that it is clear the central government can do better than any other entity.

He pointed out that the expenditure program—as differentiated from the new obligational authority program—for 1962 should not exceed the $81 billion apparently set now for fiscal year 1961. He made quite a point of the excess personnel carried by all agencies and demanded reductions wherever possible. He pointed out the very appropriate and forceful article carried in the 1 June issue of the *Wall Street Journal* written by [historian and former president of Brown University] Henry Wriston and the editorial that accompanied it. He stated that this was one of the very best pieces he had ever read and recommended it to all of us. He made comments about the desire for security on the part of our people, which seems to be a continuing and important element in their lives. He stated that in his early days at Columbia, he spoke to the freshmen class and said that if they wanted security they should pick out a state where there is a good penitentiary, commit a crime, and they would certainly achieve security. Incidentally, they would lose their independence and therein lay the moral of the story.

He expressed concern over the actions of some executive department heads who seem unable to carry the philosophy of the administration clearly to the legislative bodies on the Hill. We ought to be responsible in our responses to the committees but we ought to express the clear purpose of the administration and not operate just to satisfy Congress or the newspapers. He said that he was not opposed to the start of new programs but that we must use the money of the public intelligently and that we should remind ourselves what our duty to this nation is.

Secretary Benson agreed with everything the president had said. No other person spoke up promptly and vigorously in defense of the president's comments, and it seemed a very sober-sided group as we filed out of the cabinet room.

I got back to the office just in time to sit down with the Bell Aircraft people and listen to their complaints about their seeming inability to win a contract from us. Next a visit from Rosener of Plasmadyne who wanted to tell me what his company could do. Then at 11:30 Dan Kimball, president of Aerojet, came in to say that it was strange that it had won the 200 K rocket engine competition but had lost it to Rocketdyne. I pointed out in no uncertain terms exactly what the situation was and Dan seemed to be satisfied.[2]

At lunch, I talked with George Kistiakowsky about the future problems of the Federal Council for Science and Technology.[3] Horner, General Ostrander and I talked further about the management problems in the Rover project. It seems almost impossible to get John McCone into a room to settle this matter.

Another discussion with Corson about the problems of the study on contracting and we finally wound up the day with Richard Dimbleby of the British Broadcasting Corporation. He turned out to be a very jolly fellow over here to do two shows—one on missiles and one on space. Several of our people took part in this activity and I hope the shows will be useful ones for the British audience.

Saturday, June 4: Down to the office after packing for a trip to Dayton. I spent most of the morning clearing up the desk and then took the 12:45 plane to Dayton where I was to meet Ruth. She had been flown to Dayton by Kent Smith's aircraft and had been waiting for me for two or three hours. This never seems to bother the young lady, however, and we were glad to see each other. The Air Force had a car waiting for us and drove us out to Oxford where we had no difficulty in finding the Harrisons' home. It was a pleasant weekend we had in prospect and we were glad to see these good people—one of them a first cousin once removed of Ruth's, the other her husband. A good dinner, some good talk and then bed.

Sunday, June 5: Commencement Day at Miami University, Oxford, Ohio. We were up bright and early for we had to have breakfast this morning with President and Mrs. [John D.] Millett. Bob Harrison, professor of zoology, took us to the Millett home where we met Bishop [Hazen G.] Werner, who was to give the baccalaureate and Mr. and Mrs. [Edward W.] Nippert. He is the chairman of the Board of Regents of Miami University. Werner is the presiding bishop of the Methodist Church in the Ohio area. Breakfast table conversation was pleasant. Ruth and I had decided, somewhat against our better judgment, to attend the baccalaureate service. First, however, I had to go to the office of the president, where several of us who were to be in the platform party during the afternoon were pictured with cap and gown by several photographers. Ruth was lucky—she was

---

[2] See Chapter 6, note 6.

[3] An executive order of 13 March 1959 established this body to promote coordination and improve the planning and management of federal programs in the areas of science and technology. (*United States Government Organization Manual 1959-1960*, p. 541.)

taken on a brief tour by Mrs. Millett. They almost missed the service—perhaps they had planned it that way. The bishop spoke well with a good deal of fire but without very much substance.

We dashed back to the Harrisons' to pack and prepare for lunch at the student union building with the deans of the various schools and with Mike and Bob Harrison. Actually, we only partially packed and returned after the luncheon to get ready for the ordeal that was to start at 4:30. We put our bags into the car, which was to take us to the airport and then formed for the academic parade. There must have been 5,000 people in the stadium and with 850 candidates for graduation, it took almost 40 minutes to complete the academic procession. There was a strong wind blowing and there was the ever-present threat of rain, but providence smiled on the good people of Oxford and a pleasant commencement ceremony resulted. I was given a Doctor of Laws degree—one more pretty hood. President Millett seemed even more nervous than I about the arrangements for his commencement, but outwardly he was very calm and certainly made the people feel pleasantly relaxed and at home. Immediately following the ceremony, a Major Osver and the purchasing agent of the university drove Ruth and myself to Wright-Patterson Air Field where a Grumman Gulfstream took us aloft in our first ride in an executive-type jet job. We dropped Ruth in Cleveland and I reached home in Washington after a pleasant flight about 1:30 in the morning. Some day I must catch up on my sleep!

Monday, June 6: This day began the review of the budget being prepared for fiscal year 1962—that is, the year starting 1 July 1961. It promises to be a thorough-going review of our program. The asking price of the field organization was about $1.58 billion; the recommendation of the Washington staff directors was $1.376 billion. My guess is that we must plan on a budget of about $1.1 to $1.2 billion.[4] The morning was spent with the budget review committee as members delineated, each in his turn, the particular activities for which he had been responsible. It appears to be a well-managed job although I suspect there will be a good many headaches before we are finished. Shortly before noon, we got at the budget for the advanced research programs area. Abbott did a good job at defending his proposals, but it took all day—until 6:30 p.m.—to finish this one segment of the budget.

I had a call during the course of the morning from Estill Green, executive vice president of the Bell Telephone Laboratories. He has been recommended by Frederick Kappel, president of AT&T, as a good candidate for Horner's job. Arrangements were made for him to come down during the afternoon and I did see him at 4 o'clock. He was interested but seemed a bit afraid of the job. He is 65 years of age, has had 3 ulcers and is quite apparently not used to working at the pace at which we find we must work. He decided to give the matter further thought and let me know within a week.

---

[4] The actual appropriation for fiscal year 1962 ended up being $1.8253 billion. (*NASA Historical Data Book*, Vol. 1, p. 128.)

At 2:00, I went up to the Hill to see Senator Holland of Florida. He was attempting to get at the same problem about which the Michigan delegation visited criticism on our heads a week ago. This time, he was after me for the decision on the Saturn S-IV stage. Chrysler had told each of these delegations that their states would have the majority of the work to be done on the stage if Chrysler got the job. This was really quite an inquisition. After about an hour and 15 minutes of it, I banged the desk and suggested to Senator Holland that he might have a better way of deciding these matters and that I would be very happy to move out of my chair if he wanted it. He smiled and we parted friends a few minutes later. Apparently, one has to exhibit some strong feelings once in a while. Back home with some of the budget books to have a try at getting prepared for the next session.

Tuesday, June 7: Up at an early hour to meet Senator Sparkman at the Capitol at 8 o'clock for breakfast with one of his constituents, a Lou Jeffers of Huntsville. It appears that Jeffers is president of Hayes Aircraft Company, which has done a good bit of work for the Huntsville group in the past. It was the usual low pressure sell by a senator on behalf of one of his constituents. I was off at 11:15 in an Eastern Airlines Electra for Dallas. There was a two-hour time change, of course, and I arrived in Dallas at 1:30 central standard time.[5] There I was met by a Mr. Jackson of the newspaper and T. Carr Forrest, a consulting engineer who was to introduce me at the luncheon the next day. Neil Mallon [director of Dresser Industries, Inc.] had provided a beautiful suite for me at the Sheraton-Dallas Hotel. There were lovely flowers, a bottle of whiskey, television, radio, etc., etc. At 5 o'clock, the newspaper men came in for an interview—there had been both TV and newspaper interviewing at the airport on arrival as well.

Henry Herzig of Dresser took me over to the firm's offices, which are really very beautiful. He then drove me out to see Helen and on to Neil Mallon's home. It is a beautiful country place situated about 30 minutes out of the center of the city. About 16 top industrial executives were present for a pleasant and rewarding conversation. I broke it off early because of the two-hour change in time and was back at the hotel by 10:30.

Wednesday, June 8: Up quite early to have breakfast in time to set out for Chance Vought Aircraft with C. J. McCarthy. This was a 2-hour show with a variety of seemingly very able young men. I enjoyed it more than I thought I would. Of course, there were the usual space cadet activities but I guess this is something one must put up with these days. We drove back to the hotel where I made a speech to the Rotary Club of Dallas. George Low had prepared the speech for me in quite some detail. It was on the subject of Project Mercury and accompanied by a number of very good slides. The audience seemed appreciative and I was glad to have made the date.

After a recorded radio interview, I took off in a hurry with a group of men for a visit to Temco. I was not impressed by this group and was glad to get away from the plant. Forrest then picked me up and drove me to the airport where we had a drink before I boarded the Electra for a pleasant ride back to Washington. I worked

---

5 Texas did not observe daylight savings time until 1966.

on my speech for the commencement in Eau Claire all the way home. I was in bed by 11:30—not bad for me.

Thursday, June 9: Up very early this morning to have breakfast at 7:30 at the Statler with people from Case. I was late for the staff meeting but was just in time to make certain that several decisions were made. Sometimes these decisions get lost in the shuffle because there seems to be so much discussion. Directly following the staff meeting, Bob Bell and one of his men came in to discuss a possible case of fraud on the part of one of our employees. It is interesting to watch a security type—they all seem to see hobgoblins under each bed. I insisted that we get more information and then turn the problem over to the FBI. At the very least, we ought to operate under the guidance of the FBI in a case such as this. Actually, I am not at all certain that members of the staff aren't attempting to get rid of a man for whom they have little use and who may very well be defrauding the government in some way.

Lunch with John Hrones [vice president for academic affairs at Case] to talk over some of the problems at Case—nothing very new or startling today. At 1:30, Frank Henry of the *Baltimore Sun* came in to talk with me. It was a pleasant conversation, and I think I gave him the sort of information he wanted. At 2 o'clock, Henry Simmons of *Newsweek* came in, and this was quite the opposite. For some reason, Simmons always rubs me the wrong way, and I very nearly threw him out of the office today. He seems to delight in probing into the future and insisting upon an answer as though our entire life was made up of efforts to satisfy his curiosity. I have given instructions that he stay out of my office from this point on.

At 3 o'clock, Dryden and I went over to the State Department for a meeting with Assistant Secretary [for International Organization Affairs Francis O.] Wilcox on matters relating to the United Nations Committee on Outer Space and the possible United Nations conference on outer space. There were seven or eight members of the State Department staff—they do not seem to trust each other very much in that operation. Dryden and I made it clear that we were quite willing to accept a co-presidency of the proposed conference.[6] The Russians have proposed this and we can see no reason for being sticky about it. The State Department is concerned that this would set a precedent and that we would be faced from now on with the probability that the Russians would demand co-chairmanships and co-presidencies of a variety of activities. They may be more right than we are—at least they have to live with the problem and we don't.

I worked quite late tonight on the speech for Eau Claire. This is not an easy one for me to do. The chairs need new bottoms and I took six of them out of the

---

[6] NASA continued planning for an international conference on the peaceful uses of outer space through 30 June 1961, but the conference seems never to have occurred. (Fourth through sixth *Semiannual Report to Congress* [Apr.-Sept. 1960, p. 115; Oct.1960-Jun. 1961, p. 152; Jul.-Dec. 1961, pp. 132-133, where the conference is no longer mentioned.]) Perhaps this was because in 1959 the U.N. had established a permanent Committee on the Peaceful Uses of Outer Space to replace its ad hoc committee of that name created the year before. It first met in November 1961, and over the next three decades it adopted five conventions, agreements, and treaties plus three sets of principles in the area of space law. (Lubos Perek, "The Scientific and Technological Basis of Space Law," in *Space Law: Development and Scope*, Nandasiri Jasentuliyana, ed. [Westport, CT: Praeger, 1992], pp. 175-177. Thanks to Eilene Galloway, to whom the book is dedicated; she generously donated a copy to the NASA History Office.)

apartment so that they might be picked up by the Columbia Lighthouse for the Blind, which will do the recaning. In to bed very late once again.

Friday, June 10: Up at 6 o'clock to revise the speech for the last time and to get packed for the trip to Minneapolis and Eau Claire. Actually, I spent almost the entire day on the Hill, where I was able to talk to [Joseph] Martin and Senators [Thomas] Dodd, Lyndon Johnson, Warren Magnuson, Gordon Allott, Lister Hill, and one or two others. I think I can say that I accomplished my mission, which was to urge that the Senate exceed the amount we had requested in the president's budget to such an extent that compromise with the House would result in our getting exactly what we had asked for originally.

Having accomplished the tasks I had set for myself, I boarded Capital's flight #119 for Minneapolis. Departing at 6:45 pm., we arrived in Minneapolis at 9:45 central time. Ruth had come in an hour earlier from Chicago. We departed immediately in a Hertz car for Eau Claire. Taking off about 10 o'clock, we managed to get out of the city without too much difficulty and found that a 4-lane divided highway took us almost in to Eau Claire. About 20 minutes before arriving there, I was stopped by a state patrolman for failing to dim my lights when following his car. Apparently, one cannot have bright lights on when following a car. We managed to get away with a warning after a pleasant conversation with the patrolman. We arrived at 321 Summit Avenue to find the Midelfarts awaiting us.[7] It was about 12:15 in the morning. After an hour's conversation, we were off to bed after a very full and quite long day. I should say it was most pleasant to see Peter and Gerd again.

Saturday, June 11: Saturday was a pleasant day in Eau Claire with Pete and his family. After a pleasant breakfast we drove to Chetek where we found things very much as they were when I was a youngster. There was one very great difference—it used to take us four hours and two blow-outs to make the trip. Today it is completed on a good highway in about forty-five minutes.[8] Peter's cottage had been completely rebuilt and was somewhat smaller than it had been. Driving back to Eau Claire, we prepared for dinner with a good many of the people whom I had known as a boy. It was a pleasant evening with a reasonable amount of reminiscing and exchanging of experiences.

Sunday, June 12: Up about 9 o'clock to find that the entire household had overslept and that we had guests for breakfast coming at 9:30. We all pitched in and had a good time fixing up the fruit and other good things for breakfast. [Describes

---

[7] (Medical) Dr. Peter A. H. Midelfart (1905-1978) and his wife Gerd Gjems (1912-) were close friends of the Glennans. Dr. Midelfart had graduated from Yale in 1927, the same year as Dr. Glennan. He earned his MD from Harvard in 1931 and ran a clinic in Eau Claire that was founded by his father. When he retired, he changed his name and that of the clinic to Midelfort, having already changed his children's names to that anglicized version a couple of decades earlier. (Private correspondence from Professor H. C. Erik Midelfort, to whom we wish to express our thanks for the information.)

[8] Chetek was both a lake and a town 40 miles or so north of Eau Claire.

many of the people who came by, whom he knew as a young man.] It was really pleasant to see these people once again. We drove over to the residence of Leonard Haas, president of Wisconsin State College. Lunch was a pleasant affair. We walked over to the campus and robed for the procession to the gymnasium where the commencement was to be held. I managed to get off my speech without too much difficulty and I believe it was reasonably well received. This had not been an easy one to do—I recounted my own experiences from the time I graduated from college and then tried to draw from these experiences some thoughts and convictions that might serve, in some way, the youngsters who were graduating.

Many of the people at the commencement were familiar to me. It was a bit of a trying time to have people come up and say, "I'll bet you don't remember . . . etc., etc." However, I felt well repaid for the effort I had made.

Immediately following the reception, Ruth and I want back to the Midelfarts and took off for the airport without further delay. I had had the Convair come to pick us up because of the necessity of my being back in Washington on Monday morning. It was ready and waiting and we took off without further delay. We will be hoping that we may repay the hospitality of the Midelfarts one of these days. Ruth and I wrote a long letter to mother while enroute to Chicago. I am sure she will thoroughly enjoy hearing about our experiences in Eau Claire. We arrived in Chicago in the middle of a light rainfall and found Kitty, Sally and Frank waiting for us. We had time for a short drink and then I took off for Washington where I arrived about 12:30 midnight. Quite a weekend!

At this point, it seems worthwhile to make the observation that a return to one's boyhood home is both reassuring and discouraging. It is reassuring in the sense that one finds relatively little change—I am speaking of a relatively small town, of course—but finds signs of prosperity and good living. It is discouraging in the sense that any problems of earlier years seem still to be the significant problems of the community. I suppose this is to be expected, particularly when one has achieved a certain amount of success and mobility in his adult life.

Monday, June 13: This is the week of budget reviews and most of Monday and all of Tuesday morning were given over to this activity. I should say that I am desperately in need of some ideas as to a replacement for Dick Horner. I have been unable to find a mechanism satisfactory to the telephone company that would allow one of its better men to come to us for a year's employment. This business of having to sever connections with any company that has business with the government makes it almost impossible for a good man to come away from his industrial association. It may be that we will have to get a general officer on assignment from the military for a period of at least one year.

Tuesday, June 14: Again, budget review all morning. At noon, John Johnson and I had lunch with Dave Kendall to talk over the problem of finding a mechanism that would allow us to employ a man from industry. There seems to be no new idea although there is great desire to be helpful in the matter. At 4:45, I left for Cleveland on Capitol's flight #907. Ruth and Sally had driven in from Chicago and they met me at the airport. We had dinner at home and a quiet evening in anticipation of the activities of the morrow.

Wednesday, June 15: We reported in to Western Reserve University at 9:15 to robe for the commencement, which was to start at 9:45 in the morning. Jack Millis, president of WRU, had been in Miami on Tuesday for a speech and had difficulty getting back to Cleveland because of a strike on Eastern Airlines. He finally managed to make it by taking a late flight out of Miami to Pittsburgh and driving up from Pittsburgh. We had marched onto the stage and were awaiting the seating of all of the graduating class when Jack came in.

The citation given me was a warm and somewhat flattering one.[9] I think Western Reserve has done an excellent service for the community in awarding degrees to Kent Smith and myself. Whether or not we deserve them is something others must judge.

We went to lunch with the Western Reserve platform group at the Wade Park Manor at 1 o'clock. I spent the intervening time with Kent Smith going over some of the Case problems, then joined Ruth at the Wade Park Manor. We came away at 2 o'clock because I had to go out to Thompson-Ramo-Wooldridge to take part in the dedication ceremonies of the Colwell Engineering Center. This is a very nice building and I was happy to be part of the group assembled there. They are really fine people—Fred Crawford, Dave Wright, Ray Livingstone, Arch Colwell, Si Ramo, Dean Wooldridge, and others. I had intended to drive the station wagon back to Washington immediately following the ceremony but the Thompson people said that they had to send some materials to Washington on Thursday and offered to drive the wagon down so that I could take a plane. This was a fine solution to a serious problem for me. I got off on the 7:35 plane and managed to get a night's sleep in Washington before taking up the budget once again.

Thursday, June 16: All day was given over to budget review.

Friday, June 17: The morning was given over to discussion of the budget with the markup being fairly well decided upon. At noon, several of us went to the Sheraton-Park Hotel to attend the luncheon for [W. A. S.] Butement, who seems to be the Minister for Science and Technology in the Australian cabinet. It was a pleasant lunch but went on altogether too long. I finally left about 2:15 to return to the office. At 3:00, John Maddox of the *Manchester Guardian* and at 4:00, Serge Groussard of *Figaro* came in to interview me for articles in their respective papers. They were both very interesting, with Maddox being much the more responsible in his questions. It was a little difficult to know whether the frenchman really understood or simply put down what he wanted to after listening to my reply.

Saturday, June 18: Up early to drive to Quantico with Maury Stans. This is the weekend for the secretary's conference—an exercise held annually by the secretary of defense and the secretaries of the Army, Navy and Air Force. A very

---

[9] He was here, also, to receive an honorary D.Sc. Kent Smith was awarded an honorary LL.D. at the same graduation.

frank discussion of problems and policies takes place over a period of three days. I listened to talks by each of the joint chiefs and by the chairman of the joint chiefs, by Fred Eaton who has been in Geneva for the past several weeks in the disarmament sessions, by the commander of the North American Defense Command (NORAD), by the commander of the Pacific Fleet and the commander of the Atlantic Fleet. Actually these last three are what are known as combined commands. The discussions were excellent—some more pointed than others. Secretary Gates finished up the conference with a brief but frank statement of his reactions to his first year in office. He stated that he felt a good bit had been accomplished and enumerated some of the actions he had taken and policies he had established. He pointed out several deficiencies and then moved over for a frank statement of his concern over the lack of loyalty and discipline among the people in the Pentagon. He invited people who could not accept decisions made after they had had a hearing to resign. His statement was received with a good bit of warm and genuine applause. It is obvious that General [Laurence] Kuter [NORAD commander] and General [Thomas S.] Power [commander-in-chief of Strategic Air Command] are among those who have been less than well-disciplined in their activities relating to the need for money for their commands.

At lunch, Stans gave his usual talk. He always speaks forcefully and effectively, but it is the same story—we must control the budget. I drove back to Washington with Gordon Gray and Karl Harr. Making good connections, we took off from Butler in the Convair at 4 o'clock for St. Louis. Arriving there at about 6:30 local time, we were met by Jim McDonnell and several members of his staff. After checking in, we went to a beautiful country club for dinner where Arthur Compton joined us. It was pleasant to have a chance to talk with him again. The evening drew on until about 1 o'clock when we turned in.

Sunday, June 19: Up relatively early for breakfast and a full day at McDonnell Aircraft where the Mercury capsule is being constructed. It was an interesting day and one could not help but be impressed by the care that is being taken to do a good job. It's an expensive one all right but the gadget is really complicated.

At 3:30, we departed St. Louis and flew directly to Langley. We were all pretty tired and turned in early.

Monday, June 20: We spent the entire day with the space task group listening to a briefing on the Mercury project. The morning was given over to the manufacturing and technological problems relating to the capsule itself, and the afternoon was devoted to a discussion of the operational phases of the project. I must say that it was a perfectly splendid job. A fair amount of attention was given to the question of reliability. This had been one of the real reasons for calling this particular conference. Stopping for a few minutes after the end of the briefings, I called together a few of the project leaders and the people from Washington to talk a little bit about the reliability question. It seems to me that they are doing an excellent job in planning for the operations but there is somewhat more to be done in testing and evaluating the reliability of the capsule itself during the manufacturing process. There seems to be a great tendency to rely entirely on the actual flights—there will be many of them—for proof-testing the gear. This is not a bad idea but it may be a very expensive one, particularly if we find difficulties that might just as well have been found through tests on the ground.

Unfortunately, Nick Golovin took off in his usual "bull in a china shop" way and brought the day to an end that was not quite as pleasant as I had hoped. Surely, a person of his stature ought to know a little bit more about the methods to be used in discussions of this kind. You do not insult the people who have been working so hard to achieve the desired end product; you may question some of their premises and actions but you do not, in effect, accuse them of gross negligence. I think we were able to smooth it over a bit before we left Langley, but the day, which had started out as an excellent one, came to a rather disappointing close.

Tuesday, June 21: At 9:30, [Franklin] Floete of the General Services Administration came over to try to talk me out of the occupancy we were planning for Federal Office Building Number 6.[10] He knew he was going to lose before he started talking, but he had to go through with the exercise so that he could tell other claimants that he had made an effort. I don't think I would want his job for all the tea in China. A discussion followed with Siepert, Lacklen and others over our executive training program. We keep on writing routines and uncovering new executive training activities but don't seem to get any people into these courses. I asked that they immediately identify a promising group in each of the laboratories and headquarters and make arrangements to get them started in some of this sort of thing. Actually, what I am driving at is to have a group of younger men identified and put through a series of these "off-campus" executive training courses and through a rotational program on the job wherever possible. We are suffering very badly at the present time from the lack of well-trained management material.

At noon I joined a host of people in Secretary Brucker's office to take part in the award ceremony for Eberhard Rees of Marshall Space Flight Center. Wilbur Brucker did his usual flowery job, which always seems embarrassing to me and very insincere. It was obvious that Dr. Rees was pleased with the award and well he might be![11] Lunch with Secretary Douglas to discuss the possible availability of Richard Morse, the Army's research and development head, resulted in a stalemate. I had been talking with Dick Morse about the possibility of his taking on Dick Horner's job. Douglas states that they really need him in the Army although I am quite certain the Army doesn't believe that statement. Brucker has him effectively boxed in but Dick is not a quitter and it may be that he will get sufficient support from Tom Gates and Jim Douglas so that he may achieve some of the objectives he has set for himself. Jim Douglas said that he would talk with Dick, try to determine what ought to be done, and call me tomorrow.

---

[10] NASA Headquarters had been scattered among four different buildings along H Street near the White House until 1961 when it moved to Federal Office Building Number 6 along Maryland Avenue between 4th and 6th Streets. ("Facilities-Headquarters" folder, facilities subseries, NASA Historical Reference Collection.)

[11] He received the DOD Distinguished Civil Service Award.

At 2:30, Shelby Thompson, Bonney, Frutkin, Phillips, and others came in to talk about our posture with respect to problems arising in foreign countries where we have agreements on the establishment of tracking stations. What appear to be communist-inspired criticisms of the nature of our programs have caused some real concern in Zanzibar, Nigeria and one or two other places. We are accused of being something other than a peaceful agency—an aftermath of the U-2 incident. It was agreed that the best possible action on our part was to provide to our representatives in these particular countries sufficient information to allow them to refute vigorously any such accusations. Further, it appears desirable to have political and technical representatives of each of the countries visit the United States for an intimate glimpse at Project Mercury so they can satisfy themselves about the nature of the program.

This evening at 5:30, I hastened up to the Hill where I was to meet Polly, Sally and Ruth to attend the Congressional showing of "I Aim At The Stars." This is the film of von Braun's life, and Senators Sparkman and Hill of Alabama were holding a special showing for congressmen and their friends, followed by a reception in the north cafeteria of the new Senate Office Building. Von Braun was there and spoke briefly before the film was shown. He was his usual relaxed and seemingly good-natured self. He stated that it was a little difficult to be objective about a film of one's own life but that, given the usual discount for love interest and drama required by screen writers, he thought the film was a reasonably accurate portrayal. Ruth and I reacted rather negatively to the whole thing. Von Braun is made out to be an anti-Nazi and seems to epitomize the scientist's lack of responsibility for the end use of the products of his mind.[12] Most people seemed to like it however. The girls had a chance to meet and talk with von Braun after the film. We wound up the evening at a restaurant similar to those of the Howard Johnson chain.

Wednesday, June 22: At 9 o'clock, Horner, Phillips and Hjornevik came together to discuss our budgetary presentation to Stans later in the morning. This session with Stans at which we were joined by Dryden was not too unpleasant. I stated rather bluntly that we were asking for $1.23 billion. He listened intently and

---

[12] In fact, von Braun had joined the Nazi party on 1 May 1937 and had even become an officer (eventually a major) in the elite, quasi-military SS on 1 May 1940, but available American records, based on those in immediate post-World War II Germany, support his own assertions that he had joined both only because failure to do so would have forced him to abandon his work on rocketry and that he had engaged in no political work for either the party or the SS. He claimed that his motivation in building Army missiles was their ultimate use in space travel and scientific endeavors, but that hardly excuses him from blame for the deaths in allied countries resulting from the military uses of his V-2 rocket or for the slave labor used to produce them. (Collection of documents in folder marked "v. Braun, Nazi, SS Membership" [especially Col. Thomas J. Ford, GSC Director, Joint Chiefs of Staff, to Commanding General, U.S. Forces, European Theater, 3 March 1947; Affidavit of Membership in NSDAP of Prof. Dr. Wernher von Braun, 18 June 1947; and Security Report . . . on von Braun, Wernher, date illegible] and other materials in "Wernher von Braun," biographical files, NASA Historical Reference Collection; for further information on this score, see the sensationalistic book by Linda Hunt, *Secret Agenda: The United States Government, Nazi Scientists, and Project Paperclip, 1945 to 1990* [New York: St. Martin's Press, 1991], pp. 44, 65, 109, 120, 226 and the forthcoming study by Michael J. Neufeld on Peenemünde, which will appear as a book and in a shortened version in "The Guided Missile and the Third Reich: Peenemünde and the Forging of a Technological Revolution," in Monika Renneberg and Mark Walker, eds., *Science, Technology and National Socialism* [Cambridge: Cambridge University Press, 1993].)

finally stated, with some show of irritation, his judgment that we should have no more than we were asking for this year—$915 million. Finally, he said that he would like us to build up a budget totaling $1 billion, of which $50 million would be held in reserve to be available in the event of an emergency. I responded promptly that I could give him such a budget without working it over. We would just cut out the Saturn project and thus save a quarter of a billion dollars. I stated, too, that he could also have von Braun with such a cut and I thought that he would not very much like the result. Of course, I withdrew the suggestion, which I had made half in jest. Dryden was able to answer

The Little White House, which served as one of NASA's headquarters buildings until 1961.

Another view of the Little White House.

many of the questions placed by the director. Many of these queried the importance of space as compared with medical research or agricultural research or greater aid to education, etc. Dryden simply pointed out that these were not really comparable. If the space program of the nation were to be cancelled, the money thus saved would not be allocated to these other fields. Stans had to admit this to be the case. We wound up the session in a friendly fashion with Stans indicating that he found it pleasant to do business with us and hard to deny our requests.[13]

Immediately after lunch, several of us met on the question of providing an answer to the letter

---

[13] The ultimate NASA appropriation for FY 1962 was $1.8253 billion. (*NASA Historical Data Book*, Vol. I, p. 128.)

A third view of the Little White House at 1520 H Street, NW, showing the entire complex.

from Aerojet quarreling with our decision in the awarding of the 200 K contract. It was decided that a simple straightforward letter would be best and that such a letter could be prepared by Johnson. Obviously, Dan Kimball [director of Aerojet] is reaching for political advantage, and it is reasonably clear that he does not expect an answer different from the one he will get. At 3:30, we had a discussion on the activities of our program planning group under Homer Stewart. I am not happy about this activity because I don't believe it is quite as thoroughly organized and integrated into our total program as it should be. I'm a little at a loss as to the best method of getting at the problem, however. I wound up by asking for a statement from Stewart showing the format and intended content of his planning effort for the balance of this year. He will leave us to return to Caltech late in the year.

We were to meet Jim Perkins at the Cosmos Club for dinner. This we did and enjoyed the evening with Jim and his brother Courtland. Discussion was had of the invitation I had received that day from the Council on Higher Education in the American Republics. This is a group supported by the Carnegie Corporation and is interested in improving the quality of higher education in all of the Americas. Among the American members are Clark Kerr, Grayson Kirk, Jim Perkins, Meredith Wilson, Franklin Murphy, and one or two others.[14] Ruth and I are to go

---

[14] All of these men were university presidents or chancellors.

to San Francisco for a week at the end of February of 1961 and I am to take a month traveling in South America visiting universities there. I have agreed to undertake this task hoping that I can make the trip to South America in the Summer of 1961. Possibly, I may need the change and will take it in March of 1961, however. It sounds like a pleasant prospect.

Thursday, June 23: This morning started with a staff meeting and a series of sessions relating to the foreign program, legislative program and our internal and external public relations problems. Nothing of note except that I continue to be very much concerned about our ability to get Lyndon Johnson to deal with the legislation we want passed this session. I had stayed over today rather than going to the Greenbrier for the conference with the Case people because I felt it possible that we could get Johnson to act.[15] This turned out to be an erroneous assumption. I therefore decided I would take off for the Greenbrier early in the afternoon and ordered an airplane for that purpose. At 3 o'clock, Silverstein, Buckley, Stroud, Mengel, Goett, Horner and Dryden came in to talk about the necessity for a large radar in Alaska and for two others to be located in other parts of the world. They needed an answer immediately because the construction season in Alaska is so short. After an hour, I cut the meeting short by agreeing that they could have the Alaska station but not the other two. Strangely enough, this turned out to be a wise decision as the boys say that they really do not need the other two after taking a harder look at the problem.

I met at 4 o'clock with Hjornevik and the budget review committee to tell them what we had done with the budget mark-up. They seemed pleased and I was able to compliment them genuinely on the excellence of their task. Before taking off for the Greenbrier I decided to call Lyndon Johnson and try to talk him into making a place on the legislative calendar for our bill. When I reached Lyndon, he started right in on me stating that he had insisted in the full Appropriations Committee that $50 million be added to our appropriation request—that he had overruled the recommendation of the subcommittee, which had added only $20 million.

I responded by saying that I had taken note of this and was very grateful. I told him that he might be certain that I knew who brought this about, but I went on to say that we need now to get on with the amendments to the law. I started another sentence but was interrupted by Johnson who said, "Now why do you need to go on with that? There are only 10 days left in this session. You have or will have the money to get on with this job. Why don't you get on with it without 'them' changes in the law?" I responded, "But Lyndon, you leave us right in mid-stream. The House has approved the bill, which we will accept. We have set up an activities

---

[15] This paragraph refers to the attempt Glennan and President Eisenhower had made to amend the 1958 Space Act so as to eliminate the Space Council and repeal the Civilian-Military Liaison Committee in favor of an Aeronautics and Astronautics Coordinating Board. Johnson opposed any changes in the Space Act at least until the next administration took office, so the law remained unchanged in January 1961 when Glennan stepped down as NASA administrator. (Rosholt, *Administrative History of NASA*, pp. 170-171.)

coordinating board and have mothballed Space Council and the Civilian-Military Liaison Committee. We haven't signed the papers on the activities coordinating board as yet because we didn't think it quite proper since mention of it is made in the House version of the bill."

Lyndon broke in again to say, "Look now, doctor, you haven't a chance to get that legislation. You have no right to assume. . . ." I broke in saying, "But, I didn't assume anything. We all agreed that the Space Council was not a useful device any longer." Lyndon then stated, "That's what the president wanted. You don't have to use it if you don't want to. You are doing very well at the job as it is. I don't see any reason for giving you a new law at the present time. If I am elected president, you will get a changed law without delay." At this juncture, I said, "But I won't be among those present if you are president and I have. . . ." Lyndon broke in, "Now don't try to get me to commit myself at this early stage. I am not elected yet." I wound up the conversation by saying that I felt certain he would do a good job if elected and if I could be certain of a job with him, I might even vote for him—this last with a chuckle. Thus ended an interesting exchange.

We sat on the runway for one hour trying to get off for the Greenbrier. It was a pleasant flight even though the weather seemed to be threatening. I enjoyed seeing the gang at the Greenbrier although the fact that the employees were on strike made it seem a little less of a resort than usual.

Friday, June 24: This is another day at the Greenbrier. I sat quietly through most of it since I am merely a spectator at the moment. I did enjoy listening to the excellent discussion. I wonder if they really know how much money they are talking about?

Saturday, June 25: Saturday morning at the Greenbrier and then off to Washington with John Hrones and Ray Bolz. I had forgotten that we were to go to the Silversteins' for a party at 9 o'clock on Saturday night. We did make it but I got away at 11 o'clock—much the worse for wear.

Sunday, June 26: We were up a little bit late and then prepared for a brunch. The balance of the day was pretty well shot by the time our guests had left.

Monday, June 27: This was another day of varied activity. Bonney and Frutkin came in early to talk about the handling of Dr. Lovell's trip on 6 July.[16] We are playing this at a relatively low level but want enough material for good and effective distribution through the USIA. Gleason and Johnson came in to talk about the prospects of legislative action, and it was reasonably clear that we aren't going to make it this session. Later in the morning, a gentleman from the CIA came in to talk about matters that cannot be discussed in this diary. I am out of sympathy with the CIA's objectives.

Late in the afternoon, Dick Horner and I had a visit with General White to seek out his advice on a possible replacement for Horner should we find it impossible to get a civilian. Tommy promised that we would have our choice of

---

16 He visited NASA, the White House, and the Green Bank radio observatory in West Virginia. See entries in Chapter Eight for July 6-7.

three or four on very short notice although he fully agreed with us that we ought to bend every effort toward getting the civilian. Earlier this afternoon, I had talked with Dick Morse about this same problem again and had determined from a conversation with Jim Douglas that Morse would not be available. Morse had suggested Robert Seamans of RCA. I talked with Seamans and immediately decided to go to Boston to see him. Ruth packed the bag for me and I caught a 5:30 plane for Boston so that I was able to have dinner with Seamans.

He turned out to be a rather interesting person who had graduated from MIT in the class of 1940. He seems to have enough money so that he could take a job in government and seems also to want to take such a job. He is presently being considered for the job of scientific advisor to the armed forces in Europe. He evidenced a very real interest in our situation, and I have invited him to come to Washington next week.

Off to bed about 10:30 to be prepared for the trip back to Washington in the morning.

Tuesday, June 28: Eastern Airlines woke me at 5 o'clock to say that its plane had been canceled but that they had put me on an 8 o'clock flight on Northeast. I arrived in Washington about 10 minutes to 10—just in time to miss the Smithsonian Institute presentation of the Goddard Award to Mrs. Goddard. Further discussions with Frutkin and Horner about certain of our activities in foreign countries, and then off to the State Department to see Secretary Livingston Merchant. I was talking to him about some classified matters. He seems about as mystified as I am.

At the Federal Council on Science and Technology nothing of any great note happened. I returned a little bit early to my office to find that John McCone had canceled another meeting with me. We did have a session with Horner, Ostrander and von Braun relating to the budget for the Marshall Space Flight Center. I think we came out all right on this one.

Wednesday, June 29: This was one helluva day. I spent most of it on the Hill trying to talk with various people about the conference on our appropriation. This meeting is to be held on Thursday afternoon. The facts are that, as against our request for $915 million, the House has given us $876 million and the Senate has given us $965 million. The House cut 373 people and $4 million from our salaries and expenses account, $15 million from our construction and equipment account, and the balance from the research and development account. The Senate restored all of these cuts and added the $50 million in the research and development account. Normally, a conference of this type would result in splitting the difference. In this instance, this might mean that we would lose the House cuts in salaries and expenses and in construction and equipment while the compromise would take place only in the research and development account. This is the situation against which we must protect ourselves. I talked to Congressmen Taber, Brooks, Ostertag, and Thomas, as well as Senators Anderson, Saltonstall, and Allott. What a rat race! I hope we have made our point but I rather doubt it. They are so overworked and so much concerned about the coming recess that it is a wonder anything gets done. Most of the congressmen don't like this—all members of the House and one-third of the members of the Senate are running for their offices. This kind of an action puts them to extra expense and interferes with their campaigns for reelection. It seems obvious that Lyndon Johnson, recognizing that Congress can't finish its task in three or four days, has made this move in order that he may have a full week at Los Angeles to do a little electioneering. Smart cookie! In between visits to people on the Hill, I sandwiched discussions with a vice president of AT&T, George Best, and John Pierce of the Bell Telephone Laboratories; spoke at the dedication of the IBM Space Exhibit; had a session finally with John McCone and General Luedecke of the AEC; and finally had a session with Horner, Bonney and Thompson on their budgetary questions.

Just as we were preparing to give up for the day, Horner and Golovin came in. Obviously, the matter of reliability in the Mercury system is not moving as well as we would want it to. There is resistance from McDonnell and from within our own space task group. It is all part of a very interesting set of interactions and is a prime example of the sort of passive resistance one finds in government and on college campuses. I have decided that we will take the bull by the horns and instruct McDonnell to do as we wish at headquarters. We will pay the bill whatever it may be. The space task group will simply have to go along and take part in the analysis of the reports that are submitted. What a day!

Thursday, June 30: This morning we had a meeting with the staff. We had a discussion of the management analysis division operation, and I brought the group up-to-date on the work of the Federal Council on Science and Technology. Immediately following this session we talked with Dr. Randt about his budget. He is a very reasonable person and I think we will have no difficulty in supporting him adequately. He is going slowly enough to get really good men and seems much more concerned about quality than he is about the quantity of the group he has on board. We have been having several meetings over the past two days with respect to our problems in Zanzibar. This is a British protectorate but is one of those African nations that is being used, apparently, as a base for Communist agitation. For some months, we have had permission to install a Mercury tracking station and things had been going well. At least, so we thought. However, since the U-2 incident, the situation has been changing. In the first place, while I suppose we were aware of it, the Defense Department has pulled a fast one in attempting to install a Courier tracking station adjacent to ours. Courier is an operation involving the use of active communications satellites; is run by the Air Force; and thus has some connotations of being a military operation. Indeed, it is just such an operation.[17] In any event, there have been some demonstrations, the Communist front organization has a fairly large membership and has adopted resolutions condemning Project Mercury as a "military intelligence operation."

---

[17] The Air Force placed the Courier I-B active communications satellite in successful orbit on 4 October 1960. It had a total weight of 500 lbs., an apogee of 658 miles and a perigee of 501 miles. By October 23 when it ceased transmitting, it had transmitted 118 million words. (Emme, *Aeronautics and Astronautics . . . 1915-1960*, pp. 128-129, 150.)

It seems that the situation has been aggravated somewhat by the undisciplined comments of certain representatives of the Space Technology Laboratories who were operating in Zanzibar for the Air Force. "Ugly American" would seem to describe their activities. The result is that Courier has been withdrawn as an operating installation and we hope to be able to keep Mercury in the picture. To withdraw is to admit the effectiveness of Communist agitation. We must ride this out if we can, although we have an alternative in the chartering of an additional ship on which the equipment could be installed.[18]

At noon, Jim Gleason and I went up to the Congressional Hotel where we met with the assistants to a great many of the Republican congressmen. They call themselves the Beau Elephants and seem to be a group of very nice and relatively young men. I spoke to them on the program of NASA and answered their questions, which were for the most part very interesting ones. When asked to comment on the U-2, I simply said I would not. They took this with a smile. During the course of the afternoon, word came down to us that the conference committee had accepted a compromise that gave us exactly what we asked for last January. I am delighted with this result. I think our visits paid off, but it is clear that the action of Lyndon Johnson in insisting that the Senate vote $50 million more than we had asked provided the means for a compromise.

---

[18] This expedient did not prove necessary. Although Zanzibar did request the removal of the station, the head of state later publicly regretted the request and the station remained in Zanzibar. (Frutkin, *International Cooperation in Space*, p. 70.)

CHAPTER EIGHT

July 1960

$\mathbf{F}$**riday, July 1**: We started out with an early morning meeting with Silverstein, Horner and Dryden in an attempt to understand better what the space task group is planning to do with follow-on activities in the Mercury Project. I have some very fixed notions about this. I believe that we ought to relieve the space task group of any real responsibility for the continuation of the project beyond the demonstration of man's capability to fly in a Mercury capsule. We ought to transfer the follow-on operation to the Ames Laboratory on the West Coast. This particular research center does not have as dynamic a program as do the others in NASA. It does have some very good people and I believe that centering the manned space flight activities of NASA at that research center, including the life sciences group, would make a great deal of sense. It will allow breaking with the present pattern of operations and bringing industry more dominantly into the picture. I doubt that this will be easy to sell but it may be necessary to force the issue very soon.[1]

The boys have been talking about this activity for two days earlier this week at a meeting of the space task group held at Canaveral. It appears that the on-going activity must look nine years ahead for a manned circumlunar flight in 1969 or 1970. This program would cost at least $2 billion and probably $3 billion. Obviously, to put that sort of price tag on it at the present time is dreaming but I think it's dreaming in the right ballpark. I doubt that anyone would agree to go ahead with the program on that basis. This presents us with the usual dilemma: how much do we believe such a project should be undertaken?

In any event, I had to leave the group in mid-session to go to the White House for a meeting of the cabinet with President Eisenhower. Stans was presenting an "eyes only" paper relating to the need for planning for transition to the next administration. Apparently, this has never been done in a really orderly fashion. Eisenhower did organize a task force after he was elected, so that his cabinet and staff could get off to a running start. Harry Truman helped very little.

Obviously, defense matters, foreign affairs and intelligence matters would take the highest priority in an exercise of this kind. The president has spoken of a "legacy of thought" for his successor. He is, however, very concerned about the manner in which such a transitional operation should be accomplished. I think it might be said that this must be played in a very low key until after the election. If

---

[1] Glennan's hopes to convert Ames to a manned spaceflight center never came to fruition because the Manned Spacecraft Center was formed out of the space task group at Langley and moved to Houston in 1962, becoming the nucleus of what came to be called the Johnson Space Center. (Rosholt, *Administrative History of NASA*, p. 214.)

Nixon wins, the problems will be very much less than if the Democratic contender wins the election. Eisenhower did say that the country's interests must be paramount—that everything else must be subordinated to that concept. He wound up the discussion by suggesting that thought be given to this and that plans be made at a later date for implementation of the ideas discussed—probably after the completion of the conventions. He wound up the discussion finally by saying that there is one thing we can do for our country—we can heighten our zeal for frugality during the closing days of the administration. He spoke particularly about identifying people who were more interested in continuity of their job than in pressing forward with sound plans for governmental operations.

It was a sobering session and one that I will remember. Returning to my office, I found John Stack of Langley. He had come up to talk with Horner, Dryden and myself about his future. Stack is an interesting character—almost ready for retirement, outspoken and somewhat lacking in common sense. He is, however, one of the very best men in the aeronautical field and his imprint is to be found on many of the most exciting and productive projects undertaken by NACA. John wants to be associate director of Langley; it is obvious that he shouldn't be even though he has been considered a sort of third man on the totem pole in that organization for many years. Just how we will get out of this one I am not sure, but it points up again the very great importance that must be placed on the handling of personnel in any organization.

A lunch at the Metropolitan Club with John Corson was a rather dismal affair. The McKinsey boys are doing a fairly good job but I am not satisfied that we are getting at the heart of the organizational study; nor that we are getting what we should out of the contracting study. Dealing with consultants is difficult; their dealings with us must be equally difficult for them. I agreed to have Stans, Staats and others of the Budget Bureau meet with the organization advisory committee during the course of its next visit in Washington.

At 2 o'clock, my spirits were lifted by a meeting with Dr. Jastrow. He is an exciting young man who heads our theoretical group. He is going to do an article on "Why Space Research," which I hope we can get into the *Atlantic* or *Harpers*. He is also quite excited about the prospect of doing a television panel show at the end of the year that would bring together some of the top scientists of the country who would be talking about the impact of science and scientific discoveries on the economic development of societies. I was so interested that I called Frank Stanton of CBS late in the evening and secured his agreement to "taping" the show regardless of whether or not CBS finally releases it. We ought to have more people who are as exciting as this young man.

At 2:30, Silverstein, Buckley, Frutkin, Williams and Faber came in to talk about the Zanzibar situation.[2] It is apparent that the State Department is in a bind

---

[2] The name "Faber" does not appear in the NASA biographical files or telephone books for the period, and neither Glennan nor Arnold Frutkin could identify him or suggest another name for which "Faber" might have been a mistaken transcription in the original diary.

and that the British Foreign Office is concerned about the prospects of real trouble. They even talk about armed intervention on their part if a peaceful situation is to be maintained. Obviously, we want no part of any action of this kind, and I therefore authorized the transfer of equipment to a ship as a standby operation. We will continue to install equipment in Zanzibar on the odd chance that it will be possible for us to operate there after the present trouble dies down.

At 3 o'clock, we went into a budget session with Silverstein and Ostrander. This budgeting is really quite a game. The boys keep changing their plans; I suppose they must do this, but it is very hard to get a fix on the situation at any one time. I told them that they must prepare a budget that would come out at $1 billion and we would defer for another week the determination of action to be taken on the space flight and launch vehicles budgets for 1962.

At 5 o'clock, I went up the Hill to attend a reception being given by the House Space Committee, which had been named committee of the year by the McNaught [newspaper] Syndicate. Lots of activity, whether or not intelligent, seems to be a real criterion in this competition. Once again, it was interesting to see the manner in which the congressmen bask in the light of publicity and public relations men.

Saturday, July 2: Up early to take Sally and her friend Donna to the White House. I went to the office and read until they came by at noon when I took them to the White House mess for lunch. This evening, Laura Silverman is having dinner with the kids and then Ruth, Sally and Donna will join me in a drive to Huntlands at Middleburg, Virginia, where we are going to spend the next two days. Believe me, I am looking forward to it.

Looking back over this past week, I have become increasingly discouraged about the prospect of getting a man to replace Dick Horner. However, the visit with Bob Seamans has interesting implications, and he will visit me in Washington next Wednesday for a full day of discussions with me and other people. He is interested in the job.

The actions of Congress in attempting to get out of town before the conventions have been less than what one might expect from the responsible lawmakers of the country. They have really spent a lot of time in unproductive pursuits this spring and now stand up and blame each other for being out of town, for wasting time, for avoiding coming to grips with the issues, etc. While this is a great democracy that has withstood the tests of time for 150 years, one is struck with the conviction that some new mechanisms must be found in the near future if we are to keep abreast of the rapidly changing economic, political and international scene. I used to say that only a rich nation could afford the confusion and indecision that seem so often to characterize our national operations. I see nothing that would change my mind. What is more important, it seems clear to me that we must find a way to exert leadership as we identify more clearly our own goals. By definition, our present practice of reacting to the Soviet actions must doom us to ultimate failure.

Sunday, July 3: Up at a reasonable hour and off to Huntlands with Ruth, Sally and Donna. It was a beautiful day, not too hot but with plenty of sunshine.

Later in the day it became rather humid. Living in air-conditioned comfort with a pleasant and efficient staff, a swimming pool within yards of the house and excellent food at any time of the day, it was a pleasant weekend. Sunday was spent quietly with me attempting to put together my thoughts for the NASA-industry conference to be held late this month. I managed to complete most of it today.

Monday, July 4: Independence Day was spent in quiet splendor with much sunning around the swimming pool. This two-day respite has done me much good. We returned home about 8:00 in the evening.

Tuesday, July 5: We started off the day with a review of the Project Echo film prepared under the supervision of the Jet Propulsion Laboratory. It turns out to be a pretty good effort with only minor editing changes to be made. They have attempted to anticipate success in the next launching and have thus put the hex on the project team. Tommy Thompson came up from Langley to discuss the management problems he faces there. I hope we were somewhat helpful to him. At 2:30, Siepert and others came in to discuss the effect of the recent congressional pay raise—it was passed over the veto of the president—on NASA's excepted positions. Many of our people have had very substantial increases during the course of the last two years. Now the pay raise forces action upon us that will add anywhere from $500 to $1000 to these same salaries. In many ways, this is a really morally wrong situation, but pressure groups being what they are, we are powerless to resist. For a few, the additional raise is warranted. For most of the group, it is not.

A meeting on the NASA-industry conference took up most of the afternoon. We were listening to the papers as presented by the authors. The boys take these rehearsals very seriously. I have no judgment at the moment as to the value of the conference in the first place and the quality of the presentations themselves.

Wednesday, July 6: Bob Seamans spent this day with us. I had breakfast with him at 8 o'clock and then turned him over to Dick Horner for the balance of the morning. Lunch with Johnny Johnson to discuss some of our operating problems and to tell him about Bob Seamans, who then joined us for lunch. At 2 o'clock Professor A. C. B. Lovell—our guest from the University of Manchester in England and the director of the Jodrell Bank Observatory—came in to pay his respects. It is his station that has been so effective in tracking and communicating with our satellites and particularly with Pioneer V. He seems a pleasant chap and I think is going to enjoy his stay with us. The NASA-industry conference rehearsal took up the balance of the afternoon. At 7:30, Professor Lovell, Admiral Arleigh Burke, Mrs. Burke and George Kistiakowsky came to the apartment for dinner. It was a very pleasant evening with much good conversation. Lovell is an out-spoken character and operates very much as any of the good scientists in that he is seemingly quite apolitical.

Thursday, July 7: An early morning session to greet the fiscal officers and the administrative officers from our field establishment. They look a good lot. The staff meeting had nothing unusual to recommend it this morning. A consensus of opinion as between Dryden, Horner, Johnson and myself leads to the conclusion that we should offer Seamans the job of associate administrator. Actually, I had done this last evening just before Seamans left for Philadelphia. He is to give me an answer on Saturday morning.

At 11 o'clock, Professor Lovell, Lord Hood, chargé d'affaires of the British Embassy, [E. S.] Hiscocks, Dryden and myself visited with the president.[3] Lovell and Ike got on very well together. The president showed a great deal of interest in the activities at the Jodrell Bank Observatory and expressed his appreciation for the cooperative program that has been going on between Professor Lovell's installation and the space operating groups here in the United States. Lunch for Professor Lovell was a small affair at Lord Hood's residence. We had caused to be made a rather nice little model of the Pioneer V satellite containing a music box especially made in Switzerland for us. The box played "God Save the King," which served adequately to bind together the two countries since we also use this tune.

Dr. A. C. B. Lovell, director of the Jodrell Bank Observatory, flanked on his left by Hugh Dryden and on his right by T. Keith Glennan.

---

[3] Available records at the British Embassy do not indicate that Lord Hood was still on the staff there at this time. According to these records, Dennis Russell Hiscock [without the "s"] was the scientific advisor there in 1960. (Thanks to Marianne Hosea of the personnel office for her research in the embassy's files.) However, the Glennan files in the NASA Historical Reference Collection contain a letter from Glennan, dated 26 October 1960, addressed to Mr. E. S. Hiscocks, Scientific Attache, United Kingdom Scientific Mission, 1907 K Street, Northwest, so perhaps there were two individuals with similar names doing similar work for the British government in Washington at the same point in time.

At 4 o'clock, we held a press conference with Professor Lovell that went rather well. I left the conference to have a meeting with John Lear, science editor of the *Saturday Review*. This went well indeed. At 7:30, we had a black tie dinner for Professor Lovell at the Cosmos Club. He spoke briefly about the work of his observatory and we handed to him a beautifully engrossed scroll expressing our appreciation for his help on our problems.

Friday, July 8: This day was devoted entirely to a program review using the mechanism of the new program control center. We went through the meteorological satellite program, Project Mercury, and the Centaur program. There are a good many tough decisions to be made as these programs go forward. I have made a note of the questions that need resolution and hope to be able to get at them next week. We broke off for lunch—Horner and myself—to be with the advisory committee on organization (Kimpton Committee). Director Stans, his assistant Elmer Staats, and our budget examiner, Willis Shapley, were present as well. It was a rather disorganized luncheon but I hope the committee was satisfied with the result of the discussion.

They seem to be getting off on to some side roads but perhaps more mature reflection will bring them back onto the main street. Kimpton continues to worry about our association with the scientific community. I have told him several times that we can repair this disaffection by supporting the individual members of this so-called community in the manner to which they would like to become accustomed. Unfortunately, we don't have enough money for this. At 7:30, Ruth and I went to dinner with the Ostranders at their home. Hugh Dryden, Nick Golovin, the guest of honor, Dick Horner, and wives were present. It was a pleasant evening for all concerned. We have been very fortunate to get General Ostrander assigned to our staff.

Saturday, July 9: The morning was spent in review of more projects. The Saturn development and a further review of Mercury were put together end-to-end. Each of these involves a billion or more dollars. The Mercury Project, if carried to the conclusion the boys have suggested, will cost at least $3 billion. These are decisions a little bit tough to make. And yet, decisions must be made. The "lead time" for any one of these important projects may be as much as from five to seven years. It seems impossible to believe that the nation will be satisfied with accomplishing a Mercury flight and then stop any further attempts to undertake more difficult and longer-duration missions. I suppose that one of my problems is that I am already beginning to feel like a "short-timer." And in this situation, I tend to avoid making decisions that will affect the activities of my successor.

Sunday, July 10: This was to have been a day of rest but the presence of Dr. Lovell in this country suggested the possibility of doing a taped interview for the U.S. Information Agency. Accordingly, I met Dr. Lovell and others at the office at 11:30 and we spent about an hour in a discussion that was put on tape for use by the USIA in its foreign broadcast service. The balance of the day was spent in contemplation and work.

Monday, July 11: This was the morning when I was to take a physical examination at the Naval Medical Center at Bethesda. Actually, I have no need of this physical since I had a rather thorough one in Cleveland over the Christmas holidays last year. However, since we are just setting up this kind of an activity for our senior staff I wanted to see what kind of program was being offered by the armed services. Each of them is taking a number of our top administrative staff people and putting them through its regular routine physical examination as given to the generals and admirals. I was not impressed. Since I am writing this several days after the event took place I can say that my lack of enthusiasm seems to be valid. The report from the doctor states that everything seems to be all right except that one of the blood tests would indicate that I am a bit anemic. He does not rely particularly on this test since the second blood test was not well handled, either. The suggestion is made that I return for a further physical but I will not take the time to do this.

I have been attempting to talk with Homer Joe Stewart about the problems of finding a replacement for him. Homer Joe goes back to the staff at Caltech late this year. One of my problems is that I am unhappy about the way he has conducted the affairs of the office of program planning and evaluation. I need to get this straightened out before I set up the criteria for a replacement person. The balance of the day was given over pretty much to listening to rehearsals of the papers to be presented in the NASA-industry conference. This is a tedious and time-consuming job, but there is no substitute for practice in these matters. Some of the papers were very good, but many contained statements that give every evidence of making of NASA a space cadet organization. This will have to be corrected.

At 6:30, Silverstein, Horner and Stewart met with me to discuss tracking problems in other countries. At 7:15, Kerrigan came in to report on the program that he had prepared for the transport of members of the Nigerian government to our Wallops Island operation where they may view some of the Mercury activities.[4] Normally, there would be no problem about this but all of these gentlemen are black and it is impossible for us to take them into Virginia without giving real concern to the manner in which they may be received. In particular, we must take them down early in the morning and bring them back at night in order to avoid having to find hotel accommodations for them.

Tuesday, July 12: At 9 o'clock, B. F. Coggan of the San Diego Convair organization came in to see me. He is a person who had been recommended as a possible replacement for Dick Horner. The very welcome telephone call I received last Saturday relieved me of the necessity of talking definitively with Coggan. Bob Seamans has agreed to join us as associate administrator and will report on 1 September. Coggan was a pleasant fellow but it is obvious that he is best suited for sales and public relations efforts—not the rigorous task of running a tough and

---

4 Edward J. Kerrigan worked in the office of international programs as programs assistant and then chief of operations support. (Headquarters Telephone Directories, May 1960, p. 1, and Dec. 1960, p. 8, NASA Historical Reference Collection.)

broadly-based program. I say this in spite of the fact that for several years he has been the manager of the San Diego operations of Convair and in that job has handled a very substantial organization.

At 11 o'clock, Irving Gitlin of NBC came in with Shelby Thompson, Walt Bonney and others to discuss their proposition for a television series. Gitlin was with CBS and prepared the program submitted to us some weeks ago by that organization. Thus he had the advantage of knowing what was in his competitor's proposal. Nevertheless, he did present an interesting and exciting suggestion for our consideration. I think we are going to get out of this just what we want, although it is going to take a good bit of effort on our part. After a visit with Nick Golovin about the NASA-industry conference, I had lunch with Jim Hardie of Case. Jim has grown a good bit in stature although he has not as yet acquired very much in the way of humility. Each of us has his shortcomings—the problem is to achieve the best by taking advantage of the strengths of our people and minimizing their weaknesses. This is easily said and not so easily done. In Jim's case, he will ultimately hurt himself unless he does learn something about the necessity for humility in his day-to-day operations.

At 4 o'clock, Siepert and Wyatt came in to talk over a paper that has been in preparation for at least a year. It is intended to set forth the basic philosophy governing our operations with the Jet Propulsion Laboratory. At long last, it seems to be ready for transmittal to JPL. Later this afternoon, Miss Van Keuren, a Lewis employee who is serving temporarily as editor of the NASA-industry conference papers, came in to make a final review of my paper. She seems a very able gal.[5] Nothing much of importance happened during the balance of this day.

Wednesday, July 13: I had an 8 o'clock breakfast with John Corson of the McKinsey Company lasting about two and a half hours. I was going over with him the form of the final report that the Kimpton Committee is to present to us. There are some real differences of opinion as to the kind of organization that we ought to have but all concerned are agreed that there is need for a much better management capability than we now possess. At 11:00, Admiral Kirby Smith came in to see me. He is a retired admiral who is representing some Atlanta company seeking work with NASA. One has to see all of these people but it certainly is a waste of time.

I had hoped to get away on Friday of this week for a week's rest at the Bohemian Grove encampment.[6] With Dick Horner gone and Hugh Dryden away part of the time it has become increasingly difficult to contemplate that trip. In an attempt to salvage it, however, I called together the top staff at lunch and said that I would take the trip if I could be certain that there were no unresolved problems of major importance facing any member of the staff. This led to a five hour session in which we did come to grips with several of the problems posed in the meetings we

---

[5] Katherine M. Van Keuren worked in the reports editing branch at Lewis. (Lewis Research Center Telephone Directory, 21 Aug. 1959, pp. i, 47, NASA Historical Reference Collection.)

[6] This encampment, located outside San Francisco, brought together leaders of government, industry, and the media every summer.

held last Thursday and Friday. At the end of this session, I had agreed on the lunar and interplanetary program, had determined that the Saturn problem must have further reviews before a final decision on the scale of effort to be undertaken next year could be made, and had agreed on the Centaur program and several other problems that were bothering us.[7] Thus, I am going to take the trip.

I picked up Ruth and went to a cocktail party at the Sheraton-Carlton Hotel given by Alhaji Sir Ahmadu, premier of the northern region of Nigeria. There were many jet black people in the room and I was a little disturbed that there was so little representation from our own State Department. This was the day on which these good people had been taken to our Wallops Station, and they seemed to be very grateful for the arrangements that had been made for them.

Thursday, July 14: The staff meeting contained no unusual problems and I joined the space program council meeting at 10 o'clock. This is a meeting held every three months by Horner to bring together the directors of the laboratories and the various program offices so mutual problems can be discussed freely. I think its chief value is that of letting people get matters off their chests. It is interesting to note the differences in approach taken by the various members of the staff. Ostrander is reasonably quiet, always thoughtful and never overly determined or aggressive. In fact, he could be a little more aggressive in his own behalf. Silverstein attacks on every question and is seldom at a loss for words. One of his favorite expressions is that "he does not want to argue the question." He will state his position, then go forward to attempt to ride over all opposition. He is a very able engineer with an enormous capacity for work. It is clear that others find it difficult to work with him although many are devoted to him.

Von Braun and Pickering operate a bit as outsiders. Pickering always wants to get the record cleared down to the smallest detail and always seems to be pushing for more authority. Von Braun is persuasive and enthusiastic. He is a very able engineer and quite conservative when the chips are down. He seems to be a born leader and is able to engender very great loyalty among his subordinates. Harry Goett is a good technical man, but I think we made less than the best choice when we put him in charge of Goddard. This must be corrected by giving him a deputy who is adept in the management field. Harry is apt to be short-tempered and Abe tends to back him up in public discussions when he knows that Goett is not wholly right.

At 4:30, Frutkin came in to tell me of the success of his mission to Zanzibar. Apparently, after a diligent effort to meet with large numbers of Zanzibari, the Mercury tracking station will go forward on that island. Nevertheless, I have authorized and will continue to support an alternate station to be placed aboard a

---

[7] Presumably "the lunar and interplanetary program" referred to Ranger (discussed in previous notes), Surveyor, and Mariner, among other programs and projects. Surveyor was a series of soft-landing spacecraft to examine the lunar surface, while Mariner was an interplanetary program, with Mariner 2 becoming the first satellite to fly past another planet when on 14 December 1962 it came within 21,380 miles of Venus before continuing in orbit around the Sun.

ship. Dick Homer, probably in light of the confusion that attends his leaving the organization tomorrow, failed to provide arrangements for dinner for the group. Accordingly, Al Siepert and I took them to Costins. These are the small things that are important when you have top staff people from out of town. Left to fend for themselves, they must wonder whether or not they are really important members of the organization.

Friday, July 15: This was an all-day meeting with the space program council. I did finish up at 4 o'clock in the afternoon and Ruth picked me up at five to go to the airport. I was to have received a copy of each of the speeches to be given at the NASA-industry conference, but this package was not delivered to me. I had hoped to be able to use the five-hour trip across the country to review these papers. I suppose I will now have to do this next week while I am "relaxing" at the Grove. The flight left on time from Friendship Airport [Baltimore] and I was in San Francisco at 8:35 p.m. where I was met by Marron Kendricks. We started out at 9 o'clock for the Bohemian Grove and arrived at the gate at exactly 11 o'clock. Almost everyone had gone to bed but my tent was ready with the electric blanket turned on. I am looking forward to the next several days.

Saturday, July 16: I should have noted that the late afternoon on Friday was given over to a discussion with Abe Silverstein about the management plan for Nimbus and about the possibility of transferring these manned space flight projects beyond the present Mercury project to the Ames Research Center. He was asked to bring back his response to these matters upon my return from the Bohemian Grove.[8] Also, I had a meeting with Ostrander, Hyatt and Finger to ask for their thoughts about a study of the probable uses of a nuclear rocket propulsion system of the Rover type. I will follow up on this when I return.

That trip west yesterday at 32,000 feet was a pleasant one in some ways and not so pleasant in others. For a fair portion of the trip, we were flying over a cloud pattern that looked like a great many patches of sheep on a grey landscape. It was truly beautiful and I attempted to get one or two pictures from the cabin window. On the negative side, it has become increasingly difficult for me to travel without being interrupted by people who want to talk with me about the space business. This gives me a strange feeling of oppression that arises, I am sure, from the thought that I am unable to get away from the public nature of my office. During the course of the flight, I was accosted by one chap from the Itek

---

[8] Nimbus was an advanced meteorological satellite program approved after the first launch of Tiros in the spring of 1959. General Electric's Spacecraft Department developed and built the satellites under the direction of the Goddard Space Flight Center. A successful launch of Nimbus 1 occurred on 28 August 1964, but the satellite only operated a little under a month. Nimbus 2, launched 15 May 1966, transmitted 210,000 images over a period of 978 days. (*NASA Historical Data Book*, Vol. II, pp. 360-363.)

Corporation and by a second from the Aerojet Corporation. The first was pleasant enough but the second, after two martinis, was determined to argue with me the validity of my choice of Rocketdyne for development of the 200 K engine. I listened as politely as I could, responded pleasantly for a time and finally asked the chap to leave since I had other things about which I wanted to think.

Enough of that sort of thing—this is Saturday and I was up at 7:45 for a breakfast with Jimmy Doolittle and with General Pat Partridge and General Joe Cannon. Pat Partridge is now retired; he was the commanding general of the North American Air Defense Command. Joe Cannon is the commander of the Sixth Army in San Francisco. Immediately after breakfast, the four of us went for a three hour walk. This became a bit strenuous for General Cannon and myself and we broke off a bit ahead of Doolittle and Partridge. Lunch at the camp was a pleasant experience. Our guests today were a quartet and a five piece-orchestra. The music was quite wonderful and the food and conversation good. We wound up about 4:30. Around 6 o'clock, we went as a group to the Web Camp for cocktails. The weekend crowd has begun to arrive and the noise and confusion is increasing. Several old friends were among those present and I thoroughly enjoyed the experience. We had dinner at the "Dining Room Under the Stars." It was a large gathering, probably 1,600 in all. Apparently, there is a tradition that this opening dinner includes remarks from a member of the "old guard" and a response from one of the younger members of the Bohemian Club. The old guard consists of those members of the club who have had forty encampments to their credit.

Sunday, July 17: Up about 7:30 to meet Jimmy and Pat at breakfast. They went for a walk but did not invite me this morning. At 12:00, we listened to the organ concert at the lake side given each day by Paul Carson. Dinner at the dining circle and then a wonderful campfire program.

Monday, July 18: Up at 7:30 again and off with Pat Partridge and Jim Doolittle for a one and a half hour hike. Later in the day, I went for another walk with Jimmy Doolittle and, on my return, found much work to do. The papers for the NASA-industrial conference had come in. A good many people came in for cocktails before dinner and the campfire program tonight was instrumental and not too good. Off to bed about 12 o'clock.

Tuesday, July 19: Up at 7:30 and off for another three and a half hour hike with Pat and Jim. I came back at 12:15 to finish up the conference papers on which I had worked during the afternoon and evening yesterday and then spent 45 minutes on the phone talking to Washington to give them my comments. John Lodge, brother of Cabot Lodge and Ambassador to Spain, came to lunch today along with Preston Hotchkiss. A good discussion took place after which several of us went along to the Sun-Dodgers camp to listen in on the Socrates discussion. Plato and Socrates seem pretty distant in these times and I believe most of the audience thought this as well. The present-day controversy involving people like Oppenheimer, Strauss and Clinton Anderson could be made equally interesting— even more so.[9]

I talked with Grayson Kirk, President of Columbia University, about South America. I find it a little difficult to explain adequately my own interest in this part of the world. He listened sympathetically but suggested that it would be impossible for me to develop a real interest in Latin and South America while continuing at Case. Part of this arises out of the probability of a split in my allegiance but I would think that this could be managed. Dinner this evening was followed by a campfire show involving the Marine Corps. At least, this was the billing given to the show although the Marines only appeared to present the colors. Our luncheon quartet of several days ago stole the show.

Wednesday, July 20: Up at 7:45 and off with Pat for three and a half hours. We had a good talk about Washington, the various services and about participation in community affairs. [In the evening,] we had cocktails with the Lost Angels camp. I stayed home for dinner and worked on a speech. This being the 20th day of the month, I sent a remembrance to Ruth.

Thursday, July 21: Up at 7:15 and a four hour walk around the rim of the Grove proved almost too much for me. At the lake side today, John Lodge talked about the importance of Spain as an ally.

People are pouring in for the weekend. Lee DuBridge came to lunch and promised to come over tomorrow morning for a discussion with me. I worked on the speech again and then visited Herbert Hoover's camp which is known as Caveman's Camp. Hoover had gone to San Francisco and to Chicago for the Republican Convention but his son Herbert Hoover, Jr. and others were there. The campfire tonight was given over to the music of Jerome Kern and it was really a fine evening. After we had finished, the members of Pelican Camp gathered around the fire to hear Fred Crawford tell stories once again.

---

[9] On the Anderson-Strauss disagreements, see note 7, Chapter 5. J. Robert Oppenheimer (Ph.D. Göttingen, 1927) was an American physicist who made important contributions to quantum mechanics as related to atomic nuclei. He was director of the Los Alamos Laboratory during World War II, in charge there of developing the first atomic bomb. From 1947-1953, while he was director of the Institute for Advanced Study at Princeton, he served on the general advisory committee to the Atomic Energy Commission, while Strauss became chairman of the AEC in the latter year. Because of allegations of past associations with Communists, a security hearing in 1953 resulted in a ruling that despite his personal loyalty, his further access to military secrets be denied. This resulted in his contract as adviser to the AEC being cancelled in a widely celebrated and controversial case. Also factors in the situation were Oppenheimer's opposition on moral grounds to the building of the hydrogen bomb—a position with which Strauss vehemently disagreed—and a concern on the part of President Eisenhower that Senator Joseph McCarthy, then conducting his infamous hearings to detect Communists in the government, not give the country the impression that all scientists were disloyal. Eisenhower was successful in preventing this undesirable outcome, but in the process, Oppenheimer became a worldwide symbol of scientists who, in attempting to deal with the ethical problems that arise from scientific discovery, become the victims of witch-hunts. (See Nuel Pharr Davis, *Lawrence and Oppenheimer* [New York: Simon and Schuster, 1968]; Alice Smith and Charles Weiner, *Robert Oppenheimer: Letters and Recollections* [Cambridge, Massachusetts: Harvard University Press, 1980].)

Friday, July 22: I had a leisurely breakfast this morning and put the finishing touches on the notes for my talk. Several of us walked over to Monte Rio just to make the annual pilgrimage. The talk went fairly well—there must have been a thousand people at the lake side. There was much good comment about the speech although I did not do as well as I should have done. Fred and I visited Lee DuBridge at the "Sons of Toil" Camp and then went up to Lost Angels and Halcyon. I avoided dinner again—just too much to eat. The campfire was very fine tonight. At lunch today, General [David] Sarnoff [chairman, Radio Corporation of America] appeared and I found myself in a substantial argument with him. This time it was about his conviction that the Soviet Union was about to launch one hundred satellites carrying nuclear weapons and thus equipping themselves for a real exercise at blackmail.

Saturday, July 23: This morning I was up at 8:15 and walked around taking pictures with Fred Crawford. We had an early dinner and found good places at the Low Jinks. It was not a particularly good show but the audience was wonderful and everybody had a good time. We repaired to our camp immediately following the show, and I packed up to leave very early tomorrow morning.

Sunday, July 24: Marron Kendricks awoke me at 5:30 and after a fast ride to San Francisco, I caught an 8 o'clock plane for Los Angeles. Tom and Martha met me there and took me to the Knox home.[10] Getting into shorts and a shirt I just had a wonderful lazy day. I called on Gordon and Cora for a couple of hours and brought them up-to-date as they did me. A wonderful barbecued lamb supper, much discussion of politics and Latin America wound up the day.

Monday, July 25: This day I spent at the Jet Propulsion Laboratory. We had a good discussion and I found much to be pleased about. Certainly, the rapport between JPL and NASA is very much better than it was a year ago. At 3 o'clock, I went over to Arcadia to see Mother, Jessie and John.[11] Mother is very frail but seems just as bright as ever and we had a really good talk. I got her to tell me something about my childhood and learned much that I had not before known. Gordon came over to get me and I had dinner with him, Cora and Mary Lou and the baby.[12] This has been a really pleasant day.

Tuesday, July 26: Tom and I drove to the Burbank Airport where an Aero Commander took us to the Edwards Air Force Base for a visit with our Flight Research Center people at that station. Tom had a chance to look at the X-15 while

---

[10] Dr. Stuart C. and Rozella Knox were evidently close friends or relatives of the Glennans who lived in the Los Angeles area. On Gordon and Cora, mentioned just below in the diary, see note 12 below.

[11] Arcadia is a community in the northern part of the Los Angeles metropolitan area. John [Simmons] was Glennan's nephew. Jessie was his mother and Glennan's sister.

[12] Gordon was Glennan's brother, Cora his wife, and Mary Lou their daughter.

I visited the U-2 installation and looked over two or three of the planes.[13] We then went across the reservation to look at the big static test stands that are being erected for the testing of our 1.5-million-pound thrust engine. This is really a substantial piece of concrete. Back home about 5:30 and ready for packing for the trip back to Washington. I had a swim and then took the Knox family and Tom and Martha to the Beachcombers for dinner. The plane left on time and I arrived in Washington at 7 o'clock in the morning with very little sleep behind me.

Wednesday, July 27: Bill Detrich met me at the plane and took me directly to the apartment.[14] After a shower and a bit of breakfast with Ruth, I took off for the office where I had a date at 9:30 with [General] Quesada, representatives of Dr. York's office and of the Air Force—Dr. Courtland Perkins [Air Force assistant secretary for research and development, 1960-1961] and others—on the problem of developing a public policy and a program for the management of a supersonic transport project. I had called this meeting because of the increasing pressures for activity in this field. These pressures have been generated both by industry and by Congress.[15] And this is not to ignore the desires and pressures of large numbers of people active in government service today. I started the meeting by stating that I did not want to manage and fund this kind of an activity. Much to my surprise, this statement was greeted with a sigh of relief on the part of all concerned. Quesada was particularly happy about it and I responded by saying, "I thought you all recognized that NASA was the one government organization that is not seeking to expand its responsibilities." This brought a laugh, at least.

We agreed that the Federal Aviation Agency should take the leadership position in this activity and that the rest of us should support the activity in any way possible. FAA would plan to use the Defense Department to manage its development program and would expect research and development support from NASA. It was agreed that a small committee be set up to deal with this matter in the hopes of bringing it to the attention of the president and other elements of the government— perhaps the cabinet or the National Security Council—so that we might get instructions from the president to move ahead with planning in this field. This was

---

[13] The famous hypersonic X-15 rocket research aircraft was built in the mid-1950s under the joint auspices of the Air Force, Navy, and the NACA/NASA with the NACA as the technical director of the project when it began in 1954. The first flight occurred on 8 June 1959, and over the next decade there were 198 more flights in the aircraft, with speeds up to 4,500 miles per hour and altitudes of over 350,000 feet. (For details, see Hallion, *On the Frontier*, pp. 106-129, 329-337 and Milton O. Thompson, *At the Edge of Space: The X-15 Flight Program* [Washington, D.C.: Smithsonian Institution Press, 1992].) The U-2, of course, was the high-altitude jet reconnaissance and research aircraft in which Francis Gary Powers was flying on 1 May 1960 (see entries in diary after that date). A prototype first flew in 1955. The aircraft had a top speed of 494 miles per hour and could attain altitudes upwards of 70,000 feet. NASA used the aircraft for weather and atmospheric research. (U-2 files, NASA Historical Reference Collection.)

[14] This may have been Harry W. Detrich, Jr., who worked in the public information office and then in the office of educational programs. (NASA Headquarters telephone directories, May and December 1960, pp. 7, 2 respectively, NASA Historical Reference Collection.)

[15] See U.S. Congress, House, Committee on Science and Astronautics, *Report of the Special Investigating Subcommittee to the . . . Eighty-Sixth Congress, Second Session on Supersonic Air Transports*, 30 June 1960 (Washington, D.C.: Government Printing Office, 1960).

an interesting exercise in bringing interested parties together in the development of public policy.[16]

My desk was piled high with mail, which I have been attempting to surround. Lunch with Bob Seamans was a pleasant affair and we agreed on his program for the next month or so. At 2 o'clock, Dr. Kistiakowsky came over to talk with me about the problems involved in the communications and meteorological satellite areas. Apparently, one of the subcommittees of the President's Science Advisory Committee has been holding hearings on the communications satellite system and has become very much concerned that we are lagging in the development of public policy in this area and that the program could proceed at a very much faster pace. I told him of our activites with the Rand Corporation and asked Bob Nunn to come in to discuss the status of those researches with Dr. Kistiakowsky. None of it is very reassuring and it is clear that we will have to put some one person in charge of this particular activity. I am reminded, at this point, that no single communication satellite has flown, as yet.[17] The pressures generated by AT&T and by the military as well as by other industrial suppliers are building up quite a fire, however.

The rest of the afternoon was given over to catching up with the mail and talking with Dick Horner and others. Ruth and I joined the group at the National Press Club at a dinner for John Victory at 7:15. I was toastmaster and did as well as might be expected with the lack of sleep I had experienced. John was less voluble than usual having been warned about this on several occasions privately and publicly. It must be a bit of a wrench, however, to drop out of an activity one has had a real hand in developing over a period of 45 years. Victory was the first employee of the NACA back in 1915.[18]

Thursday, July 28: A call from Elmer Lindseth before breakfast resulted in my agreeing to meet him at 8:15 at the office in order that we might talk a bit about the work of the organizational advisory committee. They are coming down the pike toward a report framework and Elmer wants to be sure that the report will be a helpful one. There is much to be said for a continuing operation of this kind and I am hopeful that our own management analysis group will be able to take on such a task. I opened the NASA-industry program plans conference at the Department

---

[16] For a discussion of NASA's subsequent role in the effort to develop a supersonic transport, see Hallion, *On the Frontier*, pp. 177-196. For an overall history, see Mel Horowitch, *Clipped Wings: The American SST Conflict* (Cambridge, Massachusetts: MIT Press, 1982).

[17] For subsequent NASA progress in the area of communications satellites, see the unpublished paper by Leonard Jaffe, director of the communication and navigation program office (since 1963), NASA office of space science and applications, entitled "Satellite Communications: Six Years of Achievement, 1958-1964," dated 1 February 1965 (in his biographical file, NASA Historical Reference Collection.) Jaffe had joined NASA in 1959 as chief of the communications satellite program in the office of the assistant director for advanced technology. He had previously worked at Lewis for the NACA in instrumentation and automatic data.

[18] As these comments would suggest, he was retiring after 52 years of continuous government service. See the entry under his name in the biographical appendix.

of Commerce auditorium at 9 o'clock. There must have been 1,500 or 1,600 in attendance and the program seems to be off to a good start. I think our staff has done an excellent job in planning and executing this particular program.

I had lunch with Bob Nunn to talk over the steps that need to be taken with respect to the communications satellite business. It does seem probable that we should ask the president to assign to us the task of developing the basic public policy to be proposed to Congress by the administration. This is the way these things are done; if some one agency doesn't step up and seek the assignment, everyone is apt to rush in and a chaotic condition can prevail. It seems clear to me that it is our responsibility and one that we should not duck. Accordingly, I asked Bob to come up with an outline of a paper to be presented to the cabinet at an early date. This paper would request that the president assign, by executive order or otherwise, the task of developing policy to NASA. At 2 o'clock, I met with the advisory committee on organization and continued with them until 6 o'clock. There is a good bit of discussion going on about the desirability of having the individual field centers report directly to the associate administrator. Under certain circumstances I think this would be a good thing; in fact, it may be necessary to do this within the next year or so. It also seems desirable to bring together, at some point in time, the activities of the launch vehicle operations group under Ostrander and the space flight operations group under Silverstein.[19] I think we are going to get enough out of this activity to make all of us feel quite satisfied about it. Basically, the areas of management development, adequate recruiting and organizational planning, and the strengthening of the planning function throughout seem to be the principal areas of agreement. Left for future discussion and action are precise organizational changes, the urgency of which does not seem too great at the present time. Bob Seamans, Hugh and I had dinner with the organizational advisory committee at 7 o'clock. Seamans conducted himself well in a discussion period following the dinner.

Friday, July 29: Lee Atwood of North American Aviation came in to see me on short notice. He was in town for the NASA-industry conference and simply wanted to pass the time of day. He made strong remarks about the Rover program, however. Everyone wants to build this gadget but none of them wants to face up to the problems of its use. At 11:30, Barton Kreuzer of RCA came in to see me. His firm built the Tiros satellite. It wants to build the next one but I am afraid that our organization is not quite up to it. I say our organization because we have not set up our management structure sensibly for the Nimbus satellite system.[20] I am hoping to get this straightened out before many weeks have passed, however.

---

[19] These changes did not occur during Glennan's tenure as administrator but were part of the 1 November 1961 reorganization. See Rosholt, *Administrative History of NASA*, pp. 221-222.

[20] In December 1960, RCA won a contract to develop and build an advanced vidicon camera system for Nimbus, but in February 1961—under the Kennedy administration—NASA selected General Electric to build the spacecraft and do the subsystem integration, a contract on which RCA had bid. (*NASA Historical Data Book*, Vol. II, p. 361.)

We closed out the NASA-industry conference about 4 o'clock with an attendance that continued to surprise all participants. At 5:00, Dryden, Horner and Silverstein met with me to discuss the on-going program in the manned satellite field. I had proposed the possibility of moving the manned space activities to Ames Research Center at Moffett Field in California. After some long discussion, it became quite clear that there was substantial opposition to this proposal, opposition that was rational enough so that it must be reckoned with. There was substantial agreement, however, with my concern that this project would overwhelm Goddard Space Flight Center and that some alternative method of planning for it should be developed. It does appear that it could be left at Langley as a separate activity. I have asked Silverstein to explore this thought. All hands agree that the management of the project must be strengthened, although it is clear that this must be done in such a way as to avoid tampering with the morale of the Mercury group. This has been quite a day and I think some useful results have been accomplished. I decided to take the family down to Huntland for the weekend and was able to set this up without difficulty.[21] I have a great deal of work to do over the weekend.

Saturday, July 30: Up to a leisurely breakfast and off to Huntlands at 11 o'clock or thereabouts. The afternoon was spent in reading, writing and sitting in the sun. I got off to bed at 10 o'clock but did not have much sleep during the night for reasons that must be associated with this heavy organizational and operational activity facing me in the next six weeks.

Sunday, July 31: Up to a very leisurely breakfast and more work during the day. We had a good change at Huntlands and arrived back home in the early evening. George Brown certainly deserves a hearty vote of thanks for the manner in which be has made it possible for a few heavily loaded people in Washington to enjoy the pleasures of a weekend home in the Virginia countryside.

---

21 As appears in part from what follows in the diary, Huntlands was the horse farm of a man by the name of George Brown, president of Brown & Root, Inc., who permitted others besides Glennan and his family to visit the premises in the Virginia countryside on separate occasions and use them as a weekend retreat from the burdens of their work in the nation's capital. (Correspondence of T. Keith Glennan with J.D. Hunley, 24 Jul. 1993.)

CHAPTER NINE

AUGUST 1960

Monday, August 1: After I dictated the story for yesterday, a storm came up and we hastened to pack and start back to Washington. This morning is my first with the responsibility for the operating organization as well as for the activities in my own office. Dick Horner has gone west and Bob Seamans will not be on board for another month. This means that I will have my hands full but I suspect that I will learn a great many things about the organization I have not been aware of. After trying to catch up with the morning mail and make a little sense out of the thinking I have been doing over the weekend, I met with Golovin and Colonel Heaton to talk about the way we would operate during the coming month. Golovin very nicely agreed to stay on an extra week—he had been planning to leave at the end of the current week—so that I would be served for half the month by Golovin and for the second half of the month by Heaton, who reports for his assignment as special assistant to the associate administrator [Seamans] on 15 August. I think we made out well in this discussion and that I will be well served by these two gentlemen. As I may have said before in this chronicle, Golovin is an amazingly able person who completely lacks the ability to deal with people. I have found few persons able to think through a problem and to state the conclusions on paper so clearly and so convincingly. On the other hand, I have found few who can so readily arouse the antagonism of co-workers—especially those he is expected to convince regarding a proper course of action.

At 11:30, I drifted over to see Jerry Morgan at the White House to assure myself that the need for Senate action on our legislative proposals will be mentioned in the president's message to Congress. I received as much assurance as one can in a situation of this kind. Following a lunch at the Occidental for Gus Crowley —he retired a year ago [as NASA's director of aeronautical and space research]—I met with Ostrander's committee on the Juno II program. This committee had been set up some two or three months ago to review the then-ABMA program calling for the launching of four Juno II rocket systems with scientific payloads provided by our Goddard Space Flight Center but manufactured and qualified by ABMA. The Jet Propulsion Laboratory has supplied the upper stages for the Juno II—indeed, this was the first American rocket system to launch a satellite into orbit, Explorer I.[1] Unfortunately, it is really not a very good system and the parents now would rather forget it than try to make it work. However, we have a good bit of money involved

---

[1] Actually, it was the Juno I that launched Explorer I on 31 January 1958. Like Juno II (see note 2 below), it employed Sergeant rockets for its upper stages but used a modified Redstone rather than an extended Jupiter as its first stage. As the name would suggest, both launch vehicles were in the same family of rockets. (*NASA Historical Data Book*, Vol. II, pp. 46-47.)

and I am determined that we will fly these missions whether or not they all result in failure.

This particular committee, set up by Ostrander, has had the task of attempting to review the vehicle itself with the thought of improving, if possible, the chance of success. It was an excellent presentation and I think it can help somewhat in giving us greater assurance of a reasonable score with these shots. Strangely enough, the principal action recommended was that of assigning responsibility for the missions. I am not at all sure why it has not been done before, but I suspect that this is one of those things that Dick Horner left behind him during his last month or two when he was a little loath to take action in certain of these matters. At any rate, I dealt with this matter promptly and have assigned the responsibility to Marshall Space Flight Center. If these are their vehicles—they are making the payloads under scientific direction of Goddard—they have the background of experience in this particular flight article, and it must be their black eye if things fail to operate satisfactorily. I know they would prefer to get out of this responsibility, but I am determined that they accept it—willingly and with determination.[2]

At 4 o'clock, Thompson, Bonney and Phillips came in to discuss the proposals given to us by CBS and NBC regarding television programs dealing with the space exploration program of NASA. After some discussion, it became apparent that each of the proposals was reasonable and that there was little to choose between them. However, all of us agreed that CBS seems to have dealt more effectively with programs of this sort in the past and accordingly, I determined that CBS would get the nod. I managed to get away from the office about 5:30 and surprised Ruth at home. Most of the evening was given over to reading; it seems that I get satisfaction out of having discharged a task such as that of making the decisions in the television and the Juno II situations. At least, I felt very good tonight. Perhaps I ought to say that I have decided to go on a diet again, and this time I will not stop until I have reached 180 lbs. Since that goal is some 14 lbs. away, I suspect that it is going to take some weeks before the goal is attained.

Tuesday, August 2: At 8:30, Bob King [NASA director of program management] came in to discuss with me further the program management plan. Yesterday I had begun to look into this in depth for the first time and was very much surprised that we had established such a plan without a real program for making effective use of it by having the top officials in the agency review at frequent intervals the results depicted in the plan. I intend that this be done and now must

---

[2] Juno II consisted of an extended Jupiter intermediate-range ballistic missile (developed by the ABMA in the 1950s) as its first stage. This was a liquid-fueled booster powered by a Rocketdyne engine. The second, third, and fourth stages of the launch vehicle consisted respectively of 11, 3, and 1 scaled-down, clustered Sergeant solid-fueled rockets (developed by the Jet Propulsion Laboratory). Because of the separate development of the liquid and solid stages, Marshall and JPL had shared responsibility for Juno II until this decision by Glennan. As he states, the launch vehicle had had poor results, its total record being only 3 successful missions in 10 attempts. Once Marshall took over full responsibility, the record improved from 1 successful launch (Explorer 7) in 6 attempts under joint responsibility to 2 successful launches (Explorers 8 and 11) in 4 attempts. Nevertheless, NASA replaced Juno II with the Scout launch vehicle (developed by Langley Research Center) as the primary launcher for the Explorer series. (*NASA Historical Data Book*, Vol. II, pp. 46-48, 61.)

get at the job myself. These people who are paid $17,000 a year to handle problems of this kind certainly are not earning their money, in my opinion. They do not have as much initiative or as much follow-through as I would think was necessary.

At 9 o'clock, John Johnson and Bob Nunn came in to talk with me about the communications satellite problems. In the period of time since I went away for the recent holiday, a good many things have happened tending to make it important that we grab the ball in this particular program. I should say here that we have stayed entirely with the "passive" satellite program and have relied upon the Defense Department to handle the "active" communications satellite program. It appears that it has not done the job very well and that AT&T, Hughes Aircraft, IT&T and others are now becoming deeply interested in getting on with the job of putting together a communications satellite system. Some of this has leaked to the press and John Finney had a column on it in the *New York Times* within the past week. Already, we have had questions raised by congressional committees and George Kistiakowsky's President's Science Advisory Committee has asked for reasons supporting our lack of activity. I think it is important that we now take on the job of developing an active satellite system for civilian purposes. This means a re-evaluation of our relationships with the Defense Department in this particular area and a change in the agreement we made some eighteen months ago. More than this, however, it means dealing with the many difficult and complex public policy issues that are involved. The communications business in this country has always been operated by private industry, albeit an industry regulated by the government. What is the responsibility of the government for carrying on research and development in an area such as this where it is already reasonably apparent—according to our own figures—that industry can make a profit using satellite relay systems? If the government does continue to conduct R&D activities in this area, just how long should this activity continue?

I had asked Johnson and Nunn to come in because we have had a study underway on some of these problems for some six months under Johnson's general supervision. This has been carried on by the Rand Corporation and, unfortunately, is nowhere near completion. Before I left for my holiday, I had asked Bob Nunn to get active in the field and he had called in several people from Rand during my absence and had elicited from them a promise that they would have a fairly complete but preliminary report in our hands by 15 September.

For some reason John Johnson was afraid my suggestion that we prepare this story in the form of a paper for the information of the cabinet and approval by the president would result in extended delay and excessive complications. I suspect that his problem was thinking of this paper as one in which the cabinet might become involved. Actually, I had in mind a statement of the present technological developments—the state of the art—as well as a statement of the problems that face us, both domestically and internationally. Following this, I would propose that we outline a program of action the president could approve. All of this would assume an earlier discussion with the Defense Department and an agreement between

Defense, the Bureau of the Budget and ourselves as to our responsibility in the development of an active communications satellite for civilian use. In any event, I finally asked Johnny to get on with the job of developing certain elements of the statement that I wanted and I expect that he will do his usual fine job.

At 10 o'clock, Ruben Mettler, president of Space Technology Laboratories, came in, fishing for information about future work his organization might expect from NASA. I took the opportunity to discuss with him the untoward activities of some of his people in the Zanzibar incident and to question seriously the desirability of his people making speeches about the shots they were doing for NASA. I am not sure that I made any progress, but at least I tried. At 10:30, Hjornevik and [Aaron] Rosenthal [NASA director of financial management] came in to give me the operating plan for 1961. This discussion took almost two hours. In the course of it, I learned a great deal about our organizational difficulties. In this short space of two days, I am beginning to doubt the validity of the position I took with our advisory committee on organization last Thursday. It seems clear that the director of each of our principal program offices needs to cooperate at a very much more effective level or in a more effective manner with his counterparts. When one thinks that each of them is responsible for the expenditure of several hundred million dollars, this sort of thing becomes even more important. In any event, the Bureau of the Budget is withholding about $50 million from our research and development appropriation—I think they are right in doing this. In our own house, we are going to withhold some $15 million as a reserve in my hands. I suspect all of this will be expended before the end of the year.

A quick lunch with Dr. W. O. Baker of the Bell Telephone Laboratories brought me up-to-date on their interests in the communication satellite field. Bill Baker is really an interesting person. It is very difficult for me to understand his rather devious way of talking and yet he is held in the highest esteem by his colleagues and by people generally throughout the government and the research and development echelon of industry. At 1:30, I met with the Federal Council on Science and Technology. Glenn Seaborg, chancellor of the University of California at Berkeley, presented a draft paper on the need for support of science through increased attention to graduate education and research. There followed an interesting discussion covering a period of 90 minutes, at least. [Secretary of Health, Education, and Welfare] Arthur Flemming was in attendance and Jim Shannon of the National Institutes of Health was also present. I enjoyed this part of it very much indeed.

Home in a hurry, then on to a dinner at the home of Minister of the Australian Embassy Donald Monroe. He had as guests Prince Mohammed and wife of the Malayan Embassy, one or two others from the Australian Embassy and several people from the United States State Department. I was the only outsider there. Ruth and I enjoyed the evening very much. The prince and his bride were attractive and interesting people.

Wednesday, August 3: Once again, Bob King was in to discuss further the program management plan. Obviously, he has not been used to the kinds of questions I have been handing to him, and he is beginning to react in a positive manner. I will keep up the pressure because he seems to be able to take it and to get results. Nick Golovin came in and we had quite a long talk about a variety of organizational problems. Once again, I am impressed by the man's thinking ability; I do wish he could handle himself better with other people.

At 1 o'clock, luncheon at the Shoreham for 10 Russian gentlemen who are over here to look at airports and civil aviation turned out to be a rather dull affair. Pete Quesada was their host and he did his best to make it a pleasant party.

At 2:30, our people gave us a presentation on the active communications satellite—a very convincing one. I think this program will move ahead fast and it is up to me now to get on with the job of resolving problems between ourselves, the Bureau of the Budget and the Defense Department. At 4 o'clock, Jack Wooldridge of *Nation's Business* came in, complete with assistant and steno-typist, to record an interview. Fortunately, I am to have a chance to correct the transcript and am sure that I can add a few questions they didn't think of. It wasn't a bad experience. At 4:45, Dr. Mesthene of the Rand Corporation, who has been loaned to George Kistiakowsky for several months, came over to discuss the preparation of a paper entitled "Managing Important New Technologies." This was supposed to be a paper prepared jointly by the AEC and ourselves as a "legacy of thought" to be given to the new administration when it appears. It seemed to me quite an erroneous point of departure and I suggested that it might be better for us to attempt to point out how we would now organize NASA if we had the chance to begin over again. This seemed to strike a responsive chord and we will start out this way in the hopes that it will result in a satisfactory product.[3]

I dashed off to George Kistiakowsky's apartment to meet Ruth for a cocktail. Walt Whitman, MIT's chairman of the department of chemical engineering, was in attendance. He has come to Washington to serve as the scientific advisor to the secretary of state replacing Wally Brode. I am sure that he can do a very much better job than Wally; it would be hard to do one any worse.[4] That seems to have finished the work for today and I came on home with Ruth to my glass of Metrecal [a liquid diet beverage]. I hope this doesn't last forever but I am going to keep it up as diligently as I can in the hopes of getting down to that 180-pound mark.

Thursday, August 4: The staff meeting lasted for a couple of hours this morning. We reviewed the recent NASA-industry conference and discussed, at

---

[3] The apparent result of this effort was a paper entitled "Managing Major New Technologies," prepared by Walter D. Sohier, NASA assistant general counsel, dated 1 Oct. 1960. It is available in a folder marked "Federal Council on Science and Technology" in the Glennan subsection of the NASA Historical Reference Collection and ranged widely over a whole series of problems and issues having to do with organization, management, and operations. See also diary entry for 10 Sept.

[4] Whitman remained science advisor to the secretary of state through 1963. According to Allan A. Needell, who had researched the subject, Brode angered many scientists by putting foreign policy concerns ahead of purely scientific ones. This may have been what Glennan was referring to here.

some length, the plans for the United Nation's space conference, which presently is being planned for late 1961. The exhibit area of 25 to 30 thousand square feet is scheduled to cost about $2 million. It will have a very high residual value, however. It is hoped that the Century Twenty-one Exposition to be held in Seattle two years from now will make substantial use of the same exhibits. There is always the desire to have such an exhibit here in Washington as a national activity so that the citizens may see the purposes for which their money is being expended.

Lunch today consisted of Metrecal. Immediately following lunch, *Nation's Business* sent a photographer in who took several pictures in a very short time and I hope got some that might be worth publishing. Today we were hosts to a group of 60 or 70 State Department people who were being given a briefing on our program. I greeted them and turned them over to Arnold Frutkin. Then I had a further talk with Shelby Thompson and Ned Trapnell. Ned is acting as a consultant to Shelby in putting together a briefing program that we must prepare for a variety of military installations. This is almost a professional activity and is certainly one for which we do not have the necessary staff readily available. At 3:30, John Corson and Jack Young of McKinsey, two of their men and a group of our own people sat down to review the proposed report on the contracting problem faced by NASA. Rather naturally, this particular study is related to the organizational study that McKinsey is also working on. I found their efforts in this instance more definitive and somewhat more practical than the comparable efforts in the organizational study.[5]

Friday, August 5: John Johnson in at 9 o'clock to discuss with me briefly the problems Bob Seamans faces with respect to his stock holdings. We agreed that since this is not a presidential appointment and therefore not subject to Senate approval, there was no need to make public disclosure of the actions to be taken by Seamans and yet, on the other hand, I urged Johnny to discuss informally with the office of the Attorney General the plan proposed to hold Seamans harmless from any future charges of "conflict of interest." The fact that we will undoubtedly be doing business with RCA, in which Bob owns considerable stock, is bothersome, of course. A way must be found to diminish substantially the impediments that are placed in the way of employing good people from industry.

At 9:30, the Project Mercury boys came in with a Mr. Burke of McDonnell to give me the sad news on Project Mercury. Obviously, as we come closer to flying the production capsule, troubles are bound to occur. I am quite certain that every reasonable effort is being made to keep this project on whatever schedule we want to set up. It does now appear that we will be delayed a few weeks in accomplishing the sub-orbital flight but there is no reason to believe that the orbital flight will be

---

[5] On the contracting report, which McKinsey & Co. completed in October 1960, see Rosholt, *Administrative History of NASA*, pp. 154-160.

delayed by any substantial amount. Of course, it is always possible that we can catch up the time we now appear to be losing. The loss of the Atlas-Mercury shot the other day when the Atlas blew up has started the usual round of threats of investigations on the part of Congressional committees.[6] For this reason, and for this reason alone, I have asked George Low to put together a chronological history of the project with pertinent dates and schedules that have been published in the past together with the conditions under which those schedules were published. We must be ready for this kind of investigation at any time, and I propose that we take a positive stand rather than a defensive one. I had Gleason and Bonney in to talk over the possible impact of the possible change in schedule of Mercury with respect to public information and congressional problems. We agreed that there was no need for taking positive action of any kind at the moment. Certainly, if we can avoid this until after the election we will go a long way toward keeping space out of politics—an aim that I hold very strongly.

Lunch with Arthur Flemming was a pleasant affair. Apparently, he has a desire to stay on as Secretary of HEW if Nixon is elected although he has no feeling for his chances of being asked to stay on even though Nixon does win. He has no specific plans for his personal life following the ending of this administration but I know he won't be idle very long.[7] I have spent most of the day attempting to get hold of Lyndon Johnson but without success. I have been trying to impress on him the need for dealing with our administrative legislation—even to the point of flying out to Johnson City, Texas, to see him. Finally, I found his press secretary in Louisiana and the message was relayed to Lyndon. Word has come back that Lyndon will see me early next week.

Admiral William D. Irvin, chief of the Defense Communications Agency and an old submarine friend, came in to talk about his problems. He is not a "communicator" and it seems to me a rather strange choice of a man to head an agency destined to find itself in trouble almost from the start. We agreed that he might get good support and suggestions from Tim Shea and Mervin Kelly. At 4 o'clock, Johnson, Dryden and I discussed the management plan for Project Nimbus and found it wanting.[8] It is very difficult to be as rough as one should be in these cases when you know that the people concerned have done their best and

---

[6] On 29 July 1960 an Atlas launch vehicle carrying an unmanned Mercury capsule exploded about a minute after launch from Cape Canaveral. (James M. Grimwood, *Project Mercury: A Chronology* [Washington, DC: NASA SP-4001, 1963], pp. 105-106.) This was not the only setback for Project Mercury, but on 5 May 1961 Alan Shepard completed NASA's first suborbital mission, and on 20 February 1962, John Glenn carried out NASA's first orbital mission. While Mercury showed that "final launch preparations took far more time than anyone had anticipated in 1958 to ensure perfect readiness and reliability of the machines and men," for a total cost of a bit more than $400 million the project prepared the way for Gemini and Apollo. (Swenson, Grimwood, and Alexander, *This New Ocean*, pp. 352-358, 422-434, 505-511, quotation from p. 508.)

[7] A former president of Ohio Wesleyan University, he became president of the University of Oregon in 1961.

[8] On Project Nimbus, see note 7 of Chapter Eight.

don't really understand the management problem. Nevertheless, this project is almost a year old now and we are really not underway as yet. Fortunately, I think we can pull the fat out of the fire, and I plan to do this early next week.

Abe Hyatt came in at 4:15 to talk over the organizational problems facing his office. It is clear that we are not getting hold of the "nubbins" of the problem and I suppose this means that I have to get into it with both feet myself. All of this is good for me and I am sure that I will be better able to counsel wisely with Bob Seamans when he appears on the scene in September. After a visit with Bob King about certain problems in the program management office, I stopped off home to help Ruth get ready for dinner. We are having [company].

Saturday, August 6: I had promised Buck Bowie that I would see one of his in-laws who needs help finding a job. Accordingly, even though I was very tired, I got up at 8 o'clock and Ruth with me. While I was shaving, she came in to ask if I had left my money clip in the living room and had taken one of her purses from the closet and left it on a chair near the kitchen door. Having no recollection of this sort of action, I immediately started to investigate and soon found that someone had entered our apartment during the night, had taken from my bureau, at a point within three feet of my head, my wrist watch and my money clip with somewhere around $75.00 in it. They had apparently reached into the closet in the hall and taken this one purse of Ruth's, which had nothing in it of any value. How he was able to enter the apartment, then our bedroom, and take these articles without our waking is hard for me to understand. It is a fact that I had taken a sleeping pill, but this seldom ever keeps me "under" for any length of time. I called the police at once and within ten minutes two patrolmen were in the apartment. They stated that this sort of thing had happened quite frequently recently and that there was a pattern being followed by what they presume to be a single individual. During the course of the day, two more detectives came by to get further information. Fortunately, my insurance will cover a substantial portion of the loss. Since this has occurred, I have wondered what I might have done had I wakened while the man was in my room.

The visit with Buck's in-law was a pleasant one and I hope that I'll be able to help him. I then had a 3-hour session with John Corson about our organizational study. Neither of us is very satisfied with it and I think I got him started on another course of action that may give us a better grasp of the problems.

Sunday, August 7: Up at 9:00 for the usual skimpy breakfast—I weighed in at 190 stripped this morning. Then an hour and a half with the newspapers and on to the work table where I managed to get several matters well planned for the coming week. Ruth and I took a walk for an hour and half and then came back to prepare for dinner with [friends]. Another detective came in to see us today. In the course of his conversation with us, he let us know that this was the 22nd robbery that had taken place in this general area, in apartment houses like ours, within the last two months.

Monday, August 8: At 9 o'clock, Abe Silverstein, Johnnie Johnson, Hugh Dryden and I sat down to try and deal with Project Nimbus. Actually, I thought Abe did an excellent job this morning. He professed—and I believed him—that he wanted to do exactly what we did. As a matter of fact, I think he would have done more than seems reasonably possible to do. In any event, it is clear that we must have solid legal counsel on this matter of the management of Project Nimbus, and I am looking to John Johnson to give this to us. Hugh and I had lunch together and managed an interesting discussion with [Raymond J.] Steve Saulnier. Obviously, Saulnier (the chief of the president's economic advisors) is concerned about the health of the economy. Clarence Randall [special assistant to the president] came in and sat with us for a few minutes. Saulnier asked him about the level of activity in the steel industry and whether or not it could be expected to rise. Randall was not particularly optimistic. He raised questions about the administration's holding back of several hundred thousand tons in orders for steel pipe. Saulnier said that these orders had been released at the end of last week. In the debate that followed, Saulnier made clear his concern that the high cost of money [i.e., the interest rate] was deterring the developmental programs of industry. He pointed out that 5 percent money would not encourage any real industrial organization to go ahead with capital expansion. He wound up by asking whether or not we in NASA could expedite the commitment of funds for activities during the first six months of this year. This is about as close as I have been to the seat of power with respect to economic matters in this government.

At 1:30, Bob King came in to talk about the program management plan. It is obvious to me that he is not a really strong person even though Dick Horner recommended that we give him $17,500 a year. There are so many of these reasonably mediocre people in government who are paid in the range of $14,000-$18,000 a year that it sometimes makes one a little bit sick. Bob is not a total loss but he is certainly going to work for his money as long as I am here in Washington. I must say that he is willing and I believe that together we can make a reasonable operation out of the activities with which he is principally concerned.

I called Leonard Jaffe to talk about the communication satellite problem.[9] This was a good three-quarters of an hour discussion in which I found him to be quite responsive and quite responsible. He admitted having overstated some of his claims for the non-military communication satellite program. I think we came to a good understanding of what needs to be done from this point on, and I wound up by dictating a memorandum to him asking for confirmation of certain facts and figures about which we had been talking.

Ruth came downtown and we had cocktails with Dan Kimball and the Aerojet crowd in honor of Homer Joe and Mrs. Stewart. Just why they were honoring the Stewarts I do not know. Actually, there could be questions raised as to the possibility of conflict of interest here. Stewart works for us and Aerojet is one of our contractors. I think we will get by without an investigation in this case, however. I had my usual dinner of Metrecal and off to bed since I feel a little bit weary tonight.

---

[9] See note 17 of Chapter Eight for background.

Tuesday, August 9: The morning started with the usual set of discussions. I managed to be promptly on time at a television interview with Senator Wiley of Wisconsin at 10 o'clock. Wiley was brought up in Chippewa Falls, Wisconsin, just 10 miles from Eau Claire. We had a right good time together and I think made good use of the opportunity to become better acquainted. I moved from there directly to the office of Senator Stennis to urge him to speak with Lyndon Johnson about getting on with our legislation. It is interesting to note the way in which these senior senators still defer to the majority leader. Apparently, there is a code to which they all subscribe. A chairman of a committee is a real power and actually controls, pretty much, the activities of that committee. Of course, this sort of thing is not universal and very often one finds a maverick who will upset the otherwise placid tenor of their ways.

At 11:30, Hugh and I met with the president, chairman of the executive committee, chief engineer and Washington representative of Grumman Aircraft. They simply wanted to let us know that they were very much interested in the orbiting astronomical satellite project.[10] They took us to lunch at the Washington Hotel roof and we enjoyed the visit. For some reason, I was quite relaxed and really did enjoy it. Back to meet with General Ostrander, who came in for a short session with me even though he is on leave. I was discussing with him his interests and concerns about the organizational set-up. He is an able person and one for whom one must have respect. At 2:30, the vice president for engineering and the Washington representative of Boeing came in to give us the same sort of pitch about their activities. The day wound up pretty much with my going to the office of the deputy secretary of defense to discuss with him and with one of Herb York's people the desirability of our undertaking an "active" satellite program. As is always the case with Jim Douglas, we had no difficulty in coming to a reasonable agreement, and paperwork will be hammered out to support it.

Just before I went over to the Department of Defense, Dan Kimball of Aerojet came in at my request. Formerly the secretary of the navy under Truman, Dan is a strong Democrat and has some influence on the Hill. I asked him to attempt to secure action from Lyndon Johnson on our bill. Actually, these industrialists have a good bit at stake because we have been able to get the House to accept a favorable patent clause. If the bill does not pass this session, it will have to be argued all over

---

[10] This evidently is a reference to the orbiting astronomical observatories project. Langley had done a preliminary study of the concept in May of 1958. In February 1960, NASA assigned technical management of the project to Goddard Space Flight Center. In October of the same year, NASA announced plans to negotiate with Grumman for a contract to build a 1360-kilogram observatory, with that firm getting a subsequent contract for follow-on satellites and subcontracting their subsystems. The first observatory was launched successfully and entered a circular orbit on 8 April 1966, but after 22 orbits, its power system failed before it could return any data. On 7 December 1968 the second observatory was launched successfully and placed in orbit. All systems and experiments functioned as planned, and the spacecraft provided a great deal of information on ultraviolet, gamma ray, x-ray, and infrared radiation; the structure of stars; and the distribution and density of the interstellar medium. (*NASA Historical Data Book*, Vol. II, pp. 259-262.)

again and I guess I should be happy that I won't be here to participate. Tonight, we had dinner with Millard Richmond and Ray Smith of Western Electric Company in New York at a new French restaurant called Paul Young's. It was very good indeed. A pleasant evening and off to bed.

Before doing that though, let me talk a little bit about this matter of getting our bills through Congress. I have attempted for a month now to reach Lyndon Johnson. I have a continuous call in to his office to try and talk with him. I know that he is busy with other things, but I suspect that he is busy principally with keeping his political lines straight. Certainly, the first two days of the Senate have seen speeches by Lyndon attacking the administration rather than getting on with the job of passing the legislation that is needed. It will be interesting to see how long it takes for me to reach him. At this juncture, I should say that I have not had strong support from the White House, either. Perhaps this is my own fault; if it is, I intend to remedy it tomorrow.

Wednesday, August 10: This was another of those days. At 9:30, Gerald Lynch and Dr. Krause of the Ford Aeronutronics Division came in. They wanted to explain their deep interest in our activities. They are a responsible group and are already involved in the program. For some reason I did not seem to resent their discussions. At 10 o'clock, Dr. Clark Randt discussed with me various aspects of the life science program. I admire this man very much indeed. He has given up an important professional career and substantial income to join with us in attempting to push this program forward because he believes in it. Unfortunately, I am not able to give him the assurance I would like or to answer all of his questions. These relate to the necessity for his hiring a few men to get on with the planning of a life sciences center, etc. I shall try to get this matter in hand soon.[11]

At 11 o'clock, Francis Reichelderfer of the Weather Bureau came in to talk with several of us about the on-going problems in the weather satellite field. I am determined that we take a long-range view and set our sights on the development of an operating system or at least the prototype of an operating system at some specified time in the future. If we do not do this, I am sure that our own people will want to continue to research and experiment while the Weather Bureau will want to continue to talk about things rather than doing something about them. It was not easy to make this point clear and I am not sure that I succeeded. However, I am going to take the initiative in this matter.

At 12 o'clock, Dryden, Silverstein, Hyatt, Siepert, Rosenthal, King and I sat down for a quick lunch and a discussion of several problems. I am anxious that we make good use of our program management man and I am setting up formalized meetings for a review at stated intervals of the activities we have under way. There are some operating plan difficulties I hope we straightened out at this meeting

---

[11] NASA appears never to have established a life sciences center per se, but there were fairly extensive activities at both Ames Research Center and the later Johnson Space Center. See Hartman, *Adventures in Research*, pp. 321-323, 426-428, 478-485, and 500-503; and John A. Pitts, *The Human Factor: Biomedicine in the Manned Space Program to 1980* (Washington, DC: NASA SP-4213, 1985), passim.

although I think I was the one to be straightened out rather than the other participants. I broke the meeting off a little early so that I could go up on the Hill with Jim Gleason to see Senator Styles Bridges. I was trying to get him to convince Lyndon Johnson to put our legislative act on his calendar. Bridges was quite happy to see us and promised cooperation, but I doubt that anything useful will come of it. Once again, the protocol in the Senate is something to behold. Actually, I guess it is necessary for it to be this way—but it is frustrating. I dropped by the Johnson office and was happy to find that I have an appointment for Friday at 10:30 in the morning. Back to the office then for a long session with Siepert and Hjornevik. They have been doing some very good thinking on our organizational problems and presented a somewhat long-winded but excellent picture of their thinking on project management. While this is only one aspect of our organizational problem, it is a very important one. I must get them now to deal with John Corson and others so that we get some unified and positive thinking on this matter.

At 4:30, Homer Joe Stewart came in to say that he was going west on holiday and wanted to clear up some matters with me before he departs. The discussion was a fruitful one and I hope it will lead to definite progress on things Homer Joe is supposed to be doing for us. Off [for] home at 5:30 and nothing much to say about the evening except that I did some more work.

Thursday, August 11: The staff meeting this morning was a long one devoted completely to the results of an ad hoc committee study of our procurement and source evaluation board procedures. It was a really fine discussion, and I think real progress was made. The willingness of the staff to participate in these discussions is very satisfying. At 11 o'clock, G. L. Best, Harold Botkin and Dr. John Pierce of the AT&T and Bell Laboratories came in to talk with Dryden, Johnson, Frutkin, and Jaffe of NASA. Botkin had been discussing with the communications people of England, France and West Germany the possibility of cooperative satellite communications research projects. He found a willing and even enthusiastic audience. We spent almost two hours discussing both public and operating policy questions and finished up with an agreement that AT&T would provide us with an informal statement of its proposed course of action. If, indeed, the AT&T is willing to support research in this field, it is not clear that the government should do more than a minimum. On the other hand, I doubt that AT&T realizes how costly this research will be. In any event, this project must go forward and it is my task to see that it does.

We went to lunch with these good people and I returned just in time to spend an hour and a half with Shelby Thompson in a discussion of his program for handling technical information and educational assistance. He had outlined a $4 million program, which I accepted at $3 million. Now we will have to find a substantial portion of the $3 million because there was no provision in our FY 1961 budget for this activity. We can pull together the money without too much difficulty, I am sure.

Gleason came in to try to work out the strategy for our discussion with Johnson tomorrow morning. We agree that the best possible approach would be to attempt to convince Lyndon that the Republicans would make no political hay out of his refusal to act and that there was not enough real hay in it for the Democrats in case he did act but, really, we were simply trying to get on with the job he had given us. Further, I intend to press him with the statement that he, in effect, promised the president that this legislation would go through as early as last January. I doubt that I am going to get any place with it but I can do no less than try.

Off to home then for a pleasant evening with Sally and Ruth. We went over into Maryland to let Sally use the trampoline. She does quite well on it but she split her britches! Much kidding, only a part of it received by Sally with good grace. Off to bed then about 10:30. The family is going up to Atlantic City to spend a little time with Polly and to meet the MacGregors and I shall join them by air tomorrow evening.

Friday, August 12: This was really quite a day. Having been awakened rather early each morning for the last several mornings to be told that Echo I was not going to be launched, I was delighted to get the word at 6:30 that the launch had taken place. A series of telephone calls kept me advised of the situation until I reached the office at 8:15. By this time, the launch was confirmed and it appeared clear that we had an orbit. Actually, a press conference had been set up for 8:30, but the success of the transmission of the president's message from the Goldstone Station in the desert outside of Pasadena to the Bell Telephone Laboratories station at Holmdel, New Jersey, [via Echo I] was so complete that I decided to hold the reporters until such time as we had a copy of the recording. It had been phoned from Holmdel to our Goddard station where it was recorded on tape. A motorcycle escort brought it into the office and I listened to it with great glee before going to the press conference at 9:10.

While waiting for the tape to arrive, I had alerted the White House and arranged to have the privilege of taking the tape into a meeting of the National Security Council so that all present might hear it. This was done about 9:30 and the president and the members of the NSC were delighted. A little later, pictures were taken in the president's office with Dryden, Jaffe, and myself together with the president. It was a great day, coming as it did upon the heels of the success of Discoverer XIII and followed later in the day by the establishment of a new altitude record by the X-15.[12] At 11:15, I went up the Hill to see Senator Johnson along with

---

[12] With Major Robert M. White (USAF) at the controls, the X-15 established a new altitude record for a manned vehicle of 136,500 feet on 12 August 1960—more than 10,000 feet higher than the previous record for the X-2 on 7 September 1956. And the first man-made object recovered from an orbiting satellite was the 85-pound, instrumented capsule from Discoverer XIII. It was retrieved from the ocean off Hawaii after 16 orbits on 11 August 1960. (Emme, *Aeronautics and Astronautics . . . 1915-1960*, p. 126.) The Discoverer program consisted of a series of research satellites launched by the DOD. It tested components, propulsion, and guidance systems and techniques for later use in various U.S. space projects, according to a report issued at the time. (NASA, *Third Annual Report in the Fields of Aeronautics and Space*, p. 24.) Its most important mission, however, was as a reconnaissance satellite, part of the Air Force effort to develop a spy satellite to replace the U-2 in the 1960s. The purpose of the Discoverer program in general was to determine the best way to employ cameras on satellites and recover the film after it had been exposed. Discoverer XIII and XIV, the latter launched on 18 August 1960, recorded some revealing photographs of Soviet territory. (Robert A. Divine, *The Sputnik Challenge: Eisenhower's Response to the Soviet Satellite* [Oxford: Oxford University Press, 1993], pp. 11, 190.)

Jim Gleason. We spent about 15 minutes trying to convince him that our legislation should be dealt with by this rump session, but to no avail. However, I did impress upon him that he had told the president this bill would be acted upon, and I finally got out of him the promise that the staff would prepare all necessary documents and take whatever steps could be taken, so that if the opportunity presented itself, the bill might be dealt with speedily.

Back to the office to find that things continued to go well and Bonney made the suggestion that recordings be made of each of the foreign ambassadors for the purpose of transmitting them via Echo I from coast to coast. These transmissions, recorded on tape, could then be used by the United States Information Agency in its foreign broadcast activities. I called George Allen and received his hearty endorsement; the deal was made and steps are already underway to accomplish the desired result. At 12:30, Silverstein, Newell, Goett, and John Johnson came in to go over the papers on Project Nimbus. Finally, I believe we've got a document that will allow us to get the best possible management set-up under the circumstances. It is a little bit like pulling teeth but one must be both deliberate and stubborn.

At 1:00, I had lunch with Dick Harkness of NBC. We went to Jack Hunt's Raw Bar where we were greeted with open arms. It turned out that Dick Harkness had not heard of the success of Echo when I had called him earlier in the morning. He was much more excited about it than I. At 2 o'clock I visited the White House for discussions with General Persons. It appears that Leonard Hall, Nixon's campaign manager, had brought to the attention of General Persons the desire of Hughes Aircraft Company for some of NASA's money. Actually, it has a project in communications of interest to us.[13] I suppose there must be some valid reason for undertaking these excursions to bring political pressure to bear or else the activity would not be undertaken.

I broke off at 3 o'clock and started for the airport where I was able to spend an hour quietly having a drink before going to Atlantic City. The MacGregors and Ruth and Sally met me at 7:30 and we dashed on into Atlantic City because Polly's Hotel Morton dining room closed at 8:00. The girls were waiting for us—that is Polly and Kady MacGregor—and seemed genuinely glad to see us.[14] After dinner, we spent some time on the boardwalk and then retired to our motel and bed.

Saturday, August 13: Up at a leisurely hour and off to a Howard Johnson for a quiet breakfast with Ruth. We sat around in the mist that finally turned to rain and waited until noon before we went to do a bit of shopping. It had been our plan

---

[13] This was the Syncom communications satellite project on which Hughes had been working since 1958. On 16 August 1960, Hughes made a presentation to NASA and Glennan suggested the company pursue the idea further with the goal of obtaining experience in using such satellites in synchronous orbit. Syncom 1 was launched on 14 February 1963. It achieved orbit, but communications with the satellite lasted only 20 seconds. Syncom 2 and Syncom 3, launched on 26 July 1963 and 19 August 1964, were successful; thereafter, NASA transferred the Syncom system to the DOD. (*NASA Historical Data Book*, Vol. II, pp. 378-384.)

[14] As emerges below, Polly and Kady had summer jobs at the shore. The MacGregors were evidently family friends of the Glennans.

to barbecue a steak for the two girls who were to have this day off. Sally had stayed the night with Polly and about 1:30 they put in an appearance. We managed a bit of lunch of crackers and cheese and then the kids went swimming even though it was raining. There was a good pool there at the motel. We had been able to get a very good steak, and the rain let up enough so that we managed steak and corn in the best picnic fashion. All of us ate too much. One of Mac's friends had called to say that we could have tickets to the Ice Capades, so we drove on into Atlantic City and enjoyed this show very much. Back to the motel at midnight and a somewhat restless sleep. I have forgotten to say that, during the course of the day, Attorney General Bill Rogers called me to ask why I did not prepare a statement for the president to give out on the tremendous success that had attended the nation's efforts in space over this weekend. I did just this and telephoned it back to Bonney for coordination with the White House. It sounded like a good idea and maybe it will get used.

Sunday, August 14: Up again at a leisurely hour. After packing, we drove in to Atlantic City where we met the girls who were enjoying the sun on the beach. Polly and Kady had worked this morning, of course. We finally got them off the beach and into a restaurant for a quick bite of lunch after which (now almost 4:30) Ruth, Sally and I drove back to Washington. It has been a good weekend. I should have recorded the discussion last night—or rather Friday night—when Polly and Kady presented their case for leaving the Morton Hotel before their agreed-upon time of servitude had expired. They had drawn up a list of eighteen points, most of which were the usual gripes of anybody who has to work for a living. It was clear, however, that the girls were quite unhappy and that they were operating under conditions much less than desirable. Mac and I finally agreed that they might leave at the end of another two weeks but made as certain as we could that the girls understood we were not impressed by their arguments of persecution. Being a father in a situation such as this has its drawbacks; one wants to do what is right in the best sense of that term and yet one has real sympathy for the child.

Monday, August 15: The weekend news of Echo has been excellent. I have not been able to see it as yet because of the cloud cover each night, but I continue to hope that I will have a glimpse of it.

At 10 o'clock we had a review of the Saturn Project and I find myself now with the task of rationalizing an increase of $25 million in the budget we have requested for 1962. The plain facts are that von Braun and his staff oversold the project just as they had while they were with the Army. Why does government make people just a little "dishonest"? At 11:30 Wyatt and Newell came in to talk about the terms of reference I had prepared for the office of program planning and evaluation. This was resolved rather easily and they are going to give me some help on rewording the statement. At 12 o'clock, at the invitation of Secretary [of the Air Force] Dudley Sharp, I attended a White House gathering where a flag from the capsule of Discoverer was presented to the president by [General] Tommy White of the Air Force and Secretary Gates. It was a pleasant ceremony in front of the cameras and I think the president was genuinely pleased. Certainly, the sight of the capsule, which looked like a kettle drum, gave me a thrill.

At 2 o'clock, we continued with the Saturn meeting but I did not come to any conclusion. This is going to be a rough decision. Congressman Chet Holifield and Eugene and Barbara Zuckert came for dinner. It was lots of fun and they stayed until 11:30. After that, I was so stirred up from conversation and argument that I did very little sleeping.

Tuesday, August 16: This was another day! The morning was given over to presentations by the Hughes people and by the Bell Telephone Laboratory people on an "active" communications satellite. There is real pressure on the part of industry to get into this business, and it is reasonably clear that AT&T is serious about driving toward a communications system using satellites. I asked our people to develop a program for the next three or four years that would involve participation by both of the organizations we have been talking with.

After lunch with Dryden and Silverstein, I went out to the home of a Finnish sculptor named [Kalervo] Kallio to see a bust of the late General George C. Marshall. We are thinking about buying the bust to be placed at the Marshall Space Flight Center in Huntsville, Alabama. It was really a delightful experience and I think we will buy it even though the price is high.[15]

At 3:14, Norris Bradbury [professor of physics at the University of California and director of the Los Alamos Scientific Laboratory] came in to talk with Dryden, Ostrander and Finger about the Rover program. We are still trying to get John McCone into the corral on the organizational problems. Everybody has now agreed that our nomination of Harold Finger is the only sound one. McCone simply won't light long enough to deal with the Joint Congressional Committee on Atomic Energy as he should before public announcement is made. I left the office a little early—about 5:30—because I was less than full of life.

Wednesday, August 17: I started off the morning with a talk to a group of people who are going to take part in a drive to encourage the purchase of United States Savings Bonds. I guess these things have to be done, but I wish my heart were a little more in tune with an activity of this kind. At 10 o'clock, I talked with Stans and Staats of the Budget Bureau. I told them of my difficulty in presenting them with a $1 billion budget that made any sense but expressed complete willingness to talk for several hours with staff members in order that they might get a complete picture of our dilemma. I think the budget at $1.23 billion is going to be tight.

I also told Stans about the communications satellite business and our determination to go into the active communications satellite program. He seemed to agree with this as long as we coordinated properly with the Defense Department. I have worked this out with Jim Douglas as I have said earlier, and I think there is no problem on this score.

---

[15] The Kallio bust was the one used in the dedication ceremony. See photos below in the next chapter. (Ltr., T. Keith Glennan to Forrest C. Pogue, Director of the Research Center, George C. Marshall Research Foundation, 14 September 1960, Marshall Space Flight Center folder, Glennan subsection, NASA Historical Reference Collection.) Pogue, the famous biographer of Marshall, called the bust to NASA's attention.

At 11:15, I met with the president and General Persons. I was asking the president to fly to Huntsville, Alabama, to dedicate the Marshall Space Flight Center on 9 September. He has the utmost respect for George Marshall and agreed to do it but wants it to take place on 8 September. This being my birthday, who am I to kick? I told the president about our determination on the communications satellite business, and he seemed in complete agreement. I discussed with the president my problems about the legislative changes Lyndon Johnson has thus far refused to bring to the Senate committee. The president said Lyndon was not to be trusted in any sense of the word but suggested that I keep after the matter. He recalled, too, that Lyndon had said the president's desires with respect to changes in the Space Act would be dealt with in a favorable fashion.

Returning to the office, I found Admiral Bennett waiting for me. We went over to Jack Hunt's Raw Bar for lunch where Jack took good care of us again. I talked with Rawson about the problems we were facing at Case with respect to the financing of our activities in the materials field. He may be able to do something for us in this matter. At 3:00 in the afternoon, Gleason and Jerry Siegel came in to see me. The latter is a research worker and soon will be teaching at the Harvard School of Business Administration. He was a member of the staff of the Senate committee at the time of the drafting of the Space Act. I was appealing to him to do whatever was necessary to assure Lyndon that we needed the changes in the act. Siegel does not agree with me on this matter but I think I convinced him that we had no ulterior motives—simply a desire to be sure that my successor had a good, operating organizational arrangement.

At home at 4:30 to be photographed with Ruth by Bob Quinlan of the [Cleveland] *Plain Dealer*. Apparently, the paper wants to do a full-page story on Ruth with some atmosphere provided by her husband and possibly by Sally. Sally did come home in time so that we were able to have some pictures made of her in Rock Creek Park. Tomorrow, Johnnie Simmons will arrive. Cora has just called from the West Coast to tell us that Gordon has gone over to Arcadia to bring him back for the night. It's his first trip away from his mother and his first airplane ride. He should arrive "full of beans."[16]

Thursday, August 18: The staff meeting this morning went rather rapidly. Nick Golovin did a fine job of presenting the reliability program. This is suspect from the start but must be carried through to a rational and, hopefully, successful conclusion. Everyone is a little fearful of its becoming a "fetish."[17] At 11:00, three men from Westinghouse came in to tell us how much they wanted the orbiting astronomical observatory job. Lunch followed with General Ostrander. We talked

---

16 See above, entry for July 25 and notes 10-12 to Chapter Eight.

17 As originally established, the program had as its objectives to "measure the reliability of existing components, to determine what had to be done technically to increase reliability, and to devise a method for assuring that what should be done was done." There were supposed to be individually-tailored programs for each system, implemented by field centers and NASA contractors under the guidance of reliability steering committees. (Rosholt, *Administrative History of NASA*, p. 149 n. 134.)

about a good many problems, including particularly those concerned with organizational arrangements. Don is not a forceful person but is thoughtful and considerate. I doubt that he chooses men as well as he might. At 2 o'clock, the Lockheed people came in to tell me how much they wanted the orbiting astronomical observatory. Later on, Hugh and I spent two hours discussing a variety of matters relating to organizational arrangements, program decisions, etc. The agreement with the Defense Department on the active communication satellite program has been found to be acceptable by staff people in the Pentagon. It is expected that it will be signed on Monday or Tuesday of next week.

A long discussion with John Johnson regarding the policy problems that will be involved in the further prosecution of the active communication satellite program proved to be interesting. He is thoughtful and somewhat stubborn. On the way home by 5:15 to find that Ruth and Johnnie had arrived safely from the airport. He seems to be in fine shape. A pleasant dinner was followed by a separation of forces—John and Sally drove down to the Watergate Concert at which the Navy Band was playing. They stayed through only half the concert and then took a little trip looking at the Capitol, the Marine statue, the Washington Monument and the Lincoln Memorial. Ruth and I managed to see Echo pass over twice tonight. I cannot get over the wonder of this experiment—140 lbs. of a balloon, 100 feet in diameter, 1,000 miles above the surface of the earth and travelling almost 16,000 miles per hour. What a business!

Friday, August 19: Awakened to a call from Frank Phillips to tell me that the Soviets have launched something at 4:55 a.m. our time. No Russian announcement has been made as yet. We were certain they had something in orbit but not sure what it was.[18] And yesterday the Air Force launched a Courier from Cape Canaveral—it blew up two and a half minutes later—and a Discoverer XIV from Vandenburg Air Force Base. Discoverer XIV went into orbit, and an attempt was made today to recover it. Let me say right now that it was recovered, successfully, in midair![19] Thus, for once, the Soviet Union is put into the shade even though it did put two dogs into orbit this morning. What a wonderful day! Now back to the business end of this activity—it was a very full morning. George Metcalf, a vice president of General Electric came in to tell me about his company's interest in the

---

[18] This was initially called Spacecraft II and later designated Sputnik V, a satellite weighing 10,120 pounds and containing two dogs, mice, rats, flies, plants, fungi, seeds, etc. Designed to test the capsule and recovery system for development of human space flight, the satellite was recovered after 18 orbits on 20 August, having travelled 437,500 miles. (Emme, *Aeronautics and Astronautics . . . 1915-1960*, p. 149; Satellite Situation Report, 31 August 1960, NASA Historical Reference Collection.)

[19] Discoverer XIV weighed 300 pounds and was designed, according to information available publicly at the time, to gather data on propulsion, communications, orbital performance, stabilization, and recovery techniques, just like Discoverer XIII. On 19 August, as Glennan states in part, an Air Force C-119 transport recovered the satellite from midair at 10,000 feet, the first such recovery of an object from space. The Courier was designated 1-A and was a communications satellite that failed to orbit due to the premature shutdown of the first stage of the Thor-Able-Star launch vehicle. (Emme, *Aeronautics and Astronautics . . . 1915-1960*, pp. 126, 149.) On Discoverer's role as a reconnaissance satellite, see note 12.

possibility of developing a service to provide launching and data acquisition facilities for commercial use. General Electric is the company that has employed the Bechtel Corporation to search for a Pacific Ocean site suitable for launching satellites into an equatorial orbit. I called John Johnson in so that he might get some of the flavor of this expression of interest. It is obvious that there will be more questions raised about activities of this sort and we must be prepared to discuss them sensibly.

A lengthy discussion with Morris Tepper of the meteorological satellite program was most helpful to me. I am anxious that we bring together all of the organizations that make use of meteorological data so we can be as certain as possible that our satellite program makes good sense. Tepper seems to be a reasonably aggressive young man—he is a meteorologist in his own right. Naturally, each of these men likes to keep the program completely under his own control. I think we are making progress in this field, however. Bob King came in to talk about the agenda for tomorrow's program discussion and his planning for the first of the monthly meetings at which the administrator will be brought up-to-date on problems and schedule changes in each of the operating projects.

A lunch with Karl Harr was pleasant enough. Usually, he has a good many things about which he wants to give me advice—in a most helpful way, really. Today, he had nothing but praise for our program and the manner in which we are handling our relationships with the USIA, etc. At 1:30, Modarelli came in with nine different designs for a commemorative stamp I am trying to get the Post Office Department to issue in connection with our Echo Project. I have been on the phone trying to get [Postmaster General] Arthur Summerfield, but I find that he has gone to New York. Finally, I did reach him on his boat and have made a date to see one of his top aides on Monday morning.

As a followup on my Wednesday morning meeting with Stans, we had a session this afternoon with Messers. [William F.] Schaub, [Don D.] Cadle and one other from the Bureau of the Budget. I had them sit through a presentation of the Saturn problem and then told them that I was going to increase my request for FY 1962 by $20 million. This amount is somewhat less than is necessary to meet the requirements for Saturn —by about $6 million—but I think we can manage the absorption of some portion of the added requirement. I had been reluctant to give the Bureau of the Budget a program priced at $1 billion and I believe that we were able to show our guests the futility of attempting to approach the problem that way. What really will happen is yet to be determined, however. Hugh Dryden and I worked with Bob King and Al Siepert over the agenda for the Williamsburg conference. We cut it down by one day and have achieved a very much better balance and I think a better plan for the conference.[20] That ended the discussion and nearly brought the day to an end.

---

[20] See entry for 16 October 1960 below in diary.

Dinner at home and a game of backgammon with John filled in the time until Echo came across the heavens once again.

Saturday, August 20: This being the 350th monthly anniversary of our marriage, I left in my bed when I went down to work a set of records for Ruth. These were Spanish language records and might help her in her studies. A large group assembled at the office at 8:30 and we had a six-hour discussion of various elements in the program. I think these program discussions are an excellent medium for bringing everyone up-to-date and providing for a much better understanding of objectives, both at the administrative and the technical level.

Back at home about 2 o'clock with nothing to do but loaf for the rest of the day. Word has just come in that the Soviet Union brought back to earth its satellite with the two dogs, alive and well. The dispatch states that the capsule was brought back within six miles of the pre-designated spot. If this is the case, the Soviets really have achieved a significant advance. For a nation that has consistently put out propaganda by its most eminent scientists that the re-entry problem had not been solved, they are doing awfully well. As Ruth has said, it strikes us that the Soviet people—at least, their government—do not know the meaning of truthful discussion.

Sunday, August 21: Up at an early hour and off to Gettysburg to show John the battlefield. He seemed to enjoy the countryside through which we drove. It was a hot day but a pleasant one. We found a good place in the shade of some trees and among some large rocks for a "cook-out" breakfast. We did not do too much out at the battlefield but did enjoy the electric map that gave a graphic portrayal of the three-day battle described by tape recorder. Returning home in the afternoon, I attempted to get at some of my work but without much success. Edward Teller was on "Meet the Press" and did a rather outstanding job—one much more nearly objective than most of the discussions in which he engages. After some argument, I accompanied Ruth and the children to the Carter Barron outdoor amphitheater to witness and hear a performance by Victor Borge. He is really a very accomplished artist although I was not in a mood to enjoy the patter accompanying his rather amazing feats of piano magic. Home and in bed somewhat after midnight with the realization that everyone had to be up at 6:45 in the morning since Ruth and the kids were going to fly to Langley and then take a car over to Williamsburg for the afternoon.

Monday, August 22: This was a reasonable day—except for the fact that I was awakened at 4 o'clock in the morning by a reporter from the *Journal-American* in New York. I believe his name was Foley. He had a UP dispatch quoting me as saying that the U.S. far outstrips the Russians at the present time in space activity. Half asleep and very angry, I told him I would comment in the morning—that this was not a quotation from anything I had said and that I would have nothing further to say at the moment. I slammed the receiver and was not bothered until 7 o'clock in the morning when Foley called me back. Apparently, a telecast I had taped with Senator Wiley of Wisconsin had been played on Sunday in that area and, in commenting on the beneficial aspects of space research, I had said that the United States was far ahead of Russia. Actually, this is a fact. Of course, the newspaper men wanted to stretch this into a controversial statement in the face of the Saturday accomplishment of the Russians in bringing back to earth their Sputnik V with the animals. I am not sure that I accomplished my purpose with Foley in straightening him out, but at least I tried.

I visited Roy Walker, staff assistant and public relations man for Postmaster General Arthur Summerfield, to talk with him about the possibility of issuing a commemorative stamp on Echo I.[21] After a thorough discussion of the problems that face the Post Office in matters of this kind, he led me to believe that such a stamp might be issued for us. Fortunately, I had asked some of our people to suggest some designs, one of which seemed to please Walker. At least, he asked me to prepare a statement for Summerfield and one by myself explaining to the public the reason for issuing the stamp. Hopefully, this can be done—the issuance, I mean, during the month of November.[22]

George Kistiakowsky was back from his vacation, and I visited with him for three-quarters of an hour. I brought him up to date on the communications business and discussed at some length the problems of inadequate knowledge on our part about the Discoverer satellites and what the state of knowledge of other branches of the government might be on space activities, both at home and abroad. At noon, joined by Al Siepert and Al Hodgson, I was the guest of former Governor Leo Hoegh [of Iowa], who is now the director of the Office of Civil and Defense Mobilization. We flew by helicopter to the relocation center to which I have been assigned in event of serious enemy action. This was an amazing experience. I came away in much the same mood as I have from recent visits to atomic energy installations. The amount of money, thought and genius that has gone into the creation of these elements in our developing national strength is distressing when one considers that its entire purpose is defensive or destructive. Nevertheless, it must be done and I am of the opinion that this one is a good exercise.

Back at the office at 4:30 to visit with Hugh Dryden and others about some of our problems arising out of the Russian accomplishment. We are going to lay on a discussion with several of our people to try to fathom the methods of operation now being employed by the Russians. Naturally, we have little or nothing to go on except our own reasoning of the events so sketchily described by the Russians. At home about a quarter of six to wash the dishes, make the beds and prepare the apartment for the return of the prodigal's family members following their day at Langley Research Center and Williamsburg. Perhaps I'll have a chance to get some of my own work done before the evening is over.

---

21 "Roy Walker" is probably a transcription error for L. Rohe Walter, Special Assistant to the Postmaster General (Public Relations), with whom Glennan later corresponded on this issue.

22 As discussed below in the diary, the stamp was issued 15 December 1960.

Tuesday, August 23: This morning wasn't too bad. The usual early morning discussions and then at 10:30, Steve Bechtel of the Bechtel Corporation and several of his people came in to talk about their interest in making available their services to NASA. Note how I stated that—it is exactly as stated by Bechtel. This is not intended to be a derogatory statement. Certainly, his discussion of the manner in which they handle their jobs was most interesting. An extremely successful contracting firm operating in a variety of heavy construction and engineering fields, Bechtel Corporation is one of the great success stories of the land. It is really a pleasure to listen to this sort of presentation.[23]

At 11:15, Kistiakowsky, [Richard] Bissell, Dryden and I talked about a variety of problems that are classified.[24] Suffice to say that they were concerned with matters that took place during the past two weeks.

Lunch at 12:30 with Dick Harkness was a difficult one. Dick had asked me to lunch with him because he wanted to seek my cooperation in a project. Imagine my surprise when he told me that NBC was planning to do a one-hour television show on the U-2 after the election. He had a series of questions prepared by his New York office that he showed to me. They revealed a desire to get at the sort of confused issues that had never been wholly explained—and probably never will be completely explained. As may seem evident to the readers of this chronicle, I reacted by saying that I could see no reason for such a show. What was the purpose? Dick could not give me an answer. The tenor of the questions suggested that the goal was to have a show with wide audience appeal. My own conviction is that where espionage is involved—even though acknowledged espionage —nothing is gained by further discussion of an issue of this kind. Allen Dulles is to be interviewed by Harkness on this same question. It will be interesting to see what his reaction is.

Back at the office for a discussion of the necessity for the undertaking of research on environmental simulation devices. I happen to believe that we really don't know as much as we should about the technology involved. However, the staff concerned—and there are some good people involved—are convinced that the only way to acquire this technology is through actual construction of an increasingly larger and increasingly more complex series of environmental test chambers. This whole discussion was triggered by complaints of Litton Industries that we were so poorly organized that we couldn't deal properly with a proposal they have made for research of this kind. After about an hour's talk on the subject, I concluded that we had better avoid undertaking a research program for which the staff had little desire. Only one man stood out for the type of research under discussion and he was asked to prepare a proposal, a work statement, for further consideration by Dryden and myself.

---

[23] By FY 1963 Bechtel had become one of NASA's top 100 contractors. (*NASA Historical Data Book*, Vol. I, p. 211.)

[24] Glennan identified Bissell as a CIA employee who handled the U-2 program and also was involved, he believed, with the Bay of Pigs invasion of Cuba in April 1961 during the early Kennedy administration. (Comments on draft Ms. of this chapter, June 1993.)

At 4 o'clock, Bonney, [Chief of the News Division Joseph] Stein, Ostrander, Gleason and others came in for a discussion of the program for the proposed Marshall Space Flight Center dedication, which is now scheduled for 8 September. I had envisaged a very brief and closely held operation—closely held in the sense of having it completely within the confines of the Redstone Arsenal and without a great deal of fanfare. Imagine my surprise to learn that this had now grown to the point where the president would land at the Huntsville city airport and would drive in a motorcade to the arsenal. Reason—70,000 people of Huntsville would have an opportunity to see the president of the United States for the first time. Some 67 legislators would be invited, the governor would be invited to speak briefly, city and county officials would be invited, etc. Finally, since it appears these matters are well understood by people such as Jim Hagerty and others in the public information and press relations operations of the White House I gave in, regretfully.

At 5 o'clock, Gilruth, Silverstein, Low, North, Mengel, Dryden, Kistiakowsky and I discussed the Mercury project and the possible implications of the recent Russian success in recovering their dogs. An hour of discussion brought forth no new light on the subject, but we did reaffirm our determination to proceed with our program with a strong sense of urgency but without hysteria. Coming home a little bit late, I found things in good shape for a dinner with Congressman and Mrs. Albert Thomas. It may be remembered that Albert is the Chairman of the subcommittee of the House Appropriations Committee that deals with our money bill. I cooked a steak on the grill and it turned out to be excellent. Ruth had one of her usual fine dinners, which elicited much favorable comment from the Thomases. It was a good evening and I continue to have some admiration for the congressman.

Wednesday, August 24: This day started off with an hour and a half of photography. Yousuf Karsh of Ottawa, Canada, came in to make a series of color and black and white photographs of me, presumably for use on the cover of the Sunday magazine, *Parade*. If it is like any of the other of these accumulations of material to be presented in magazines, this will gather dust in the files. But it was an interesting experience. It is obvious that Karsh is an artist of great ability. At luncheon with the people of the National Geographic Society, he was described as the successor to the mantle of [Edward] Steichen and probably greater than Steichen ever was. I found myself at complete ease with him; this speaks volumes as compared with the experiences I have had with other cameramen or photographers.

Off to the Hill for a filmed interview with Congressman Riehlman of New York. This went very well and allowed me to get to the next appointment a little bit early. This was with Congressman Lindsay of New York, who is a freshman congressman. It was a matter of discussing with Lindsay the space program of NASA and the United States, and I found it enjoyable. Not very stimulating though! Lunch at the National Geographic [Society] with Dryden and Thomas McKnew [executive vice president and secretary of the Society] was a pleasant experience. We did a short tour of the building afterwards and found the usual progressive professional activities being carried out.

Back at the office at 1:45 for a meeting with Hodgson, Jack Young and one other man from McKinsey and Co. We are coming down the home stretch on the report of the advisory committee on organization and I wanted to get some ideas across to Young before the report gets too far crystalized. It was a useful discussion. At 3 o'clock, Dr. Si Ramo of Thompson-Ramo-Wooldridge came in for a pleasant 60-minute conversation. Si was doing a fine job of giving me the soft sell on three projects in which I have really great interest. It was a good discussion and I think I may have aroused his interest in coming to Washington—provided Nixon is elected.

In between times today, I have been over to the White House twice to try and get settled the important matters relating to the president's visit to Huntsville. This is a very complicated procedure and requires coordination at every turn of the road. What is most important at the present time is to secure a final sign-off from the president to the effect that he is actually going to make the trip. It appears that he will; I hope I can get this settled tomorrow.

Thursday, August 25: This morning we had a staff meeting that was quite uneventful. I had planned to be at the meeting of the Naval Research Advisory Committee (NRAC) at the David Taylor Model Basin (DTMB) for the entire day. This did not work out. There was much to be done about the trip to Huntsville and during the course of the day, it was decided and announced that the president would be there on 8 September—my birthday. Now comes the rest of the mechanism necessary to a trip of this sort. It is really quite an operation. Arrangements must be made to take along the White House correspondents, give them ample time to reach Huntsville before the president arrives, provide them with a special location for their cameras and writing tables, provide trucks to allow them to precede the president to the various places on the reservation where he will stop for inspection. How he stands it, I will never know.

Finally, at about 12 o'clock, I dashed off to the DTMB for lunch. I stayed through an hour and a half of discussion in the afternoon and then came back to the office to try and clear the desk once more. At 6:45, I boarded the USS *Sequoia* for dinner with the NRAC crowd. It was a pleasant evening with Arleigh Burke assuming the role of moderator of a long and serious discussion. Home at 11:30 to find that Ruth had finished most of the packing and was really pretty well prepared for the morrow.

Friday, August 26: Up at 6:30 and helped with the breakfast while Ruth finished up the last bit of packing and making of sandwiches for lunch, etc. They had planned to be off at 8 o'clock and actually made it about 8:05.[25] Arriving at the office, I called General Persons to alert him to the necessity for asking Mrs. Marshall [if she would be willing to travel to Huntsville for the ceremony of dedicating the space flight center to her husband]. He begged off and suggested that I do it instead

---

[25] They apparently went to Cleveland. See the entry for 2 September.

in order not to put too much pressure on Mrs. Marshall of the type that would occur were the invitation to come from the White House from such an old friend of the general's. So be it. I reached Mrs. Marshall's daughter on Nantucket Island late in the evening and will get an answer from Mrs. Marshall next week. It begins to look as though Ruth will not have to make the trip.

I did get over to the NRAC meeting this morning, which started out with Admiral "Red" Raborn [director of the Polaris program] giving us a briefing on the Polaris weapon system. He is a positive, optimistic and extremely confident man. Perhaps he ought to be! The rest of the morning was given over to an executive committee meeting, and I must say I have considerable respect for the way this particular advisory committee works. Hugh and I had lunch at the White House mess and then back for a meeting at 2 o'clock with several of the people from the office of launch vehicle programs and the office of space flight programs. I am concerned about the Saturn project and the necessity for pushing it ahead of a variety of other activities. If, indeed, the spacecraft will be ready for the Saturn, we ought to move ahead without hesitation. On the other hand, it appears that the Saturn is really destined principally for manned space flight and some of the more difficult space science missions. I am not at all sure that these craft will be ready to be flown on the Saturn. As a result of the discussion, I asked to have such a group established for the purpose of integrating the planning on Saturn, man-in-space beyond Project

Later photo of the Saturn I launch vehicle at Huntsville, shown against a background of earlier rockets to show its much larger size, hence its greater thrust.

Mercury, and the other activities requiring thrust of the sort we should get from Project Saturn.

At 4 o'clock, some of our friends from "down the street" came in to talk about the Russian landing of the space capsule with the dogs. This is still an enigma to most of us but there are some reasonably good theories about the variety of the circumstances surrounding this flight. Back home at 6 o'clock to an empty apartment and the "thrill" of making the beds, washing the dishes, etc. Oh well, it won't be for too long a period, anyway. During the course of the day, it became quite apparent that Congress is not going to stay in session for many more days. This means that our legislative amendments will not be handled this session and thus a great deal of work goes down the drain. My successor will certainly have quite a task to convince Congress that it should take up this bill in the early days of the next session. There has been so much work done on the part of the House of Representatives, however, that it may be that he can get it through that body without too much trouble. Surely, the patent clause will cause some debate once more and it may well be that we will lose it on this second run. This might be a good time for me to say something about the political campaign. I have tried to see Dick Nixon for several days now to bring him up to date on our program and to talk with him about the necessity for the legislative amendment actions I have been trying to force upon Lyndon Johnson. All this without any success; the vice president is very busy, of course. It appears to me as though Kennedy and Johnson are not doing too well with this session of Congress and that their stock has drifted downward while Nixon's has begun to rise. This is an impression but one that seems to me to suggest the Democrats are going to have to work pretty hard on this one. My discussions with Lyndon Johnson continue to give me respect for the man as a politician, a recognition of his great desire for power, an appreciation of and for his willingness to put the important needs of the country before his own desires and generally a concern over what he would do were he to be elected president. I question seriously whether he would make a very good president even though I think he is a better man than Kennedy.

Tomorrow, I am off to the West Coast for a couple of days. I look forward to next week since it will be the beginning of a new era at NASA. Bob Seamans will be on board!

Saturday, August 27: Up at a reasonable hour and over to Walter Reed hospital to see Herb York. At age 39, he had himself a heart attack—a coronary thrombosis—and then found himself in a good deal of trouble with reactions to antibiotics. He is now on the mend but will be out of action for at least two more months. He seemed very cheerful and resigned to his enforced idleness. I suspect he will do a fair amount of work from his telephone at home. Off to the airport and the flight to the West Coast. I had refrained from taking any cocktails—simply had a glass of wine with my meal—and felt fine when I arrived. Gordon and Cora joined me at the Knoxes for dinner with Tom and Martha, Rozella and one other. It was an exceedingly pleasant evening and the chicken was just right. As of tonight, no indications of a new Glennan [meaning his first grandchild]!

Sunday, August 28: A reasonably good night's sleep and up at 9 o'clock to talk with Gordon and Cora. At 11:00 I dashed off to Los Feliz to be with Tom and Martha and the Knoxes for the rest of the day. Tom and I indulged in a two-hour discussion that might have been termed an argument. He was defending the government's practice of employing organizations such as the Rand Corporation to do operational analysis and long range studies. I was not really arguing that point—rather I was trying to convince him that Rand should not be allowed to out-bid governmental organizations for good people. His reply was a pragmatic one: if the government couldn't handle the situation within its own regulations, it should not be deprived of the good offices of an organization such as Rand. I think I must admit to carrying the argument to some lengths without real conviction but I am convinced that an answer to this problem must one day be had. Perhaps these university and non-profit institutions are a necessary part of the present scene. Perhaps the government can never provide the right kind of an atmosphere for the people whose advice it must have. If this be the case, then Rand and others should have a brilliant future. With no responsibility for their decisions or their recommendations, they can go on dreaming and recommending and perhaps influencing the course of history by a measurable amount. Not a bad job if you can get it!

Monday, August 29: Up at 7 o'clock and a pleasant breakfast with Gordon and Cora before Tom picked me up at a quarter of nine. We drove over to the University of Southern California campus where the Aerospace Corporation, the Ballistic Missile Division of the Air Force, and the Space Technology Laboratories were sponsoring a ballistic missiles and space symposium for the fifth year in a row. I had appeared on this program last year and had talked rather bluntly about the desirability of taking a more realistic view of our shortcomings in the space business. Today, I am the luncheon speaker and I have chosen to avoid discussing program matters and to talk about philosophy and the reasons behind the program we presently are following. The audience was very responsive and I felt well repaid for the effort.

Immediately after lunch, Tommie drove me down to our West Coast operations office and I called on Bob Kamm. He has been deathly ill and certainly looks it. He lost some fifty pounds and is slowly regaining his strength. It appears that he will be back at the office in another three or four weeks. Dick Horner came by and we drove together over to Jimmie Doolittle's for a cocktail with a good many friends from the Ballistic Missile Division and the Space Technology Laboratories. Dick then took me to dinner at the Miramar and finally deposited me at the airport at 9:30 where I waited until midnight for my plane.

Tuesday, August 30: We arrived on time this morning but without very much sleep. I drove on home and went to bed for a couple of hours after which I felt reasonably well refreshed. There wasn't much on my schedule for the day but I did manage to make a United Giver's Fund luncheon for which I received a very kind letter from Bob Anderson who is chairman of the UGF drive for this year. The afternoon was spent in discussions with a variety of people—nothing of very great interest. Home at 5:30 and my usual diet dinner, which didn't enthuse me very much.

Wednesday, August 31: The last day of the month and only four and a half months remaining in this maelstrom! I had several meetings during the day—none of them very exciting. At 3 o'clock I answered a good many questions put to me by Mr. Fred Blumenthal of *Parade* magazine. It appears that I am to have the cover spot on 23 October, and they want some sort of a question and answer story to go with the physiognomy.[26] During the course of the day I became very much concerned about the inability of some of our people to understand problems relating to control mechanisms such as report systems, etc. It does appear that our technically-minded people need a good bit of assistance in this area. I guess one shouldn't get too excited about it but I mean to have a whack at the problem before too many days have passed. Back home at 5:30 to another diet dinner and then some reading.

---

[26] The interview did appear in *Parade* on 23 October 1960. (Glennan biographical file, Glennan subsection, NASA Historical Reference Collection.)

CHAPTER TEN

SEPTEMBER 1960

Thursday, September 1: This was a red letter day. I swore in Bob Seamans at exactly 8:30. The staff meeting was long and thoughtful with a good many policy problems discussed. Among them were the relationships between ourselves and the Air Force on a variety of matters—including prominently the relationship between Discoverer and Mercury. I must lay on a conference with General Schriever about this matter.

I took Bob Seamans to lunch at the White House Mess. I was able to introduce him to several of my friends over there and get him a reasonably good start. During this portion of the day I talked again with Allen Dulles and found that Dick Harkness had been in touch with him about the same U-2 program that had

Dr. Robert C. Seamans being sworn in as associate administrator of NASA by Administrator T. Keith Glennan on 1 September 1960.

been proposed to me. Apparently, Allen had the same response but there seems to be a leak in the dike someplace. It may be late but I think we must try to plug that leak.

During the morning staff meeting, Randt had made some pointed comments about the necessity for getting on with the life science program. I arranged to have him come in early in the afternoon, and we sat down with Seamans and one or two others to try to beat out these problems. Much of his trouble stems from his lack of knowledge of the budgetary process in government. I think we laid on a program that will make him reasonably happy. It is really nothing more nor less than scheduling a series of activities that should have been undertaken anyway. He is an excellent man—very enthusiastic and full of the kind of energy that we so badly need. Later in the afternoon, Seamans met with me, John Johnson and Bob Nunn to set up the program for the communications satellite of the future—the operating satellite system. Bob Nunn will report directly to me as special assistant and will oversee all facets of this program. There is so much in the way of public policy, international relations, public information policy, possible development of legislation, etc., that someone must pull this together. Bob is enthusiastic about the program and should do a good job for us.

At the end of the day, [my friend Millard] Richmond called me and said that the strike on the Pennsylvania Railroad had done him out of a ticket to New York. He was on an airplane but had to give the ticket up to his boss who had had a train reservation. Accordingly, Rich took Bob Seamans and me to dinner. Both being New Englanders, Rich and Bob seemed to have a reasonably good time. Rich had with him a copy of the *Congressional Record* and pointed out to me a memorandum put in the record by my good friend, Lyndon Johnson. This is a statement commenting on the three important portions of the legislation we have been trying to get Lyndon to do something about. The upshot of it was that there was no urgency about these matters, that we were doing pretty much what we wanted to do anyway. A neat maneuver and a matter of privilege only to the senator. Back home to read a transcript of the Joint Committee on Atomic Energy hearing on Monday last. This was a secret document although I wish it weren't. Senator Clinton Anderson, another of my good friends, really beat John McCone about the head. I just don't understand why anyone has to take the kind of chatter, accusation, half-truths and insinuations that persons like Anderson heap upon persons like the atomic energy commissioners. Statements made about the integrity and abilities of men of whom Anderson has scant, if any, knowledge are really exceedingly damaging to their reputations. Much of the backing for the senator's statement comes from the members of the committee's staff who, as I have said before, wield a great deal of power. Basically, the committee is determined that it will maintain jurisdiction over the entire Rover program. I'm sorry I am not going to be around to dispute that with them because I think they would have a bit of difficulty beating me to it in this case. In any event, it's a shabby story and one that ought to be removed from the record.

Friday, September 2: The morning was given over to a variety of activities, the most important of which was a meeting with Johnson, Bonney and Mel Day. We were discussing the desirability of having a "public service" television film prepared on Project Mercury by one of the television networks. It may be remembered that I had decided the Columbia Broadcasting System should have the right to make a series of programs for us next year. The National Broadcasting Corporation, disturbed at having lost this particular privilege, has served notice that it wants the rights to a film on Mercury. It seems clear to me that we should consider this a matter of serious public information and that, regardless of cost, we ought to do this program ourselves. Accordingly, I decided that we would give it to no one of the networks; rather, we will make a film that will become public property. This seemed to please everyone on the staff. I went over to the White House and talked further with General Goodpaster and Jim Hagerty about the NBC program, which is to be filmed for distribution after the election. It is to be remembered that Dick Harkness talked to me about this activity, which centers around the U-2 incident. Clearly, Jim Hagerty has stepped out of bounds on this one—I hope some ground may be recovered. In any event, Allen Dulles and I will not participate. The afternoon was rather routine and I left for Cleveland on Capitol Flight 207 where I found Dr. Clark Randt as a fellow passenger. We had an opportunity to mend the fences that had been breached in the morning and it was, all-in-all, a pleasant flight.

Ruth and Polly met me and I had just a minute or two to board John [Simmons]'s jet to say goodbye to him before he flew off for Los Angeles. He seemed very pleased with himself and certainly grateful for the activities he had indulged in during the past two weeks. [Later,] shortly after retiring, I heard a call from below the stairs, "Anyone home?" It turned out to be Kitty and Frank [Borchert] who had driven straight through from Chicago. It was fun to see them again and to see them looking so well and happy. Sally came home shortly thereafter and the house settled down about midnight. At 1:30 in the morning, the phone rang and it was Tom who, in answer to my query, "Well?" said, "Hi, Dad, it's a boy!" Kitty had heard the phone ring and was up and at them as she wanted to talk to Tom. Apparently everything went well on the West Coast although Martha was in labor about nine hours. The household settled down about 2:30—with a sigh.

Saturday, September 3: Up early for an 8:30 breakfast with [old friends]. At 10, I went to the Case campus to meet with Kent [Smith] and [others].

Sunday, September 4: We had a rather late breakfast and then went over to the Country Club to have lunch. Returning to the house about 3 o'clock, we found that Kitty and Frank were just back from a wedding and were in need of some rest. Shortly thereafter we drove on down to the campus so that Kitty might see the new library. Back home at 5 o'clock, we put the leg of lamb on the grill and had a late dinner.

Monday, September 5: [Discusses various matters of Case business.] We put the roast on the barbecue and managed to have our Monday dinner about 2:30

in the afternoon. Soon thereafter, Kitty and Frank left for Chicago and I packed for the trip back to Washington. It was a pleasant flight although I had little opportunity to contemplate the pleasures of the weekend. A former secretary for Admiral Bennett was on board and took up all of my thinking and reading time.

Tuesday, September 6: This was a rather ordinary day with a briefing at 10 o'clock for the office of launch vehicle programs delineating the study contract program for the coming years. It is clear that the Saturn project is going to be an exceedingly expensive one. I am anxious to receive the results of the integrated study our people are doing in an attempt to bring together the Saturn program and the spacecraft that will use these large vehicles. Dick Horner took Bob Seamans, myself and two or three others to dinner at the Shoreham. We watched Echo cross the heavens twice during the course of the evening. What a thrill!

Wednesday, September 7: This was a full day. Much discussion has taken place with the White House over the last two days as we have attempted to complete arrangements for the president's trip to Huntsville tomorrow to dedicate the Marshall Space Flight Center. Really, if I had had any idea of the problems involved, I doubt that I would have proposed that he make this trip. Questions about who is going to ride with whom seem more important than almost anything else. Discussions with Al Siepert about management training were productive, and I think we are well on the way toward the solution of one of our problems—the training of project managers.

Lunch with Bob Seamans at the White House again and then an afternoon with Jack Young of the McKinsey organization as we worked over the report it is preparing on our procurement activities.

In the midst of the afternoon session, I was called out to discuss a problem that has arisen in our Western Operations Office in Los Angeles. It has to do with the dismissal of a secretary who had apparently become involved with the director of the office. She was about to prefer charges against our organization for dismissing her even though the case seemed pretty solid against her. I decided that we would fight it out and that the man involved would have to answer for his actions at a later date. George Kistiakowsky called me and we had cocktails at his apartment and then went out to a French restaurant for dinner. I talked to Ruth twice—it is rather lonesome here in Washington without her. We have set up a couple of dinner engagements for next week in which I hope Polly will be involved.

Thursday, September 8: Up at 6:30 to greet the day on which I was born some 55 years ago. This has turned out to be a memorable day for me. I left the apartment at 7:30 having put out the laundry and arrived at the Military Air Transport terminal at 8 o'clock. There I joined Hugh Milton, under secretary of the Army and General Lemnitzer, chief of staff of the Army and soon to be chairman of the Joint Chiefs of Staff. We had a cup of coffee and watched the honor guard dispose itself about the *Columbine*. At about 8:20, we were told that the helicopter had departed the White House grounds with the president, and we then boarded the plane. We had barely taken our seats when the helicopter appeared over one wing and it seemed that almost immediately we were underway. Apparently, the president boarded the plane through the forward hatch and the four motors on the *Columbine* started in unison. This is quite unlike any commercial flight where the motors are started one at a time.

On board with us were Bryce Harlow, Jim Hagerty, Dr. Snyder, Ann Whitman (the president's secretary), Dr. Kistiakowsky and three or four secret service men. One or two additional White House aides were visible. Soon after we were airborne, Bryce Harlow asked me to help him write an ending for the president's speech. This was to provide him a few words that would introduce General Marshall's widow and prepare the stage for the unveiling of the granite bust of General Marshall that we had purchased. We were invited into the president's cabin to see a 14-minute film put together by the Air Force. Believe me, that organization never loses an opportunity to put its propaganda before the people who can help it. It was a film in which the Air Force took credit for practically all of the recent launches, including Echo. Certainly, the Air Force is to be credited with an assist in many of these activities but the blatant nature of its propaganda is a little bit disturbing to me. Following the film, [Hugh] Milton, Lemnitzer and I sat with the president for an hour—oh yes, Kistiakowsky was there too—talking about a variety of matters. He stated he had thought that as soon as the nominees for the presidency were certain, he could relax a bit and avoid the speaking circuit. Instead,

President Dwight D. Eisenhower descending from the presidential aircraft, Columbine, *at Redstone Arsenal airstrip for the ceremony dedicating the George C. Marshall Space Flight Center on 8 September 1960. Administrator T. Keith Glennan is behind him in the door of the airplane.*

From left to right, T. Keith Glennan, Wernher von Braun, President Eisenhower, and Katherine [Mrs. George C.] Marshall at the dedication ceremony.

he has had a plethora of speaking invitations, most of which he feels he cannot refuse. Interspersed have been many dedications of the type he is participating in today. He is such a human guy! We landed about 10 minutes early on the strip at Redstone Arsenal. Colonel William Draper, the president's pilot, is really the captain of the ship, and he is a very great confidante of the president. He pretty much runs the show for the president while he is aloft.

We were met at the arsenal strip by Governor [John Malcolm] Patterson of Alabama, General Schomburg, the commanding general of Redstone Arsenal, Dr. von Braun and others. A motor cavalcade was formed immediately with the president riding in a convertible with the top down. The Ford Motor Company had provided white Fords with NASA decals for the entire party. I rode in the third car along with von Braun and Milton. Arriving at the administration building, the president drove into a courtyard while the rest of us moved directly to the platform. I was seated directly to the president's right in the front row. Just a country boy! After a brief pause, the president came out and the Army band played "Hail to the Chief." There was some sort of a snafu since the band was apparently expected to play an additional number but somehow or other the instructions were mixed. In any event, von Braun picked up the baton as master of ceremonies and introduced the minister, Reverend Wade, plus the congressmen in attendance and several of the distinguished guests on the platform. He then introduced me and I made the shortest speech on the record in introducing the president. I said, very simply but with great fervor, "Ladies and gentlemen, the president of the United States."

His speech was received well although it was not the type that would occasion great outbursts of enthusiasm and applause. The unveiling of General Marshall's bust by the president and Mrs. Marshall was a rather touching ceremony, and I then handed a bouquet of roses to the president who handed them to Mrs. Marshall. After the benediction and the playing of the national anthem, we returned to our car and the cavalcade started through the NASA area. We made a stop at the assembly building and the president was given a brief but quite thorough and exciting story on the Saturn vehicle, which was being assembled there. He asked intelligent and penetrating questions and received thoughtful and satisfying answers. I suppose that von Braun, the president and I were photographed three or four hundred times at this point. We then boarded the cars again and drove to the test area where the president put on a hard hat and was shown the "battleship" version of the Saturn installed on the stand. He wanted to go up to the firing floor and several of us accompanied him there. Photographers were everywhere in evidence. Again, the president showed substantial interest in the activity and in the giant rocket system that someday may fly.

At this point, we boarded the cars to drive back to the airstrip where the president was to embark for his flight back to Washington. On this leg of the trip, I rode with him. It was interesting to note that, as he drove along the

Wernher von Braun explains the Saturn launch vehicle to President Eisenhower and Administrator Glennan following the dedication ceremony at the Marshall Center.

roadway, he had a wave and a word of greeting or farewell for the people standing there. This was genuine on his part, not a perfunctory operation. At the airstrip, he thoughtfully went over to say goodbye to Mrs. Marshall and then was accosted— I can think of no better word—by Senator Wiley of Wisconsin, a Republican and a member of the Senate Space Committee. Wiley "thumbed" a ride back to Washington with the president—something that he had been trying to do on the trip down and had managed to achieve on the trip back. After the president was airborne, we returned to the officer's club where the top NASA staff, the top Army people and several of the citizens of the community had lunch together. I suppose we missed some people—I didn't see the mayor in attendance—but it was a pleasant party and I made a few remarks to close the luncheon. These were directed principally to an expression of appreciation for the assistance given us by the Army and for the very pleasant relationships that are now developing between the Army and the Marshall Space Flight Center people.

We then boarded our cars and went back to our own Convair for the trip back to Washington. We had three or four congressmen aboard and several of their aides. We arrived at the airport in Washington at 6:30 and I am now at home at about 9 o'clock—ready for bed. One or two comments about the events of the day may be appropriate. When I proposed to the president that he undertake this task, I had indicated the ease with which he could land on the Arsenal's airstrip and make his talk and then get away without further delay. As I may have stated earlier in this chronicle, Jim Hagerty and others had immediately laid on a motor trip through Huntsville, etc. The president had vetoed this. On the aircraft, however, he expressed himself as being disappointed that he hadn't realized that the city of Huntsville had a population of some 70,000 and that he might well have started out an hour or two earlier so that he might ride through the city to see and be seen by the populace. He consoled himself by the fact that he had three engagements in Washington on his return and a dinner and speech in the evening. Apparently, he resists this sort of thing just as I do and then, when the party is in progress, he sort of wishes that he had gone the whole way and exposed himself to some of the inconveniences of the occasion in order that the people might be better satisfied.

On the aircraft coming back, a story was told that is quite appropriate, it seems to me. Edward Teller is credited with having attempted to answer a question from a member of an audience in a manner typical of Edward's quick and sharp wit. Someone had asked, apparently, "What good is all of this space business?" Teller replied without hesitation, "When Columbus set out on his voyage from Spain and Portugal, he was attempting to improve trade relations with China, India and the Far East. We have not yet accomplished his objective, some 350 years later, but just look at the by-products!"

On arrival at the apartment, I found a wonderful letter from Tom telling about the birth of Keith III and some of his thoughts on family relationships. A most satisfying letter that I had to read to Ruth over the phone. How I wish she were here tonight. What a birthday!

Friday, September 9: This turned out to be a very full day. Dick Horner, Bob Seamans and Don Heaton came in at 8:30 to discuss the problem of deciding on the proper location for the next phase of man in space. I had been very much concerned about this and am not satisfied to make the decision on the evidence thus far presented. I guess that is the real trouble—there has been little evidence presented. It does look as though Langley Research Center might be the proper location. In fact, that is where the present space task group is located, and it would mean a minor disruption to our plans to continue them at that location. At 9 o'clock or shortly thereafter, we started out on a first attempt at a program analysis session. This ran through 12:30. There are many disappointments in the scheduling as we look ahead. Certainly, much of this can be rationalized as part of any research and development program. On the other hand, we are so visable that we must make what appear to be promises in the matter of dates for the accomplishment of program objectives. These rise up to smite us at a later date. Obviously, too, we have not yet evolved a mechanism for the initial approval of development programs. They may be approved in a conference in the hall, through a memorandum or through some oral discussion that is never reduced to writing. I have asked Seamans to get on this job immediately.

We stayed in for lunch and Heaton and Seamans were joined by Abbott in a discussion of the selection of a man to head the study to be made of the site for manned space flight programs. There was much discussion of the probable effect on the morale of the present Mercury group if such a study were handled imprudently. I happen to believe this can be handled properly and am determined to go ahead with the study. We decided to ask Bruce Lundin of the Lewis Research Center to come in on Monday for a discussion of his willingness to serve as chairman of this study group.[1]

At 2 o'clock, representatives of a company referred to us by Congressman John McCormack came in. The appointment had been made for two people and five appeared. Once again I found myself in a rather bad mood about this situation since I believe it is a reflection on the company concerned when it feels that it must secure an appointment through some congressman. I think we did help these people somewhat.

At 3:30, Silverstein and Seamans came in to discuss further the manned space flight center location and the method that I propose to use in resolving this issue. Abe is really a logical thinker and a very strong proponent of his own ideas. Fortunately, they are usually right. He stated that be has no problem at all in deciding the proper site for the operation; his real problem is in working out the proper organization. That is a brash statement in face of the fact that we must support our conclusions for a congressional committee one of these days. I am determined that we will not build another new center. I really find myself quite much in agreement with Abe but believe that we must go through with this study. We agreed on Lundin as the chairman, and he will come down on Monday to discuss the matter with us.

Bob Nunn came in to talk about his work in setting up the communications satellite program and the other than technical phases of that operation. He will serve

---

[1] As noted below, Lundin did serve as chair of the study group, which was organized on 17 September and submitted its report on 14 October 1960. It recommended that the "manned space flight activity" not be moved until at least the fall of 1961 to avoid delaying Project Mercury. It did not make a firm recommendation for a location of such an activity but generally favored Ames over Langley, while pointing out that such a move would have to be justified politically in terms of savings in time, cost, or efficiency. ("Report of Special Working Group on Location of Manned Space Flight Activity," 14 October 1960, filed with Lundin biographical files, NASA Historical Reference Collection.) As discussed elsewhere in the diary, the center eventually was located in Houston, Texas. (See Dethlof, "*Suddenly Tommorrow Came*".)

very much as a special assistant to me and will attempt to deal with public policy, legislative and international aspects of the program, while keeping fully abreast of the technical developments. We are awaiting from Silverstein a development program that should cover the next several years. John Johnson and Jim Gleason came in to talk about the desirability of "calling" Clinton Anderson on the derogatory remarks he made in the *Record* the other day about Harry Finger and about NASA generally. We finally agreed that this matter ought to be talked over with Senator Hickenlooper, a Republican on the Joint Committee on Atomic Energy (JCAE), merely to indicate our concern and to enlist his active support should the matter ever be raised again. One does not take on lightly this senator from New Mexico. Lewis Strauss did this to his sorrow. In many ways, Clinton Anderson approximates McCarthy in his tactics, although very much of his operation is cloaked in the secrecy of executive sessions of the JCAE. I suspect this is the wise course of action, but it is difficult for me to remain silent in a situation of this kind where one of my people is being seriously attacked.

I closed up the desk at 5:30 and came on home to do a bit of shopping. As I entered the apartment, the phone rang and Mrs. Charyk was calling to see whether or not I could have dinner with them. I accepted quickly and hoped to forget some of these problems for the rest of the day. [Not to end on a negative note, however,] one of the very bright spots of the day was the receipt of a personal letter from the president thanking me and NASA for the day at Huntsville and sending me birthday greetings. I shall treasure this one. It is signed D. E., which is the signature he uses on his most personal correspondence.

Saturday, September 10: This has been a dull day. Getting up at a leisurely hour, I had the Falcon washed and then went to the office for an hour-and-a-half discussion with Walter Sohier. He is preparing for submission to the National Security Council, a paper for me entitled "Managing Major New Technologies." He has done an amazing job of writing but is all too inclined to make a social science exercise out of it. Back at the apartment this afternoon for spasmodic attempts at cleaning up—both the apartment and some of my work. As usual, I had put off until tomorrow quite a number of things. Perhaps it is just as well to get a little rest, but I hope to begin planning an automobile trip to the West Coast, which Ruth and I will take we hope, immediately following the inauguration next 20 January.

Enough for another week; all-in-all, it has been a wonderful one. Beginning with the news of the birth of the first grandchild, running through an early celebration of my birthday and the first wedding anniversary for Kitty and Frank, continuing with the Huntsville expedition with the president, which was climaxed by a very welcome call from Kitty, ending with a quiet day at home. What more could one want?!

Sunday, September 11: A dull day with intermittent rain and without Ruth and Polly! I polished the Falcon and worked at various jobs around the apartment. I made a cake and cleared out my closet. Otherwise, just another day and not a very good one, at that.

Monday, September 12: This was a really full day with a variety of visits from people in Republic Aviation, Bendix Aviation, the Westinghouse organization and two or three others. They were all attempting to convince me of the deep interest each maintains in winning the orbiting astronomical observatory project. I think practically all of the organizations that have made proposals have come in to see me personally.

At 11 o'clock, Bruce Lundin came in with Seamans, Abbott and Heaton to talk about the task of selecting a home base for the follow-on man-in-space projects.[2] Abe Silverstein was not invited but he managed to barge in and had with him a paper that solved the entire problem. I must say that he is an aggressive and resourceful person. Actually, he will follow instructions but he would much rather follow his own lead—as who wouldn't? Lundin agreed to come back on Thursday with a proposal for carrying out the task we had asked him to do. He strikes me as a very good man.

Lunch at the Cosmos Club with Alan Waterman dealt with the preparation of speeches for the convocation to be held at Yale on 7 October. I am to be on a panel with Waterman, Bill Bundy [then working for the CIA], and [Pulitzer Prize-winning poet] Archibald MacLeish.[3]   The luncheon was not very productive of ideas. During the rest of the day, I had discussions on the communications satellite program, the Rover program and the briefing we are to give the Strategic Air Command people at Omaha later this month. I tried to get away early because Ruth and Polly are coming down from Cleveland today. I was only partially successful.

Tuesday, September 13: At 9:15, Bonney came in to talk about a job offer he had received from the Aerospace Corporation. Here I faced one of the tough ones. Bonney is not a particularly strong individual —in fact, he is much of a baby and certainly is not a good manager of men. He has made a great many very good friends in the aviation industry, largely through his willingness to serve them adequately. I encouraged him to look carefully at this job offer, stating to him that he would not be replaced at NASA, that I was not fully satisfied with his work, that he had done a good but not outstanding job. He was hurt, of course, but it seemed to me only proper that I should tell him frankly how I felt about his work.

We started out the United Givers' Fund program with a meeting of the chairman and vice chairman at which I attempted to instill a little fire. I must say

---

[2] This was probably a reference to the various projects making up the Apollo program. It was named (at the suggestion of Abe Silverstein, who followed the precedent established by Mercury of naming spaceflight projects for mythological gods and heroes) on 25 July 1960. Gemini—the two-person earth-orbital rendezvous that came between Mercury and Apollo—was not approved until December 1961, although there were already plans for the rendezvous itself. Thus, this could also be a reference to the Gemini concept if not the project itself, or even to Gemini and Apollo together, both still in the process of formulation. (*NASA Historical Data Book*, Vol. II, pp. 155, 180; Rosholt, *Administrative History of NASA*, p. 238; Wells, Whiteley, and Karegeannes, *Origins of NASA Names*, p. 99.) As to the home for these projects, see note 1 above and the index under "Manned Spacecraft Center."

[3] All of these men except Waterman were Yale graduates.

I had a rather cold audience. I do think, however, that they will catch fire as the program moves along. At 2 o'clock, Barton Kreuzer of RCA came in to tell me of the interest of his company in our projects. At 2:30, Hjornevik and Lacklen came in to talk about the necessity for getting a training program for project managers underway. I heartily agree with this; in fact I have been urging it.[4] At 3 o'clock, Jim Gleason and I went up to talk with Senator Hickenlooper about our relationships with the Atomic Energy Commission and particularly with Senator Clinton Anderson. I am still burning about the statements made by Clint Anderson about Finger. Hickenlooper says that Jim Ramey, the Director of the JCAE staff, is the culprit in all of these matters. We discussed a letter I had drafted to be sent to Anderson. After some revision, I am going to send the letter. I doubt that it will do any good but at least there will be an answer in the record asking Anderson to "put up or shut up."[5]

At 4 o'clock, noting that John Finney of the *New York Times* was in the halls, I asked for some guesses as to his interests for the day. It appears that he may have news of the Aeronautics and Astronautics Coordinating Board activities. On the odd chance that he is going to break it in the *New York Times* tomorrow, I authorized the release of a statement we have had prepared for the last several days. I won't let them get ahead of us if I can prevent it.[6] At 4:45, John Johnson came in and talked with me for several minutes about my problems with Bonney. He agreed that I had done the proper and probably the helpful thing in talking to Bonney as I had. John says that he is absolutely astounded at the poor quality of writing that comes out of our public information office. Ruth and I went to the Newton Steers for dinner.[7] It was one of the oddest functions I have attended. We had thought that there might be one or two other couples present but it turned out that there were at least fifteen or twenty. The host wandered in about twenty minutes after we arrived. His wife is the stepsister of Jackie Kennedy. We got into a little argument after

---

[4] On this issue, see Rosholt, *Administrative History of NASA*, pp. 144-146.

[5] The letter to Senator Anderson, dated 15 September 1960, read in part: "After the several satisfactory meetings and conversations you and I have had on the subject of the management of the Rover Project and Mr. Finger's selection to head that Project, and in the light of the favorable statements made to you by those having responsible concern for the AEC portion of the Project, your personal comments on the ability and character of Mr. Finger are hard for me to understand.... Predictions of failure of a project which has great national significance, at the very inception of a new management arrangement, can hardly be construed as encouraging in view of the difficult responsibilities that Mr. Finger and others directly involved will carry." The full text of the letter is in the Glennan subseries, "Chronological - September 1960" file, NASA Historical Reference Collection.

[6] On 13 September 1960, NASA and the DOD announced the creation of this board "to review planning, avoid duplication, coordinate activities of common interest, identify problems requiring solution either by NASA or the Department of Defense and insure a steady exchange of information." The co-chairmen of the board were Hugh Dryden for NASA and Herbert York for the DOD. (Emme, *Aeronautics and Astronautics . . . 1915-1960*, p. 127.)

[7] Newton Ivan Steers, Jr. was a Yale graduate and was president and general manager of Atomics, Physics, & Science Fund, Inc. of Washington. He was married to Nina G. Auchincloss.

dinner at our table relating to the amount of competition there presently seems to be in the oil industry. Bob Wilson, formerly of the Standard Oil Company of Indiana, became quite irate at some of the accusations made by poorly informed persons. Quite an evening.

Wednesday, September 14: At 10 o'clock, Morris Tepper and Colonel Keaton came in to discuss setting up a meeting with the Department of Defense and the Weather Bureau on the on-going meteorological satellite program. I am determined that we make certain the users of the information we may acquire be prepared to make good use of that information.[8] The rest of the day had its usual quota of discussions with members of the staff.

Thursday, September 15: The staff meeting was routine. The rest of the morning was given over to a session with Milton Ames and [Herman] Kurzweg, a new man who is to be in charge of our aerodynamics and flight mechanics work. Also, George Clement of the Rand Corporation came in to talk about contracting problems with NASA. At 2 o'clock, Bruce Lundin came back in with his thoughts about the way he would tackle the job we have given him. It looks to me as though he is quite well on the road to doing a good job, although it is clear that we should delay it for perhaps a week. Interestingly enough, Bruce used to work for Abe Silverstein and understands quite well the way Abe arrives at his decisions. He does not seem to stand in awe of his former boss.

At 3:30, George Best, a vice president of the AT&T, and Bill Baker, research vice president of the Bell Telephone Laboratory, came in to give me a letter indicating the interests of the telephone company in the communication satellite field. They stated they were prepared to spend as much as $30 million for three satellite flights. They will spend a great deal more than this if success attends their early efforts. This is the first real break in getting support from an industrial organization using its own funds. The Bell Laboratories have been doing this in a small way in connection with Project Echo, but this move brings new life into the communications picture.

At 5:30, I picked up George Kistiakowsky and took him to the apartment for drinks. After that, he took us to a little French restaurant for dinner where Ruth had snails for the first time. I think she rather liked them. It was a good evening.

---

[8] An interagency meeting on the establishment of an operational meteorological satellite system took place on 10 October 1960. The departments of commerce and defense, the FAA, and NASA had representatives there, but Colonel Keaton was not among the DOD attendees. At the meeting, Glennan stated NASA's intention to cooperate with the other agencies involved in meteorological research and weather forecasting as it executed its meteorological satellite program. The representatives of other agencies expressed their interests and needs in the meteorological arena. The meeting concluded that all concerned agencies were very much interested in using the data acquired during the research and development of meteorological satellites, that not only R&D but operational programs needed to be pursued, and that "an accelerated research effort" should be begun to assure an ability to handle and distribute the data that would result from the program. (Minutes of the meeting, part of a package on meteorological satellites in "M - Official - Miscellaneous," Glennan subseries, NASA Historical Reference Collection.)

Friday, September 16: This was the start of a weekend. At 11 o'clock, I boarded the *Wall Street Journal* plane to fly to northern Minnesota for a two day conference—a "Seminar on Science and the News." This was sponsored by the University of Minnesota, the Mayo Clinic, the *Minneapolis Star and Tribune*, Carleton College and by the National Science Foundation. The purpose was to bring together leading editors and outstanding scientists in an effort to improve the understanding on the part of each of the other's interests and problems. I was very tired but it was good fun and I think a highly useful exercise. There were three nationally known science reporters in attendance—Alton Blakeslee, Earl Ubell and Victor Cohn. Among the managing editors were Turner Catledge of the *New York Times*, Wilbur Elston of the *Minneapolis Star and Tribune*, Bill Steven of the same paper, William Randolph Hearst, Jr. of the Hearst newspapers, Norman Isaacs of the *Louisville Times*, Barney Kilgore of the *Wall Street Journal*, Ben McKelway of the *Washington Star*, Walker Stone of the Scripps-Howard newspapers and Russ Wiggins of the *Washington Post*.

Among the scientists present were Athel Spilhaus of the University of Minnesota; Larry Gould, president of Carleton College; Warren Weaver, vice president of the Sloan Foundation; Roger Revelle of the University of California oceanographic outfit [the Scripps Institution of Oceanography]; Harold Urey, [professor of chemistry-at-large, University of California]; Edward Teller; Wendell M. Stanley, biochemist of the University of California; and Harlow Shapley, professor emeritus of astronomy at Harvard. Just how I was included in this group is not clear to me, but I was grateful for the chance of being present.

The papers given by most of the scientists were really quite exciting. It was clear that they would like to run the editor's business for him and tell him how to write headlines and how to report stories about science. The editors were a little bashful at first but on the second day managed to state their case rather forcibly. All-in-all, it was a good exchange of views. Edward Teller interrupted almost as much as Kilgore but did it with a smile. Of all the scientists present, I thought he did the most obvious political job. This is characteristic of Edward. I left late on Sunday for Minneapolis so that I might get the plane to Texas the following morning.

Monday, September 19: Up early and off to Texas via Chicago and Braniff Airlines. I was met at the airport and allowed to retire to my room for a bit of rest and work. My hosts, the Texas Oil and Gas Association, wanted me to have dinner with them. After attending several cocktail parties, we did just that, but I was able to beg off early in the evening—about 10:30—and get to bed.

Thursday, September 20: After several telephone conversations with reporters, I managed to give my paper, which was reasonably well received. I think it was one of the strangest groups I have seen. All of them are oil men or gas men, and they are thinking more of self protection than they are of almost anything else. This is a very crude accusation; but certainly this was the impression that I gained from listening to a part of their meeting. Remembering that this speech was given at the request of Bob Anderson of the Treasury, I think I must be a little prejudiced in these statements. Catching the plane at 2 o'clock, I reached home a little after 9 o'clock, where Ruth met me.

Wednesday, September 21: Bob King came in early to discuss several of the meetings he is setting up for me.[9] The Williamsburg session looks to be an interesting one and will be held on 16 October. I hope Ruth will go down to the Inn with me. At 11 o'clock I had Johnson, Nunn and Dryden in to talk about the desirability of my going up to Holmdel, New Jersey, for a demonstration planned by the Bell Telephone Laboratories for the Federal Communication Commission. The advice I received was that this would seem to be an unwise trip since it would indicate, unnecessarily strongly, approval of AT&T as a competitor in the communication satellite business. I was not pleased with the discussion and I am afraid I made my displeasure rather obvious. AT&T is going to be in the business and if we are going to take leadership in getting this program off the ground, it seems to me that we have to take a positive rather than a negative viewpoint in matters of this kind.

I made a recording this morning for the Voice of America. These half-hour tapes are not too bad since you are allowed to talk quite naturally and editing is done by good people at the Voice of America. A report on the Saturn and Saturn application problems was delayed until next week. At 4 o'clock, I called in Bob Nunn to tell him that I was going to Holmdel and we got the program laid on for the next day.

Thursday, September 22: Up early and off to Holmdel in a special air missions plane—an Aero Commander. Jim Fisk met me at the Monmouth County Airport and I had a chance to talk to him frankly about some of the concerns expressed by our people over the development of policy in this communication satellite field. I secured from Jim an agreement that Bell Telephone Laboratories would participate in whatever satellite programs NASA might undertake. The FCC commissioners arrived about 9:30 and a picture was taken of the six of them along with me. This picture was developed, printed and then put on a facsimile transmitter for transmission by wire to Stumpneck, Maryland—the Naval Research Laboratory transmitter is located there—where it was radioed to the Echo satellite balloon and bounced back to the receiving dish at Holmdel. The resulting picture, I have a copy, is an excellent one. What a demonstration! Believe me, I was glad that I had made the trip.

Friday, September 23: The morning was given over to a meeting with the chairmen of our research advisory committees. This was simply an opportunity for an exchange of ideas and a chance for me to say how much we appreciated the work of these committees. After lunch with the chairmen and a fairly large group of our people, I met with John Corson to talk over the next meeting of the Kimpton committee and gave him a copy of Frank Borchert's experience record. Corson is going to have someone in the Chicago office interview Frank. Ruth and I attended a reception given by General Twining in honor of General and Mrs. Lemnitzer at

---

[9] On King, see also the diary entries for 2 August, 3 August, etc. (See index for specific pp.)

the Bolling Air Force Base officers' club. General Twining is retiring as chairman of the Joint Chiefs of Staff and is being replaced by Lem. Both are good men but I think that Lemnitzer is the stronger of the two. We left rather early; neither of us likes this kind of activity.

Saturday, September 24: I spent practically all of the day at the office in a review of the Thor-Delta program and of the scientific satellite program. These reviews are very worthwhile but they do take up another of the few Saturdays I have for myself.

Sunday, September 25: Working at home the entire day on the speech for the Yale convocation, I finally gave it up and went over to call on Herb York with Ruth about 4 o'clock. After that, we had dinner with [friends].

Monday, September 26: I finally decided to attend an early morning breakfast at the Statler as a guest of the Republican National Finance Committee. All they wanted was some money—$100 a ticket for a Party dinner on the 29th of September. Quite a pitch! At 10 o'clock, I had a session with Seamans, Hjornevik and Phillips on staffing for the headquarters. I agreed to add 120 persons even though Congress had expressed its displeasure at the size of our headquarters staff. We simply have to have the people to manage this program—it makes no sense to be spending $900 million without adequate management. As a matter of fact, I think I have been short-sighted in fighting to keep this staff as small as I have.[10]

At 11:30, I went over to the Pentagon to hear a briefing on a proposed aerospace plane. What a fantastic proposition![11] The purpose of the session was to give me this briefing before I heard about the project from General [Thomas S.] Power at Strategic Air Command headquarters in Omaha. I think it is clear that people should be thinking in expansive and expensive terms when concerned with the defense of the country; it is not clear that they have to be 20 or 30 years ahead in their efforts to keep the Air Force ahead of the other services. At 3 o'clock, Bob Bell came in to give me a briefing on some of our internal security problems. It was really quite amazing to see what has to be done to be as certain as possible that we are avoiding, as well as we can, the possibility of security and loyalty problems. I think our program is a rational one—one that could be supported by almost anyone.

---

[10] During the course of the entire year (1960), NASA Headquarters increased by 204 employees, a growth of 45 percent, while NASA as a whole grew from 9,567 to 16,042 people, a 68 percent growth rate—a large part of which resulted from the establishment of the Marshall Space Flight Center and the growth of Goddard. (Rosholt, *Administrative History of NASA*, p. 139.)

[11] On 31 October 1960 the Air Force announced that it was considering proposals for an "aerospace plane" capable of gathering tons of oxygen from the upper atmosphere before entering space, and then reentering the earth's atmosphere for landing as an airplane. (Emme, *Aeronautics and Astronautics... 1915-1960*, p. 129.) Apparently this was not a reference to Dyna-Soar because the DOD had endorsed it and the Air Force negotiated a letter contract for step I of the program already in April 1960. (Clarence J. Geiger, "History of the X-20A Dyna-Soar" [October 1963], p. xii, available in the "Dyna-Soar" files, aeronautical subseries of the NASA Historical Reference Collection.) It seems, rather, that this was one of the single-stage-to-orbit systems studied by the Air Force between 1960 and 1962. In the latter year, the USAF Scientific Advisory Board recommended that such systems be deemphasized in favor of two-stage systems. (*The Hypersonic Revolution: Eight Case Studies in the History of Hypersonic Technology*, ed. Richard P. Hallion [Wright-Patterson AFB: Aeronautics System Division, 1987], p. 752.)

At 4:30, Bob Jastrow and one or two others came in to give us a briefing on the results of the Echo project to date. The effect of radiation from the sun—streams of protons—in moving the big balloon about is almost unbelievable. Much as I am thrilled by these things, I get an even greater thrill from watching people like Bob Jastrow in action. How I would like to have him on the Case campus along with several of his bright young men. Nunn, Johnson and Frutkin came in to talk with me about a speech I hope to give in Oregon on 12 October. I want to make it a policy statement on the communication satellite program. I am not sure how government policy is made in these areas; I'm going to try and find out. Ruth and I attended a reception at the Canadian Office Building—the vice chairman of the Canadian Defense Research Scientific Board was the guest of honor—and then had dinner with [advertising executive, automobile manufacturer, and financier] Ward Canaday at the 1925 F Street club. It is always fun to be with Ward and I think Ruth enjoyed the evening as well.

Tuesday, September 27: At 8 o'clock I repaired to the office of the dentist. Fortunately, I needed only to have my teeth cleaned. The rest of the morning was given over to a variety of tasks—I was to go out to Goddard but did not make it—and then Dryden and I had lunch with Kistiakowsky. I left at 3 o'clock for Omaha where we were to brief General Power [commander-in-chief, Strategic Air Command] and his staff on the morrow.[12] Arriving at Omaha in time for dinner, we had a fine steak and I got to bed very early.

Wednesday, September 28: The briefing went well and Tommy Power seemed pleased. He took the stage after we had finished and made clear his conviction that the president is all wrong in talking about space for peaceful purposes. Tommy would go all-out for a strong military space program and everything else that would be necessary to make this country absolutely invincible from a military standpoint. He admits the problems of politics but would seemingly override them. I got on the plane and left at noon so that I could get back home with some prospect of getting a little rest. Actually, I worked the entire five hours on the flight home and think I was very much better off than if I had waited for the balance of the group who came home on an Air Force plane.

Thursday, September 29: This was a really rough day. We had a long staff meeting and later in the morning met with Dr. John Keyston, vice chairman, Canadian Defense Research Board, with whom we had cocktails on Monday night. He was just interested in paying a courtesy call and making certain that there were no communications difficulties between NASA and the Canadians. Another visit from the Chance Vought Corp. representatives. Do they never get enough of this

---

[12] The briefing dealt with NASA's programs for the exploration and use of space. At its outset, Glennan stated: "While our program is centered around the conduct of research and development for peaceful purposes—as the Space Act puts it—we recognize that the results of our effort will be of interest and of use to the military departments. Indeed, the law requires that the results of our programs be made available fully to the military services." ("Remarks at SAC briefing, OMAHA, 9/28/60," Glennan subseries, NASA Historical Reference Collection.)

kind of thing? Because of the bad weather, I took a train for New York at 12:45 and arrived just in time to get over to the cocktail hour before the dinner. This was a really wonderful aggregation of top industrial people brought together by the A. & H. Kroeger Organization. Les Worthington, president of U. S. Steel Corporation, and I were the principal speakers. There were several good friends from Cleveland there and I think I did a reasonably good job. Ruth had gone up in the morning and had accomplished quite a bit of shopping as well as a visit to the Guggenheim Museum. She had dinner with Mrs. Kroeger and then attended the theater. We met together at the Barclay and prepared for bed when I happened to look out the window and noticed that it was really foggy. A call to the Weather Bureau indicated that there was little chance of a flight getting off in the early morning. Because I have a real tough schedule for Friday, I decided that we would go back by train. Accordingly, we dressed and repaired to the Pennsylvania Station where we were able to get a bedroom to Washington. What a life!

Friday, September 30: Our train came in on time and Ruth and I had breakfast at the Statler after which she went home. The entire morning was spent with the heads of all of our laboratories in a review of the Saturn program and the applications that will use the Saturn. This is a real problem program. The ten-year period looks as though it would cost at least $5 billion—probably closer to $10 billion.[13] What are the criteria one would use in making a decision on a problem of this kind? I broke away from the morning's session to have lunch with the Kimpton committee at the Metropolitan Club. This was its last meeting and I had a chance to thank the members for their help. They asked a good many pointed questions of me. I think their report is going to be very helpful to us.

Back to the meeting at the office, for other and more mundane matters. At 4 o'clock, Nunn and Jaffe came in to talk about the communications satellite program. Nunn has prepared an excellent paper but Jaffe has not done what I had requested. I wanted an outline of a possible development program with some estimate of the cost. Instead, he wrote a paper that seemed to be aimed at convincing me—as if I needed to be convinced—that we should go ahead with a development program. Sometimes it would be a good thing if engineers just continued to be engineers!

---

[13] See note 9 of chapter Two, which shows that the $10 billion estimate was very close to the mark.

CHAPTER ELEVEN

OCTOBER 1960

Saturday, October 1: Down to the office at 9:30 for a meeting with Bill Hines of the *Washington Star*. He wanted a round-up on the [initial] two years of NASA's operations. I had Bob Jastrow come in and we managed to get through it without too much difficulty. Jimmy Doolittle came in for a few minutes; he had very little to say but we did talk about the possibility of Si Ramo as a candidate to follow me in this particular job after 20 January next. At 11 o'clock, I listened to a briefing on the problems of the Centaur program and agreed that we ought to increase the spending rate by $10 million in the current fiscal year. Where do we get the money?[1] I don't know—we simply have to get it. Finally, I got home about 3 o'clock and set out for a long walk with Ruth. After a dinner of Metrecal, I got into bed and I believe that almost thirteen hours passed before I arose again.

Sunday, October 2: Work occupied the entire day until 5:30—I am now dictating the finish of this day's effort—as we prepare for dinner for the Seamanses and the Harknesses. I may add something to this later on.

It was a good evening with most of the conversation carried by the Harknesses and ourselves. Bob and Jean Seamans are really quite delightful people and have all of the attributes necessary to make them a relied[-upon] part of the Washington community. Apparently, her father was an official in the State Department for a number of years. She seems gracious and at ease. He has made a very excellent impression at the office. He seems to know how to ask significant questions and does not back away from requiring a good performance or from making decisions.

Before going on with this chronicle, I think it might be good to recall that we are in the midst of one of the most significant and, at the same time, peculiar meetings of the United Nations. Khrushchev has been alternately beating the Western powers over the head and seeming to be somewhat conciliatory. At all times, he has been wooing the neutrals and the nations that have just been granted membership in the U.N. I am reminded of the comments of a British psychologist named Sargent who believes that Khrushchev is following Pavlovian theories in his handling of international affairs. Pavlov was the great Russian psychologist who

---

[1] According to the *NASA Pocket Statistics* for December 1961, research and development (R&D) funding for the Centaur increased from $4 million in FY 1959 to $36.64 million in FY 1960 and $62.58 million in FY 1961. Direct R&D obligations for the Centaur reached their high point in 1964 with $108.3 million and totalled $543.3 million by 1968, when the annual R&D obligations had fallen to $100,000. (*NASA Historical Data Book*, Vol. I, p. 152.)

was able to create interesting behavior patterns in dogs by alternately tormenting them and treating them with great kindness. He was able to get conditioned reactions with consistency. Sargent believes that Khrushchev is applying this psychology to the large issues presently facing the Western world. For our part, we have been waiting with some anxiety for the announcement of a new Soviet success in the space business. As a matter of fact, we entered upon a positive campaign four or five weeks ago to anticipate such announcements—particularly in the field of manned space flight. I believe we have achieved our goal but we are left at the church by the fact that the Russians have not, as yet, undertaken any startling space experiment. Why? I just wish we knew.

I have been saying that there are several reasons for anticipation of a Russian effort. First, we all concede that the Soviets must have the means to do very nearly what they want to do in space. I say this in spite of the fact that I am not at all sure that they have any better understanding of the hazards in human space flight than we have. I am not one of those who believes that the Soviets will recklessly put a man into space—I don't think they could stand the horrified criticism of the rest of the world were they to do this. On the other hand, there is no guarantee that they have not already put a man into space and left him there. There seems to be some evidence that the May 15 shot was just such a shot.[2] In addition, the Soviets have quite frequently chosen anniversary dates or significant public occasions for the launching of one of their space shots. We convinced ourselves that Khrushchev's arrival at the U.N. would be one such occasion, that the third anniversary of the 4 October 1957 launching of Sputnik I would be another, and that the early part of October—being a favorable time for a minimum energy trajectory to Mars—would be still another. There is yet time for that last effort—the Mars shot.[3] On 4 October, by coincidence, the Army's Courier satellite was placed in orbit and seems to be operating satisfactorily.[4] On that same day, after having tried for two weeks to launch our first complete Scout launch vehicle, we made a successful shot carrying a small payload package for the Air Force in a "piggy-back" fashion.[5] Thus we celebrated the third anniversary of the real beginning of the space age.

---

[2] On 15 May 1960 the U.S.S.R. had launched Spacecraft I into orbit. It weighed 10,000 pounds and marked the first successful effort to orbit a vehicle large enough to contain a human passenger, although efforts to recover the space capsule were not successful. The cabin contained a "dummy space man" and tested life support systems. (Emme, *Aeronautics and Astronautics ... 1915-1960*, pp. 123, 147.) Of course, on 12 April 1961 the Soviets succeeded in placing Cosmonaut Yuri Gagarin in orbit on Vostok I and returning him safely to earth. (Roger Bilstein, *Orders of Magnitude: A History of the NACA and NASA, 1915-1990* [Washington, D.C.: NASA SP-4406, 1989], p. 57.)

[3] In fact, the Soviets seem not to have launched any satellites in October 1960.

[4] This was a 500-pound communications satellite to test the feasibility of global communications using delayed repeater transmitters. The satellite successfully transmitted and received signals. (Emme, *Aeronautics and Astronautics ... 1915-1960*, p. 150.)

[5] This was the second complete Scout launch and the first to reach its predicted 3,500-mile altitude and 5,800-mile impact range. The launch occurred at Wallops. The first complete launch from Wallops on 1 July 1960 was marred by failure of the fourth stage to operate properly. (*Ibid.*, pp. 124, 128.)

On another note, I spent $5.00 to buy General Medaris' book the other day. I have not read the entire book but I did go over the last three or four chapters, which deal with the fight between the Army and the rest of the Pentagon and the discussions with NASA over the Army's position in the space business. Medaris makes a hero out of Roy Johnson and says that Brucker is the best secretary the Army has ever had. These two characterizations are enough in themselves to damn the rest of the book. This was not enough, however; Medaris went on to characterize Herb York as a second-rate scientist. Nothing could be further from the truth. I think my statement would be joined in by all responsible people active in the fields of science and technology.[6]

It is interesting to reflect on the impact made on the nation by books such as this written by generals who became so obsessed with their own viewpoints that they lost perspective and finally left the army. Clearly, there is room for argument on matters such as these, and public debate ought to be encouraged. Character assassination is quite another thing, however. In this particular instance, it appears that everyone is out of step except Medaris and those who agree with him.

Monday, October 3: We started the day with a dry run on the Rover program in preparation for a meeting with the commissioners of the AEC on Tuesday, 11 October. I think the boys have done a good job in bringing this program into focus and in laying out the activities for the next five to seven years. It does appear that we have a bear by the tail. Above all, I intend that we not be caught in the same situation as the boys who have been spending hundreds of millions over the past ten years on the problems of nuclear propulsion for aircraft. As of the present, under the determined pounding of Clinton Anderson and the Joint Congressional Committee on Atomic Energy, we are headed full tilt for that path. The meeting with the commissioners will be an interesting one.

Later in the morning, Bill Littlewood, chairman of our committee on aircraft operating problems and a vice president of American Airlines, came in to pay his respects and talk about the work of his committee. Much to my surprise, having belabored NASA for doing too little research in support of aeronautics, Littlewood wanted to have his committee authorized to deal with operating problems in spacecraft. I pointed out this seeming inconsistency, but he blandly asserted that there was no real relationship between the two.

Lunch at the Carlton with Nat Finney of the *Buffalo Courier-Journal*. Nat is a responsible newspaperman and was attempting to put together a piece for the 80th anniversary of his newspaper. I find it easy to talk with him and hope that his piece will be worthwhile. This was an instance in which Walter Bonney had let me down since he had agreed to prepare a statement for Finney some days earlier.

At 2:30, we gathered in the USO auditorium for the second annual NASA service award ceremony. The lady civil service commissioner, Mrs. [Barbara

---

6 The book in question was entitled *Countdown for Decision* (New York: Putnam, 1960).

Bates] Gunderson, spoke briefly and well.[7] She is an attractive and bright person who must bring to the deliberations of that body many new points of view. These awards ceremonies seem to mean a good bit to the participants. At 4:30, I left for Cleveland via Capital Airlines. After dinner at Stouffers with Sally and Ruth, I found myself sufficiently tired to retire without any further delay.

Tuesday, October 4: This is a day to remember. [Relates the events in connection with a $1.6 million grant to Case by the Olin Foundation for a center devoted to the materials sciences and his flight back to Washington, D.C.]

Wednesday, October 5: I am resuming dictation on this journal today, 22 October, after having been so completely busy and so much occupied with traveling that I have had to forego my daily dictation. For some unknown reason, I had thought that I would have more free time after Congress had adjourned. This has not turned out to be the case. In fact, I seem to be busier than ever and it does not seem to me that I am getting as much done as I must accomplish before leaving this job. However that may be, I will just have to keep at it. After yesterday's exciting event, this day seems a bit of a letdown—not because it wasn't a busy one. At 8:45, Dryden, Seamans, and several others came to talk about trajectory problems on some of our shots. The Atlas-Able appears to have dropped a piece of metal somewhere in south Africa, probably near Pretoria. This will be an increasingly difficult problem since it is almost impossible to avoid overflying the territory of another country in these shots. I have determined that we must discuss this with the State Department and convince it that we are taking every possible care to avoid this sort of difficulty but that we must be prepared for a problem of this sort from time to time.

Abe Silverstein brought in the new assistant director of the Goddard Space Flight Center—Eugene Wasielewski. He seems very much of an extrovert and comes well recommended. Apparently his most recent job has been with Curtiss-Wright although he has worked for NACA in the past. At 9:30, a presentation by the source evaluation board on the orbiting astronomical laboratory. The competition has been tough and almost all of the companies that proposed on this project have sent their presidents or vice presidents in to see me. Grumman won the competition and, from all present indications, will do a good job. I think the contract is in the neighborhood of $23 million.[8] We then had a briefing on the proposed changes in the construction budget for the Jet Propulsion Laboratory. It has developed a fine master plan and it would appear that over a period of several years we can really improve working conditions there. I approved this $5-million project and complimented all concerned.

During the noon hour I had the first of two flu shots and then we gathered at the control center for a briefing by the Central Intelligence Agency on all activities

---

[7] She was the vice chair of the U.S. Civil Service Commission, which was chaired by Roger W. Jones.

[8] This figure was, in fact, the contract estimate.

that have taken place in the last year. Dryden and I then gathered for some discussions of organizational changes and came into reasonable agreement as to what ought to be done. A skull session was then held with Silverstein and Tepper to prepare for the discussions we are going to have early next week with other agencies relating to the program for meteorological satellite. Finally, I wound up the day with Elmer Staats telling him of the probability of a supplemental budget for Centaur and for the communications satellite business and informing him of the large sums of money that must be spent if we are to carry on with man in space beyond the present Mercury project. Elmer did not seem surprised or concerned; I don't know what this means.

During the course of the day, I was approached by a member of the McKinsey organization who asked whether or not I would be willing to talk about accepting the post of vice president of General Aniline and Film Corporation and general manager of its Ansco Division with the probability that within a year or two I would be president of the parent organization. I quickly said, "No!" This is a bit of irony for it was from Ansco that I moved to Case back in 1947. We wound up the day with dinner at the Cosmos Club with Dr. Clark Goodman [a nuclear physicist who formerly worked for the AEC] and his wife. I left early so I could get home to do a little bit more work.

Thursday, October 6: Staff meeting this morning was not unusual but I rushed it through so I could attend the meeting of the National Security Council at 10 o'clock. As is often the case these days, the part of the meeting in which I was interested was deferred. Immediately after lunch I flew to Hartford. Congressman Daddario and my host, Frank Williams of the Connecticut General Insurance Company and the general chairman of the Yale fund drive, met me at the airport. The usual press conference was held—not a bad one this time—and then I had dinner at Frank Williams' house with several of the top people from United Aircraft and from the life insurance company. At 8 o'clock, I spoke to about 400 members of the staff of Connecticut General and had a really good audience. After I spoke for 45 minutes, they kept me for another similar period of questioning. All in all, it was a very good evening.

Friday, October 7: Up at 7 o'clock and off to a breakfast with Congressman Daddario, hosted by [Raymond A.] Gibson of the Hartford Electric Light and Power Company. About thirty of the top business leaders and professional men in the community were present. It was a good session with lots of interesting questions and I hope some reasonable answers. Daddario then drove me to New Haven where I met with Charlie O'Hearn, assistant to President [Alfred Whitney] Griswold. An hour's discussion with Bill Hinkley and members of the survey committee of the engineering school, appointed by President Griswold, revealed just how inadequate the engineering program at Yale seems to be. The school has had very little in the way of aggressive leadership for twenty years and is in need of this examination. I invited them to visit Case and they said they would. Griswold is said to be anxious to drop engineering from the Yale curriculum. Nothing could be further from the truth, but he does want a good school—one that is as modern in its concepts as is the rest of Yale. He needs help.

After lunch at Mory's, I took part in a panel discussion before the alumni group that had come together to be told the aims and campaign plans for the

$69 million fund-raising effort. Bill Bundy led off and moderated the panel. Alan Waterman then spoke about the significance of science in the national scene, and I followed with a thoughtful discussion of the NASA program and the problem we have in convincing Congress and the executive branch of the importance of going ahead on a broad and costly space exploration program. All of this was intended to indicate the need for really well-prepared men and women who can deal effectively with problems of this kind on the national scene. Leadership—that is the word. The panel was completed by Archibald MacLeish. I had never heard him speak before but was so much impressed by the beauty of his language that I missed pretty much what he had to say. He received a great ovation. A question and answer period gave me an opportunity to give Whit Griswold some backing on his engineering study. My plane had come to New Haven to pick me up, and Alan and I flew back to Washington reaching home just in time to turn on the television for the second of the "Great Debates." On this occasion, it seemed that Nixon had regained his confidence and was on the attack. It is so difficult to be objective in these matters that one hesitates to designate a winner. It did seem to me that Kennedy was as glib as ever and that Nixon appeared the more mature, a person who recognized the awful burdens of the office.

Saturday, October 8: The entire day was given over to a review of all of our projects. Mercury has slipped some more and some of the other projects have had to be delayed a bit. We are bunching a number of shots in December—so many that I authorized delaying one into the early part of January. Ruth returned from Cleveland and things began to be a little more normal again.

Sunday, October 9: The day was given over to work as I completed the script for my speech in Portland, Oregon, next Wednesday.

Monday, October 10: At 9:30, we had a meeting with representatives of the Federal Aviation Agency, the Department of Commerce and the Weather Bureau, the Department of Defense and our own meteorological people for the purpose of looking at the problems that face us in developing a meteorological satellite system that will be operational within the next four or five years. It is so obvious that the Weather Bureau is poorly prepared to take on the research necessary to deal with this very difficult problem, one wants to step in and help. Unfortunately, it doesn't appear as though we'll be able to do very much because we don't have too much of the necessary experience. The Defense Department will move very solidly in this area, I am sure. At 2:30, a group of the boys came together to discuss policy matters with respect to the meteorological satellite program and the communications satellite program. We reviewed my speech for Portland very carefully because it contains the first policy statement in the communications field. It is apt to cause a good bit of discussion in Congress and perhaps in the executive branch.

Ruth and I went to a white tie dinner with Admiral Burke and then to the Navy Ball. All of the Joint Chiefs of Staff were present as were the ambassador from Sweden and his wife. The ball was a big and noisy affair, and we stayed only long enough to be polite. I asked Mrs. Burke whether she ever had an evening to herself. She said they came so seldom she couldn't remember them. This is typical of the load placed on the leaders of our military departments. Admiral Burke has held his job now for five years.

Tuesday, October 11: John McCone of the AEC came over for a meeting on the nuclear rocket and nuclear power programs. He and the commissioners had already had a briefing prepared by Harold Finger indicating the very great cost of the Rover program and the high probability that we would not be able to make a test flight in 1965. I asked that we all listen to the same briefing again so we would be talking from the same script. I think the facts as presented are straightforward and that they cannot be brushed aside. John McCone, under the prodding of the Joint Congressional Committee on Atomic Energy, has been attempting to push this program forward, although he is concerned over the use to be made of such a rocket system. I am concerned about whether or not anyone will ever let us use the rocket with the possibility of radioactive hazards that undoubtedly will be present. In any event, we agreed that we should pursue an aggressive research program but should not launch into an all-out hardware program until some of these uncertainties have been dealt with. Louis Schreiber of du Pont came in to see me. He had been a graduate student during the time I was at Yale and had served as assistant to Ruth's father. Thus, it has been more than 30 year since I had seen him. He wanted me to deliver a speech in Wilmington, which I was unfortunately unable to accept.

At 3:45, I left Baltimore for Portland, Oregon, traveling with Admiral Bennett. Lately we have been admonished by the Bureau of the Budget to travel tourist class in the jets. I had succumbed to this order to the extent of taking the short flights tourist class, but I ride first class on the longer flights. Thus, I found myself riding from Baltimore to Chicago tourist while Bennett rode first class. We did get together on the Chicago-Portland leg of the trip.

Wednesday, October 12: Governor Mark Hatfield of Oregon had asked me to be the principal speaker at a luncheon on this day. It turns out that he is a young and very attractive man with a great deal of energy and evidence of leadership capabilities. Oregon has not been an industrial state, and he is determined that a broader industrial base be developed. This was a meeting of the development committee for the state and he had asked the Army, Navy and Air Force to tell a group of 200 businessmen of the requirements those services would have for research and development activities in the future. While the morning session was on, I met with the governor and listened to a briefing by some of his state development people. They have set aside a very large tract of land in north-central Oregon along the Columbia River for an industrial development. This has been done with the thought that it would make an excellent site for missile or space vehicle testing. The whole pitch is at the so-called space age industries. Unfortunately, they are about ten or twelve years late and it isn't at all clear that the remote site in a remote part of the nation is a useful one. Nevertheless, it may be possible to encourage industry to develop subsidiary plants in that general area. Certainly, if enthusiasm means anything, they will improve their situation substantially.

The luncheon speech went well and was widely reported throughout the nation. I was talking about policy matters with respect to the communications satellite business. I pointed out that communications had always been an operation for private industry in this country and I saw no reason for changing that in the event satellites became part of the system. I proposed that the government provide launch vehicles and launching services at cost to those companies, such as AT&T, willing to pursue their own development and pay the costs. On the other hand, I pointed out that we had an obligation to keep the United States moving ahead in all fields of space research and that we must therefore support technically attractive proposals made by industry at our request or voluntarily. Now we will wait for the congressional investigation!

A visit to a couple of local plants finished out the day and brought me back to Portland for dinner with a group of bankers and businessmen. I then flew to Seattle where I expected to take off at 10 o'clock on a Northwest Airlines jet for New York. Unfortunately, the flight engineers had gone on strike and I found myself taking a United Airlines jet at 11:45 p.m. with only three hours of sleep before I had to change planes in Chicago. I did arrive in Washington at 9:45 a.m. on schedule.

Thursday, October 13: Arriving at Friendship Airport without much sleep, I made tracks immediately for the White House where I was to attend another meeting of the National Security Council. For the first time, my car telephone came in handy as I kept in touch with developments between the airport and the White House. As luck would have it, after I waited thirty minutes, the item on the National Security Council agenda in which I was interested was canceled. Immediately after lunch, I reviewed the Portland situation with Bob Nunn and Homer Joe Stewart. We must move rapidly on this communications business before it gets out of hand. At 3 o'clock, I met with the president. I had called in from Portland the day before in order to get this date because of the probability that the president was going to be away from Washington for two weeks. I found him tired and preoccupied. I simply brought him up-to-date on our activities, told him of my speech in Portland and what meaning it had for the executive branch and then indicated that we were going to require additional money in the way of a supplemental. He had no comment to make on this matter. I told him something of the costs that appear to be involved in Project Apollo, the follow on to Project Mercury. He expressed himself once more as having little interest in the manned aspects of space research. He was cordial enough but it was obvious that he was not at his best today.

I went directly home and went to bed since I appear to be coming down with a cold. This would not be strange after the night I had spent returning from Portland. Several pills helped me to get through the night and seemed to arrest the cold.

Friday, October 14: At 9:30, we had a presentation by Bruce Lundin and his study committee giving the results of the investigation of the proposed location for Project Apollo. They have done a thorough and objective job but find themselves as a committee of four split right down the middle. Thus, I have the task of making the decision myself, although I now have much better information on which to base such a choice.

I did decide, at Bob Seamans' urging, that the life sciences center should be located at Ames Research Center in California. This decision has been delayed in the hopes that we could make a sensible decision with respect to the location for Project Apollo. It does appear, however, that to locate our life sciences group in the Washington area is to locate it within a disease-oriented medical group. In the San Francisco Bay area we find a number of medical schools and research institutes where forward-looking basic research is being carried out. It, therefore, seems very desirable to have our life sciences group situated in that kind of an intellectual climate. I know that Clark Randt will be happy to have this decision made.

Saturday, October 15: Up early and off to the airport where Ruth and I took a TWA ship to Cincinnati. Actually, the field was fog-bound and we were about an hour late in departing. We were met in Cincinnati by Neil McElroy, former secretary of defense and now chairman of the board of Procter and Gamble. He took us to his home in the Hyde Park section of Cincinnati. This seems to be the older section of the city. The home is a large, rambling but comfortable house. Camilla [his wife] seemed glad to see us and we had a pleasant luncheon before taking off for the football game between Wichita and the University of Cincinnati. Unfortunately, it began to rain soon after the game started and we did not bother to stay through more than five or ten minutes of it.

A relaxing afternoon ensued and then we dressed for dinner. I was to speak to the Commercial Club of Cincinnati and Mrs. McElroy had several ladies in for dinner with Ruth. The speech went well—a summary of the space program and some indication of the direction we will take in the future. It was the same general talk that I've given to several groups in places like Cleveland and New York. After the dinner, we returned to the house and spent another hour and a half with the ladies.

Sunday, October 16: Up at a leisurely hour and breakfast with the McElroy family, which by this time included Malcolm, the youngest son, and Barbara and her husband Dave. This latter couple has a cute little baby who seems to be deficient in heart development. They have practically despaired of her life. A trip to the Mayo Clinic in Rochester, Minnesota, is in prospect. Packing for the trip to Williamsburg, we then drove out to the home of [Bayard Livingston] Kilgour, president of the Cincinnati and Suburban Telephone Company, where we had a quick drink before driving on to the country home of the Harrisons, friends of the McElroys. This was sort of a day in the country with an interesting group of people but we were able to stay only a short time. Back to the airport where we took one of the Procter and Gamble Grumman Gulfstreams for a two hour flight to Williamsburg, Virginia. Arriving at Williamsburg about 4:30, we found we had preceded the balance of the group from Washington—most of our top staff—by only a few minutes.

This is to be the beginning of a three-day conference with representatives from headquarters and from all of the field centers. The agenda concerns itself principally with reports of activities undertaken in the last six months and a look at the future from both a programming and policy standpoint. Ruth joined us for a reception and dinner, after which [Maurice] Stans spoke to the group. I had asked him to come down principally to let him be seen by our people and to provide an opportunity for him to know some of our top staff. I think it was a very successful move—both sides seemed genuinely grateful for the opportunity to meet each other.

Monday, October 17: The conference got underway at 9 o'clock with Bob Seamans taking over in good fashion. A series of reports by various elements of the headquarters organization took up the morning. Larry Kimpton and John Corson had flown in to give us a report on the management advisory committee study. This turned out to be interesting and a little controversial. Since the report is not as yet completed, I kept cautioning the staff that it must not expect to accept the report as given or even as written. It is a series of recommendations to us and we will accept or reject these recommendations as seems best for all concerned.

In the middle of the afternoon, immediately following the completion of Kimpton's report, I succumbed to a feeling of extreme nausea. Having set up a cocktail party for part of the staff, I was concerned about keeping that engagement. I lay down for awhile and was able to be up just long enough to greet the guests when I made it back to the bed. Ruth took good care of the party. I found myself with a temperature slightly in excess of 100°, and the usual stomach evacuation process followed. Fortunately, we had scheduled nothing for this particular evening.

Tuesday, October 18: I seemed to be reasonably recovered by breakfast time and went back to the conference at 9 o'clock. The first order of the day was a panel discussion on the subject, "Where should NASA be headed?" This was an exceedingly useful discussion and might well have taken up the time of the entire conference. All elements of the program were discussed briefly by a panel of five men, followed by comments from the floor. Emphasis seemed to be on taking the bold approach, but there was more conservatism displayed than I had expected. There were some useful suggestions, particularly by General Ostrander, about improved relationships with the military. During the morning coffee break, I spoke rather casually to Clark Randt about the decision to establish the life sciences center at Ames on the West Coast. Much to my amazement, he expressed himself as being exceedingly unhappy about everything he had heard and stated that he could not continue [working for NASA any] longer. I'm afraid that I was not very cordial in my reply for I felt that he was acting rather like a petulant youngster. I simply said that if he couldn't make up his mind to carry the ball, perhaps he had better leave. Immediately after our reconvening, I put Bob Seamans to work to gather Randt in for a conference in my room after the evening session.

The afternoon session produced nothing startling, and we were off to the King's Arms for dinner after picking up [Major] General and Mrs. [William B.] Keese of the Air Force. He is to talk to us tomorrow. My paper on "The Transition to a New Administration" took about an hour of discussion. It seemed to go rather well with some considerable expression of regret that I seemed to make it so positive that I was leaving the organization after the inauguration on 20 January. There was a rather spirited debate over the desirability of undertaking the preparation of a second budget on the odd chance that a new administration would want to spend more money than we will be able to get from the current administration. This was decided in the negative since a great many people would have to be involved and it seemed wholly impossible to keep such an exercise away from Congress. Political dynamite is contained in any operation of this kind, and it is tremendously important that we avoid providing anyone with material for a congressional attack. The discussion gave me an opportunity to remind everyone that a key characteristic of our operation had been the integrity with which we approached all of our planning and operations, both internally and externally. I stated that we were not going to change that policy as long as I was the head of the organization.

After the evening session, Randt came in with Dryden and Seamans, and we debated for about an hour without getting much of anywhere in satisfying Randt's concerns. He offered to resign immediately or at least by 1 December. I pointed out the extent to which we had been delayed in giving him the support he wants and the reasons for those delays. Obviously, Randt has attempted to go about his activities without understanding as clearly as he might the business of developing relationships that would be productive for him. The fact that none of us are medical doctors or have had any experience in the life sciences makes it most difficult to achieve the degree of understanding necessary in this situation. I finally broke up the meeting by saying that we had better sleep on it.

Wednesday, October 19: The session this morning was given over to a discussion of management problems—particularly the development of better project managers and project management techniques. A good discussion ensued, although it is quite clear that we have a long way to go before we get complete understanding and enthusiastic cooperation from all concerned. At 11:30, we turned the meeting over to the Air Force people, who described their study program. They spend about $5 million a year but are able to acquire projects costing another $10 to $15 million by playing on the eagerness of the contracting fraternity to get into the business. Part of this is legitimate, since all production contracts carry some money to be expended by the contractor on advanced research activities, including studies of this kind. I thought the subject well presented and it did demonstrate the need for close association between planning groups and long-range study groups of this kind.

I summarized the conference immediately after lunch and we then took off for Washington where we arrived about 7 o'clock in the evening. All-in-all, it has been a good three days, and the feeling of good fellowship and better understanding that seems to pervade a meeting of this kind must surely pay off in the long run.

Thursday, October 20: We started off the day with a review of the briefing we are going to give the State Department on the "over-flight" problem. It is a good briefing and I agreed that it ought to be given at the working level so we could determine the best approach to the higher levels of the State Department in our search for a solution to this knotty policy problem. I had lunch with Lieutenant John Davey of the Naval War College at Newport, Rhode Island, as I discussed with him the possibility of his interest in becoming the director of our office of public information. Actually, he is a little too young and inexperienced for the job, but he is an attractive and hard-driving man. I did have him meet most of the rest of the top staff so that we might be able to counsel together on his candidacy.[9]

Late in the afternoon Wernher von Braun and others gave me a briefing on their five-year funding plan for the Saturn vehicle. It turns out that it will cost about $1.8 billion. I am determined that these figures be known to all who may have anything to do with our budgetary operations. The briefing was a good one and I think these sessions incline the staff to recognize the deep concern it must have, along with me, for the validity of our planning.

Friday, October 21: Yesterday was the 351st monthly anniversary of our wedding and I brought home some Sweetheart roses for the queen of the household. How long I will be able to keep this up, I don't know. We started out the day with an 8:30 meeting with Jim Freeman of AT&T. He was delivering to me a memorandum from Jim Fisk detailing the plans for a communications satellite experiment by the Bell Laboratories. They seemed determined to get into this business and to pay their own way. This I applaud but I do find it a little difficult to deal with them on a piecemeal basis. I have asked Jim Fisk to give me the complete program plan by the laboratory so that we can do our planning in a sensible fashion.

At 9 o'clock, John Corson sat down with members of our business administration staff to review the first draft of the report on contracting. This will be a useful document. At 11, Clark Randt came in to express his regret at having blown up. Apparently, he will stay on with us—at least, he told Seamans that he would stay on for some period of time to see that the life sciences center at Ames is well-started. I gathered that he would continue to stay with us if the support he needs is forthcoming.[10] At 11:30, Peter Murray of the Wright Air Development Division came in to have a visit and lunch. We had asked him to consider seriously

---

[9] Davey apparently never came to work for NASA. On 15 November 1960 Shelby Thompson assumed the acting directorship of NASA's office of public information in addition to his job as director of the office of technical information and educational programs. O. B. Lloyd became director of public information on 10 February 1961 after having worked for United Press International from 1946-1959 and then joined the Washington staff of Senator Lyndon B. Johnson, who of course became vice president in January 1961. (See the biographical files of both Thompson and Lloyd in the NASA Historical Reference Collection.) For an interesting perspective on the history of public relations in NASA from its beginnings through 1986, see Bruce V. Lewenstein, "NASA and the Public Understanding of Space Science," *Journal of the British Interplanetary Society*, Vol. 46 (1993): 251-254.

[10] NASA did establish a small research facility for life sciences at Ames in February 1961, but plans for a larger facility never materialized. Randt resigned from NASA effective 1 April 1961, and the headquarters office was realigned in a reorganization on 1 November 1961, becoming a subordinate organization called bioscience programs within the office of space sciences when it had been an independent office with its director reporting to the associate administrator. (Rosholt, *Administrative History of NASA*, pp. 127, 344ff.)

joining us as deputy to General Ostrander in the launch vehicle division. It is probable that Ostrander will stay with us somewhat less than a year, and thus we are looking for a good man to replace him. Murray seems to be such a man.[11]

At 1 o'clock, we started out on a rehearsal of the briefing to be given the President's Science Advisory Committee next week. This is intended to be a full disclosure of all of the programs that will use the Saturn and other high-thrust vehicles. It's the program that will run to $6 billion and maybe much more. I broke out of the meeting for a few minutes with Gil[bert Harrison] Clee [director] of the McKinsey Company. Gil had been active in the McKinsey study for Metro in Cleveland under my direction. The talk was an interesting one although I am not completely certain of its ultimate implications. Clee is a political scientist interested in management problems and economics problems in governmental areas of all kinds and sizes. Presently, he states that McKinsey has more business than it needs and is financially well-off. He and others in the McKinsey organization are now of the belief that their major clients are going to find the foreign field an increasingly difficult one within which to work. At the same time, he believes that our future is inextricably tied up with the development of less fortunate nations as effective economic and social units. He was offering me the opportunity to join McKinsey in any manner that I saw fit—no mention of money except to say that there would be no difficulty over that—to work at this sort of a problem at my own pace and in my own way. I told Clee about the discussions Ruth and I have been having with Jim Perkins of the Carnegie Corporation. All of this seemed to fit together rather well and it was agreed that we should have further discussions as my mind became more clear on these matters. Perhaps something will come of this interest in the foreign operations of the government before too many years have passed.

I waited around until 5:30 for John Rubel of the Defense Department so we might talk about the communications satellite problem. He is deputy to Herb York, who is not yet back on the job. Rubel is a good man although somewhat inexperienced in the in-fighting that must take place in government. We agreed on the need for meshing the DOD program with our own so that a clear delineation of fields of responsibilities would be evident to all concerned.

It has turned quite cold and Ruth and I found it chilly as we walked out a bit after dinner. Returning to the apartment, we listened to the fourth of the "Great Debates." Both of us felt that Kennedy was less self-assured than in the earlier debates. (This seems to be a subject on which one believes what he wants to believe for neither of us has been able to talk with anyone since the debate who would agree with us on this particular matter.) Whatever may be said about the winner of each of the debates, it seems clear that Kennedy has made an over-all gain out of this exercise. No longer is it possible for anyone to say that he is too young and immature for the task he will face as compared with Nixon. They stack up together as very able, well-informed and determined young men. I continue to believe that Kennedy does not recognize the difficulties of carrying out the kind of a program he is promising, nor does he seem to give evidence of understanding the economic problems into which he will drive the nation if he is able to carry out the liberal programs he has been discussing. I still expect to vote for Nixon.

Saturday, October 22: This was a relatively lazy day. I have been feeling a bit numb in the head these past several days so decided to give up any work for this day. The snow tires were put on the car, anti-freeze installed in the radiator, and then I watched the Notre Dame-Northwestern football game. At 7 o'clock, Ruth and I joined John and Mrs. Corson for dinner. Among the guests were Larry Henderson, vice president of the Rand Corporation; Jim Webb, former director of the Bureau of the Budget under Truman; Bob Calkins, director of the Brookings Institution; and Elmer Staats, deputy director of the Bureau of the Budget. Each was accompanied by his wife, of course. It was a good evening with some useful and pleasant discussion. A couple of the ladies seemed very much upset about the work of Albert Schweitzer. Just why they had this feeling is not clear but it soon became evident that the only one in the room who knew very much about Mr. Schweitzer was Ruth. I think we won the argument but I think the opposition remained unconvinced.[12]

Sunday, October 23: Up at 9:30 for a leisurely breakfast and then down to the office for two and a half hours with Homer Joe Stewart. He has been attempting to complete a revision of the 10 year plan. I found little with which to quarrel and so we agreed to reproduce this draft for discussion purposes. The budget for FY 1965 looks to be $350 million higher than it was in last year's version. Most of this is occasioned by the Saturn speed-up and because costs just seem to climb. The rest of the day was lazy and hopefully refreshing. Believe me, I am going to need a long rest when I finish this job.

Monday, October 24: Before entering anything more in this chronicle today, let me say something about some of the events of the past two or three weeks. It will be remembered that I made a speech in Portland setting forth my ideas on the policies government ought to follow in connection with the development of communications satellites for peaceful and commercial purposes. This was well received by the bulk of the press and I had no questions raised by anyone in the administration although I was reminded that the cabinet would want a briefing on

---

12 Webb, of course, became Glennan's successor as NASA administrator. As is well-known, Schweitzer (1875-1965) was a physician of humanitarian sentiment from Alsace as well as a theologian, philosopher, and musical scholar. He had set up a medical mission in Lambaréné, French Equatorial Africa (later Gabon) and preached a reverence for life that formed the basis for his humanitarian efforts. He received the Nobel Peace Prize for 1952, and his address on the occasion, published in English translation (from the German) as *The Problem of Peace in the World Today* (1954) had a worldwide circulation.

this particular problem. During the course of the past week, a letter came down from Lyndon Johnson on behalf of the Senate Space Committee asking a great many questions about this matter. These included a request for delineation of areas of responsibility in the active-repeater communications satellite field as between NASA and the Department of Defense. Further explanation of some of the phraseology used by me in the speech was requested. Questions were raised as to the nature of the decisions to be made with respect to support of privately-proposed satellite projects and the source of funding for these projects. Included was a question on the nature and means to be employed for international regulation in this field.

All of these are reasonable and good questions and I can just hope that they are meant to be helpful in clarifying for Congress this rather complex field of public policy. More likely than not, however, there will be an attempt to embarrass the administration in this matter. As an example, the question is asked, "To what extent does the proposal to make vehicles, launching and tracking facilities and technical services available to industry on a cost-reimbursable basis represent official executive branch policy? Has this been approved by higher authority?" We must do our best to answer these questions responsibly and effectively. It is the old story—no one makes progress without sticking out his neck. I don't expect that we have much to worry about in this situation, but I am reminded again of the effectiveness of the congressional committee staff operation in matters of this kind.

On Friday night, a letter came down from Lyndon Johnson asking for a summary report on Project Mercury with a discussion of any noteworthy problems that have arisen in the past few months and the steps being taken to meet them. Further, Lyndon wants a detailed schedule of all launches planned by NASA between now and the end of the calendar year with a clear indication as to whether any of these launches will involve live passengers—animal or human. Surely he cannot want this information for anything other than political purposes—perhaps to reassure himself that the administration is not going to pull a fast one on him and make a significant launch just before the election. As it stands, we will make one launch before the election—unless it is delayed—but it is a relatively routine scientific satellite on an old-style launch vehicle.[13] Thus is the political field being served in this political year.

Now, back to the work of the day. This was a rather full day with a meeting starting out at 9 o'clock with Hjornevik and Rosenthal on the budget. The boys have let this business get out of hand this year and we are in somewhat of a mess with the Budget Bureau. I hope we can retrieve our position before it becomes time for the final session with the director of the Bureau of the Budget but that remains to be

[13] This, presumably, refers to the launch of Explorer VIII by a four-stage Juno II launch vehicle on 3 November 1960. The satellite contained instrumentation for detailed measurements of the ionosphere, which it provided until 27 December 1960, when the transmitter ceased functioning. (Emme, *Aeronautics and Astronautics . . . 1915-1960*, pp. 130, 150.) On Juno II, see note 2 of Chapter Nine.

seen. At 10:30, a group came in to discuss policy matters in advance of a session I am to have with a writer for the *Saturday Evening Post* on Wednesday next. Mr. Spencer has seized this opportunity to write a think piece on the communications satellite business and wants information on policy formulation and operating aspects of this particular project. We spent an hour and a half on the subject. This would seem to indicate the rather complete and significant concern we have in attempting to unravel these particular policy issues.

I stayed at my desk for lunch so I could get on with some of the reading that I have not been able to do. A variety of people came in to see me but it wasn't until 2 o'clock that I was able to visit with John McKnight of the USIA. I have been trying to interest him in becoming the director of our office of public information but without success. At the same stand but one hour later, I visited with a Mr. Constantine who might be induced to come with us to handle our executive development problems. He seems a very sincere fellow and I think we must make every effort to get him to accept our offer.[14] About 4 o'clock, I became a little bit concerned about the fact that the White House knows less about this communications satellite business than it should. I was able to reach Jerry Persons and spent the better part of an hour bringing him up-to-date on my activities. There was no criticism made of our posture in this business and it was agreed that I should have a paper prepared for the cabinet by 11 November. I guess the important thing is to get on with the job—at least that's the way I'm going to play it.

At 6:30, I went to the British Embassy for a cocktail party in honor of the groups of Britishers who have come to this country to talk over communications satellite problems. It was not a particularly pleasant operation—I don't seem to enjoy these parties for I really have very little interest in getting to know the people concerned. It isn't like being part of an organization where you feel that your life is going to be inextricably involved with a particular group of people for a long time. Back home and a quick supper followed by an attempt to understand some of the budgetary materials I had brought home with me. Not being too successful, I turned to other reading material before going to bed.

Tuesday, October 25: The day started with a visit from Finley Carter, president of the Stanford Research Institute. He was bringing to me a proposal given him some days ago by a man who suggests that the first use of communications satellites should be to link all of the U.N. capitols together in one great communications network. The suggestion is a naive one but does give an indication of the international value of some of these activities. I settled with Finlay the nature of the speech I will give in San Francisco next week. It shouldn't be too bad. [A while back] I asked John Rubel, deputy director of defense research and engineering, to accompany our top staff to the west coast where we will meet on Friday with the top people in the Ballistic Missile Division of the Air Force. Apparently, Rubel's being

---

[14] No one by the name of Constantine appears in the NASA Headquarters telephone directories for the end of 1960 and the beginning of 1961. On the public information position, see note 9.

in the office of the secretary of defense makes him suspect among his military colleagues. After much cogitation, I called him and rescinded my invitation telling him simply that it was the belief of members of the staff that Air Force personnel would not speak as freely to us about their plans and problems were he to be present. This is one hell of a way to run a railroad!

A visit from a Bureau of the Budget management review team at 11:30 occupied about 45 minutes of discussion. Periodically, these reviews attempt to make certain that agencies are using modern management methods and are working on programs to improve their management. In some ways, this is a thankless task but I do believe the present effort might produce some useful results. Today's meeting of the Federal Council on Science and Technology was interesting enough. Most of the time was given over to a discussion of the report we have all been putting together for the National Security Council. The first draft, which is an attempt to condense into 31 pages the material contained in 350 draft pages, is less than satisfactory for any purpose. More importantly, should this paper fall into the hands of a new Democratic administration, it would really cause a sensation. Actually, it is an attempt to identify areas of importance for study in the years ahead, but it could be shown to be a statement of the failure of this administration in most of the areas identified. Since there was considerable disagreement as to the present format, it was suggested that a revision be made. I countered by suggesting strongly that Gordon Gray of the National Security Council be asked to read it and indicate whether he wanted it revised. My best estimate on this one is that it ought to be put in the furnace. We will see what happens.

Back to the office to clean up the desk and get ready for the dinner in honor of Walt Whitman. During this interval, we settled on the complement of people to make the trip to the West Coast on Friday and on the agenda for that meeting. The dinner for Walt Whitman given in his honor by Det Bronk, president of the National Academy of Sciences, was indeed a delightful affair. The "Great Hall" of the National Academy's building is a very high-ceilinged and domed room about 50 feet on a side. A square table—that is, tables formed so as to make a hollow square, seemed quite festive. The food was excellent and beautifully served, and the wines were very good. The program was limited to several very short and amusing talks given as a sort of welcoming ceremony for Whitman, who is the new science advisor to the secretary of state. All-in-all, it was one of the best such affairs that I have ever attended. Back home at 11:15 and off to bed with a sleeping pill.

Wednesday, October 26: It was a little hard to get up this morning. Probably a combination of sleeping pill and just being tired out, but I finally made it. A British commission of some twelve people came in to spend the day with us to talk over the commercial satellite research development field. Obviously, they are somewhat bewildered by the present turn of events and are not quite sure of whether the AT&T or NASA is really running this program. I hope I cleared [up] their minds on this matter. They are to make a visit to a good many places in the country, and then will come back to us early in November for a final discussion. We need very much to have them work with us since we need a terminal on that side of the ocean. At 10 o'clock, we started a full day's meeting with the space panel of the President's Science Advisory Committee. This was an exceedingly good session and the briefings had been very well prepared. Actually, Pat Hyland, president of Hughes Aircraft, said that this was the best organized [session] and best example of planning that he had seen in 15 years in the government. While this is probably an overstatement, it was satisfying to think that our efforts had been appreciated.

The briefing attempted to set forth all the programs that will use very large boosters, beginning with the Saturn class. It turns out that within the next ten years, we will spend at least $7 billion on vehicle development and on payloads and flight operations using this class of vehicles. There is really not much use doing this unless we are aiming at placing a man on the moon—a feat that everyone agrees will one day be accomplished. If we are to accomplish it, it appears that it will not be until some time in the mid-1970s, and our very wild estimates today would indicate a total cost for that particular exploit ranging from $14 to $35 billion. These are not decisions to be taken lightly.[15] During the course of the morning, Bob McKinney came in to visit with me briefly. He has just completed a report on atomic energy matters in Western Europe for the Joint Congressional Committee on Atomic Energy. He is an out-and-out Democrat and manages to know where some few of the bodies are buried. It is not clear to me, however, just how much weight he does pull within party circles. I worked with him on the McKinney panel on peaceful use of atomic energy back in 1955 and came to like him very much. He told me today that John McCone was complaining we were pricing the Rover program so high that it would naturally be killed. If this is a true report, it is about time that I had another argument with John McCone. I guess I really don't understand this man.

At 12:30, we entertained the British group for lunch at the Mayflower— a pleasant affair. I left the space panel briefing a little bit early to visit with Donald Douglas, president of Douglas Aircraft. He brought me a gold-plated model of the Thor-Agena launch vehicle. This is a replica of the vehicle that was used for the 100th launching of a Thor rocket—the one that carried Discoverer XV into orbit a few weeks ago.[16] Douglas also wanted to know how his firm could get more business. At 5:15, Louis Dunn came in to ask about the same problem. His firm [Ramo-Wooldridge's Space Technology Laboratories] has been identified with systems management for so long that it wants very badly to have a "hardware"

---

[15] As everyone knows, Neil A. Armstrong and Edwin E. Aldrin, Jr. landed on the moon on 20 July 1969 as part of the Apollo 11 mission, but this feat would not have been achieved so soon without President Kennedy's decision in May 1961 to land Americans on the moon within a decade.

[16] A Thor-Agena launch vehicle delivered Discoverer XV—a DOD satellite—into polar orbit on 13 September 1960. Its purpose was stated at the time as being to gather data on propulsion, communications, orbital performance, stabilization, and recovery techniques. Rough seas precluded recovery of the capsule from the ocean south of Hawaii on 15 September 1960. (Emme, *Aeronautics and Astronautics . . . 1915-1960*, pp. 127, 149.) For the then-classified goals of the Discoverer program, see note 10, Chapter Nine.

contract. I told him if Atlas-Able V was a success, I doubted he would have to worry much about additional business.[17] Off to the Washington Hotel to look in briefly at a cocktail party for Walt Bonney given by the Air Force Association.

Thursday, October 27: In yesterday's chronicle, I should have said that we have been "stewing" for the last two or three days over the proper kind of response to make to Lyndon Johnson on his request for information about our flight program and about the Mercury Project. We have decided to play this in a low key, offering to give him a much more complete study of the Mercury Project within the next two or three weeks. This will place the report in his hands after the election and thus prevent him from doing anything with it as a piece of campaign material. The schedules we are giving him will relieve his mind as to any possibility that the administration is planning some startling space shot as a clincher on the campaign arguments over the character and quality of the nation's space program. As I said to George Kistiakowsky today, "I'm not sure whether we are political scientists or scientific politicians."

The staff meeting today was a rather good one. Ray Einhorn, head of our audit division, gave a 30-minute discussion of the activities of his relatively small group of 17 people. It is an internal auditing organization—that is, it takes various aspects of our work and audits the methods, procedures, accuracy and validity of any particular area. The General Accounting Office, the watchdog over the executive branch that is responsible to Congress, probably has three or four times that number of auditors constantly working in our various offices and field centers. The discussion ran on so long that I had only a few minutes to visit with Henry Herzig of Dresser Industries in Dallas. He wanted the usual information about contracting, but I am afraid I was not as cordial as I might have been.

At 11 o'clock, Jim Fisk, president of Bell Laboratories and George Best, vice president of AT&T, came in to talk about the interests of the laboratories and the telephone company in communications satellites. This is the third or fourth such meeting, but this time I had asked Jim to set out in writing a program the telephone company would propose to undertake. This came to us on Monday and we were able to get it priced so that we could discuss rather frankly and factually the costs and probable schedules if the program were to be undertaken. Assisting me were Bob Nunn, Leonard Jaffe, Hugh Dryden, Bob Seamans and one or two others. It continues to be clear that the telephone company wants to go ahead with a program of this kind, but it is just as clear that it really doesn't understand the complexity, the costs, and the long lead time necessary to acquire a place on one of the firing

---

[17] The Able designation referred to a series of upper stages built by Space Technology Laboratories and derived from the Vanguard launch vehicle. There were four launches of the Able upper stages used in conjunction with the Atlas first stage, the last two of which bore the Able V designation. All four were unsuccessful, with the last one occurring on 15 December 1960. As a result, the Atlas-Able vehicle was retired without a successful launch. There had also been three earlier failures of the Thor-Able combination, but this vehicle also enjoyed some successes, including the launches of Pioneer V and Tiros I. See *NASA Historical Data Book*, Vol. II, pp. 34-35, 72-73; U.S. Congress, House Committee on Science and Astronautics, *A Chronology of Missile and Astronautic Events*, prepared by Dr. Charles S. Sheldon, II (Washington, D.C.: GPO, 1961), pp. 137, 164.

schedules and on the launch vehicle required. We continued our discussion through lunch and agreed to set up working groups in each organization that might meet together on occasion to be helpful to one another as the telephone company attempts to revise its thinking. We made it very clear to them that others were involved in this program and we proposed to move forward without further delay. This does not mean that their shots will be delayed themselves—rather, they will fit into the ongoing program at some proper point in the future.

Back at 1:30 for a presentation on the need for another Atlas launch pad at the Atlantic Missile Range. This is a critical situation. We have no back-up pad for Mercury or Centaur. The Air Force and NASA will use these pads jointly, and we agreed, therefore, to split the cost and to modify existing facilities so that we would have emergency capabilities available for each program. At 2 o'clock, Steven Spencer of the *Saturday Evening Post* came in to talk with me. Apparently, he has been writing an article on the communications satellite and had already talked to a great many of our people and some of the contractors such as telephone company people. I chatted with him for a time and found him a reasonable person although every once in a while he would strike out on some tack that gave evidence of a background of newspaper reporting. It will be interesting to see how his article comes out.

At 2:30, Siepert and Hjornevik came in to review the headquarters budget for FY 1962 as well as some of the allocations yet to be made in the FY 1961 budget. This was not too bad and I think we did a reasonable job for all concerned. Of course, no one will be satisfied. Immediately following this discussion, the group going to the coast tomorrow came in and we reviewed the agenda we will follow in our discussions with the Air Force people at the Ballistic Missile Division in Los Angeles. I got away as quickly as I could to prepare for the white tie dinner being given by the premier of Malaya for Secretary of State Christian Herter. This was not a particularly large party and Ruth and I found ourselves knowing only a relatively few people who were present. I took the wife of George Allen, head of the USIA, as my dinner partner. I doubt that the conversation could have been characterized as scintillating. Ruth was squired by General Lemnitzer, chairman of the Joint Chiefs of Staff and a really fine person. Fortunately, Jim Douglas, deputy secretary of defense and Mrs. Lemnitzer were seated across the table from me and thus I had a pretty good evening. After dinner, the premier read a short tribute to the United States in relatively good English and toasted the president. A response by Chris Herter and a toast to the Supreme Being in Malaya ended the formal part of the festivities. We stood around for a time having coffee and a liqueur and I had a chance to talk very briefly with Chris Herter.

Our Atlas-Able V, it will be remembered, failed to go into orbit.[18] The second stage actually broke up over Africa and some pieces apparently landed in or near Pretoria, the capitol of the Union of South Africa. It appears that these are pieces of the second stage and it would appear, also, that they did not burn up

---

[18] This was the intended launch of a Pioneer lunar probe on 25 September 1960.

because the stage had not achieved a sufficiently high velocity to cause vaporization of the material. We have been somewhat apprehensive of international incidents of this kind, and I spoke to Chris in an apologetic way. He said very promptly, "Go right ahead with your program. We will take care of any of the problems that arise. Don't worry about it at all." I was very much pleased at this statement but must admit to being somewhat startled by hearing Chris speak so positively.

Friday, October 28: Up at 6:15 in order to be on my way by 6:45 for a date at Andrews Air Force Base at 7:30. Eight or nine of us came together with General Schriever, commander of the Air Research and Development Command, and four or five of his people for a trip to Los Angeles where we planned a one-day conference. A lot of additional Air Force people went along for the ride—at least, they were not involved in our activities. We were in an Air Force 707 and took off promptly at 8 o'clock, arriving in Los Angeles at 9:15 Pacific Coast time—there is a four hour time change because we are still on daylight saving time in the east. We went right into conference and stayed with it throughout the day. I had wanted a briefing on the Discoverer series so that we might learn something of the problems they had encountered and the solutions they had found for these problems. Further there were other programs about which our top people have little or no knowledge and I wanted to get as much of this information as we might be able to get from the Air Force. Then there were several "interface" problems between NASA and the Air Force that I think must be talked out at the highest level. Actually, two or three of these evaporated almost immediately when they were brought up, by reason of the fact that both General Schriever and I could speak with finality in suggesting solutions. There just is no question but that these two organizations must learn to live with each other and to exchange information and cooperate effectively if we are to have any peace from the congressional committees. Even though we are able to work together, it is not clear that we will have that kind of peace, but at least we can try.

At 4:15, we gave up and headed back for the aircraft. We were airborne at 5:05 and touched down at Andrews Air Force Base near Washington at 1:45 on Saturday morning. Thus we had flown almost 6,000 miles, worked for four hours on the aircraft on the way out and somewhat more than eight hours in Los Angeles—all in one day. It is a crazy world!

Saturday, October 29: Polly had come down from Swarthmore yesterday afternoon so that I was pleased to see her this morning. She helped to make breakfast and we have been very leisurely in our activities today. I am still recovering from the lack of sleep and the pace of yesterday. We did go to *Sunrise at Campobello*, which I thought was quite good but for which I was not quite in the mood.[19] Tonight,

---

[19] Presumably this was the Warner Brothers movie written and produced by Dore Schary but possibly his play, upon which the movie was based. Both covered the story of Franklin Delano Roosevelt and his family during 1921-1924.

I am taking Polly to Costin's for roast beef. Then I hope to get into bed and get some more sleep.

Sunday, October 30: It was a good dinner and I think Polly enjoyed it. This day started out as a pleasant one—a day of rest—but it wound up as a day of work. Polly was studying a good bit and seemed anxious to convince us that this was a necessary part of her life on the weekends. I worked some in the morning and then helped Ruth prepare for a brunch at which we had Mr. and Mrs. Richard Scammon as our guests. Dick is a member of the staff of the Governmental Affairs Institute, a non-profit institution dealing in the matters its name would indicate.[20] Dick's particular specialty is election research. He has been an observer at elections in Israel, West Germany, Russia, etc. He is a Democrat—through and through—always on the attack. It was a right pleasant luncheon—one of Ruth's best—and the conversation was brisk and interesting. He was convinced that Kennedy would win although not by a very large margin.[21]

They left at 2:30 and Bonney came in at 3 o'clock to tell me that Astronaut Donald ["Deke"] Slayton was to have a gall bladder taken out within the week. We agreed on a very casual press release in order to play down the incident. At 4:30, Dr. Mesthene came in to have a drink and to talk over the report he has been preparing for the National Security Council. This is a compendium made up of ten or twelve reports prepared by agencies such as our own. I think ours has been judged the best of the lot but that is scant praise, in my opinion. Mesthene is a discouraged social scientist, a member of the staff of Rand and seemingly a good man.

Thus it goes—a day of rest becomes a day of relatively frenzied activity!

Monday, October 31: At 8:15, I received a quick briefing on the new things that have been happening at Huntsville and Cape Canaveral in anticipation of my trip to those locations tomorrow and Wednesday. Immediately thereafter, we started in on a program review of Project Mercury. This took all day and turned up lots of problems, but it is quite apparent that we have a good crew on this project and they are on top of those problems. Some of them may not respond to the kind of treatment we are able to give them but it will not be for want of trying. It appears that we may be able to fire two shots on 7 November—the day before election.[22] One of these will be a test shot from Wallops Island and the other a Mercury-Redstone

---

[20] Presumably this was Richard Montgomery Scammon.

[21] This prediction, of course, was accurate. Although Kennedy won handily in the electoral college by 303 to 219 electoral votes, the popular vote gave him only a very narrow victory over Vice President Nixon, 34, 221, 531 to 34, 108, 474, by one tally.

[22] There apparently were no launches on 7 November, but on 8 November a Little Joe 5, the first of a series carrying McDonnell production spacecraft for Project Mercury, launched from Wallops to check out the spacecraft in an abort that simulated the most severe anticipated launch conditions. Following a normal initial launch, the escape rocket motor ignited prematurely and the spacecraft failed to detach from the launch vehicle. It was destroyed upon impact. (James M. Grimwood, *Project Mercury: A Chronology* [Washington, D.C.: NASA SP-4001, 1963], p. 117.)

shot from Cape Canaveral.[23] Already, we are being accused of scheduling these two shots in order to influence the election—how silly can they be!

I had lunch with General Ostrander to settle my plans about Abe Hyatt. We then went back to the Mercury Project briefing which continued until 4:45 when Dryden, Seamans, Silverstein and I decided that we had better get going on the communications satellite program. Getting going, in this case, means determining that the Space Technology Laboratories will serve as our systems contractor and will manage the spacecraft contractor to be selected after a competition. At 5:30, Dr. J. M. English of UCLA came in to tell me about the course his university is preparing for engineering executives. Really what he wanted to do was to find out whether or not NASA would be interested in providing funds for UCLA. The weather has turned bad. We may have difficulty in making our schedule for tomorrow's trip. After much telephoning, it developed that the IBM plane would take off from New York at 7 o'clock in the morning, thus delaying us on the departure from Washington, but it is the best that can be done. I reached Fred Crawford and Ward Canaday and arranged to have breakfast with them.

---

[23] The Mercury-Redstone flight test at the Atlantic Missile Range was attempted on 21 November but had to be terminated before liftoff because of faulty ground-support circuitry. The launch occurred on 19 December 1960, when a modified Redstone booster launched an unmanned spacecraft in a suborbital trajectory. It impacted 235 miles downrange, having reached an altitude of 165 miles and a speed of almost 4,200 miles per hour. (Emme, *Aeronautics and Astronautics . . . 1915-1960*, pp. 131, 134.)

CHAPTER TWELVE

NOVEMBER 1960

$\mathbf{T}$uesday, November 1: I was up early to get Ruth packed so that she might take off for Cleveland without delay. I then picked up Ward [Canaday] and we had breakfast with Fred [Crawford] at the Carlton. When we arrived at Butler at 8:45, the IBM plane was ready and we took off immediately for the trip to Huntsville.[1] Fred talked over the arrangements for my return to Case while we were enroute. I think everything is in order and I have no complaints. About an hour before we reached Huntsville, I gathered everyone together to tell them something about our organization and what they might expect to see during the course of the next two days. Included in the party were the following: Tom Watson, president of IBM; Dick Watson, president of IBM International; A. L. Williams, executive vice president of IBM; Fred Crawford, chairman of the executive committee of TRW; Chuck Percy, president of Bell & Howell; Keith Funston, president of the New York Stock Exchange; Henry Heald, president of the Ford Foundation; Ward Canaday, president of Overland Corporation. We arrived at Huntsville at 11 o'clock their time and were met by von Braun. A briefing to the group detailed the progress on Saturn, and then we had lunch with General Schomburg. I had invited him to speak to the group if he desired, and he certainly made use of the opportunity. He told a very engaging story about the Nike-Zeus system.[2] He had the group eating out of his hands and most certainly, it would have ordered the expenditure of $10 billion without very much more nudging. An interesting example of how little some of our best people know about the facts of life in this crazy game.

---

[1] As Glennan stated in a letter, "The purpose of the visit is that of acquainting very important and responsible men with some of our activities in the space field." The visit was not only to Huntsville but to the launch facilities at Cape Canaveral. (T. Keith Glennan to Major General August Schomburg, Commanding General U.S. Army Ordnance Missile Command, 26 October 1960, Glennan subsection, NASA Historical Reference Collection.)

[2] Nike-Zeus was the first U.S. anti-missile missile. It was part of a complex antimissile system, on which the Western Electric Company had been working for the Army since 1955, that came to be called Nike-X in 1963. Nike-Zeus successfully intercepted an ICBM fired from Vandenberg AFB in 1962. When in September 1967 Secretary of Defense Robert F. McNamara decided to deploy a limited anti-ballistic missile defense system for the U.S. employing elements of the Nike-X, it was called Sentinel. And in 1969 when it was modified further, the name changed to Safeguard. (Fred S. Hoffman, "Space Missile-Killer Gets Go-Ahead," *Washington Post*, 29 September 1965 in "Army Nike-X" folder, NASA Historical Reference Collection; *A History of Engineering and Science in the Bell System: National Service in War and Peace (1925-1975)*, M. D. Fagen, ed. [Bell Telephone Laboratories, 1978], pp. 394, 410, 434, 462-463.)

After lunch, we went to the fabrication shop, the guidance and control laboratories, the computer center and finally to the test stand where the Saturn is installed. They are about two weeks away from a first test firing of the final flight configuration, but they did have a single engine mounted on another test stand that they were going to fire. This was timed very properly so that our group was able to witness a firing of 150 seconds. All were impressed! We were off at 5 o'clock for Cape Canaveral, having dinner aboard and arriving at the Cape at 10 o'clock. Our folks met us as did the commanding general [of the Air Force Missile Test Center and the Atlantic Missile Range, Major General Leighton I. "Lee"] Davis. We were taken to the Holiday Inn and immediately went to bed, although Fred and I took a half hour walk.

Wednesday, November 2: We were up at 6:30, had a quick breakfast and then off to the base to hear a very good briefing on the activities of the Atlantic Missile Range. Following this, we took the half-hour trip to the Cape where Dr. Debus and his staff gave us a briefing about the NASA activities at the Cape. After lunch with General Davis, we visited the Mercury capsule hanger, the Mercury control center, the Saturn pad and block house and one or two other spots. By this time the party had become very much intrigued by the business and, learning that there was a shot to be made at midnight, five of them decided to stay over. This was really a very good thing but Henry Heald and I managed to get a plane back at 5 o'clock from Orlando. Before departing the Cape, let me say that the Mercury control center is really a fine piece of design and logic. Displayed before your eyes is a map of the world with the orbits drawn on it and a moving Mercury capsule to indicate exactly where the capsule is at all times. As many as fifteen or twenty consoles are positioned in various parts of the room so that individuals may monitor various bits of information coming back from the tracking stations around the world and from the capsule itself. The immensity of this project really begins to be impressively demonstrated when one looks at a system of this sort. I should say that we are having strike troubles at the Saturn pad and have been held up six weeks through a jurisdictional argument. Very soon, we must take positive action to clear this. There was a time, not many years ago, when there was no trouble with unions at the Cape. Now it is an old operation and it is ridden with the same jurisdictional and wage disputes as any other operation.

Thursday, November 3: Up early to learn that Juno II went off at the appointed hour of midnight and is successfully in orbit with all channels working.[3] Once again, it appears to be a demonstrated fact that we have successes at the Cape when I am not present. If this would insure success, I would never go near the place.

---

[3] This was the launch of Explorer VIII by a four-stage Juno II launch vehicle. The Juno II consisted of an extended Jupiter missile as its first stage plus 2nd-4th upper stages consisting, respectively, of 11, 3, and 1 scaled-down Sergeant missiles. Built at Marshall and designed to study the ionosphere, Explorer VIII measured the influx rate of micrometeoroids and discovered layers of helium in the upper atmosphere. (*NASA Historical Data Book*, Vol. II, pp. 46-47, 238.)

At 10:20, I took off for Houston in a DC-7. There were only twelve people aboard and I had the front compartment all to myself. This was fine because I was able to work all the way to Houston. I was met at the airport by the acting president [of Rice University], Carey Croneis, Dean [of engineering, LeVan] Griffis and Howard Thompson. We repaired immediately to the campus where I spent two hours with Dean Griffis and several of the people who are working on research projects financed by NASA. I had an opportunity to drive about the campus, which has changed a great deal in the past twelve years. Our work is being supervised by Franz Brotzen, formerly of Case. I gave him his Ph. D. several years ago. He is apparently an excellent man, and I wish we had him back at Case in our center for materials sciences. The academic atmosphere at Rice is something to behold— indeed, it is very stimulating.

I drove on out to George Brown's home where I had a cup of coffee with Alice before dressing for the dinner with the Rice

At 12:23 a.m. EST on 3 November 1960, a Juno II launch vehicle lifted off its launching pad with the 90-pound Explorer VIII aboard.

University Associates.[4] They have a fine organization, copied after Case Associates, although their principal membership comes from individuals. I think I did reasonably well with the speech although I had real competition—Dick Nixon came to town on short notice. There were more than 200 at our dinner and most everyone was cordial and gracious. Back home with George and Alice and off to bed because I had to get up very early in the morning to take my plane to Dallas.

---

[4] George Brown was an engineer and corporate executive who was chairman of the board of trustees at Rice.

Friday, November 4: At 5:30 I was up and shaving for breakfast with George Brown. I talked with him about several of the Ford Foundation programs and promised to send him information about them. His chauffeur failed to appear, so George dressed very quickly and drove me to the airport. At Dallas, Neil Mallon, chairman of the board of Dresser Industries, met me. He has been trying to get me to agree to go on his board of directors. This I cannot do but I agreed to take a further look at the matter. After picking up a couple of boxes of candy, I took off for Washington where we arrived on time, about 2 o'clock in the afternoon. At 3 o'clock the British communications group came in after a two weeks trip around the United States. The members were very gracious and grateful for all that we had done for them but gave us very little in the way of solid information.

We then had a discussion of the means of acquiring the additional pad at Canaveral. The Air Force has decided it cannot take the bigger part of the costs so we have got to find $20 million while it finds $14 million. I think the involved parties are realistic about this and not attempting to get the better of us.

I was so very tired that I went directly home and managed to miss the party to say "goodbye" to Mr. Hiscocks of the British mission to this country.[5] I went to bed early.

Saturday, November 5: We spent all morning at the office working over the budget. This was in preparation for a discussion to be held on Monday with the budget wrecking crew consisting of Shapley, Schaub, and Cadle.[6] They're really a fine group of people and more often than not are clearly on our side.

I went home without lunch having asked Admiral Bennett to come over to watch the Army-Syracuse game with me. We had several drinks during the afternoon, which did us no harm. Homer Joe Stewart came in about 4:30 with the first copy of the new version of the 10-year plan. They left about 5 o'clock and I managed to put together a steak, a salad and some coffee before going off to bed.

Sunday, November 6: This was a day full of work, although I took a little time to watch the "pro" football game. I did it with due regard for atmosphere as I made myself a couple of hot dog sandwiches.

The [presidential] campaign is drawing to a close. We [at NASA] have felt very little in the way of political involvement except for blasts by Johnson and Anfuso accusing us of setting up the two Mercury shots for 7 November. Wouldn't it be funny if the shots went off on schedule and failed to achieve their objective? I suppose then I would hear from the Nixon camp. In any event, I am asking no one about this and we will go ahead as planned. This campaign has been an amazing one. How two men can stand the pace that has been set for so long a period of weeks I

---

[5] See note 3 of Chapter 8.

[6] William F. Schaub was chief of the military division at the Bureau of the Budget. Shapley (see biographical appendix) worked under him, and presumably Cadle did also. He may have been the Don D. Cadle who later went to work for NASA in the resource programming directorate.

just will never understand. Clearly, someone has to find a new answer to this problem. They can't say something new at every whistle stop, and with the television and radio, everyone of the whistle stops is reported to the entire country. Most people to whom we have talked recently have become quite fatigued and out of sorts with the whole operation. I think we still cling to a hope that Nixon will turn the trick. Certainly, he seems to have been closing fast these last few days and has gained a good bit—probably due, at least in part, to help from Ike.

Monday, November 7: The day started out with an 8:30 meeting of the National Security Council. I was there for the first item on the agenda, which was a briefing by John Rubel of the office of the secretary of defense. John gave an excellent picture of the space activities under the various elements of the Department of Defense. He brought into his discussion certain of the activities of NASA—all in an appropriate fashion. Finally, he discussed the effectiveness of our Aeronautics and Astronautics Coordinating Board. At this point, John McCone wrote a note and passed it to me. I kept it and am quoting it here: "In view of the joint NASA/DOD (ARPA [Advanced Research Projects Agency]) Management Committee, why not consider a merger of NASA, DOD (advanced projects—more than ARPA), AEC (provide full utilization of national labs), Science Advisor to the President into one agency." This is the second time that he has made a suggestion of this kind. It may be that there is more to the suggestion than I presently am able to see. In any event, John is losing no opportunity to push his ideas forward. I think he is a devoted person, but he operates in a peculiar and individualistic manner.

Immediately following the NSC meeting, I joined a group with the Bureau of the Budget and we spent the entire day in going over our program with the budget examiners. They are excellent people, able and tolerant but wise to the ways of an agency on the make. Getting back to the office a little late, I called Shelby Thompson in to discuss further with him the problem of filling the post being vacated by Walter Bonney. Shelby suggested that he might serve as an acting director while retaining his present post. This would serve to bridge the period during which a new administrator would be appointed and would give the new man an opportunity to appoint his own director of public information. This seems like an excellent idea and I believe I will buy it.

Home to a quick supper and then some work and off to bed to listen to the closing statements of each of the candidates. Once again, I'm afraid my prejudices show. It seemed to me that Ike and Lodge and Dick Nixon performed exceedingly well while Kennedy attempted to make a bit of a show of his final appearance. Well, tomorrow will tell!

Tuesday, November 8: Election Day! I was in early and called Shelby Thompson in to tell him that I was going ahead with the proposal that he had made the day before. I then talked to Joe Stein and told him that he would not be appointed to the directorship. We moved immediately into a program review that occupied us through lunch. At 3:00 p.m., I was off to San Francisco, arriving on time to have my first visit with my grandson, Keith III. He's a cute little fellow, healthy and happy and quite apparently, all boy. I didn't quite know how to act—it's been a long time since I had a two month old baby in my hands. Tom and Martha took me to Rickey's for dinner and we had an opportunity to get the first election returns before they left me for the night. It begins to look bad for Nixon. I was off to bed at 11 o'clock, which is 2:00 a.m. Washington time.

Wednesday, November 9: It seems to be all over. The *New York Times* conceded very early that Kennedy was the president. It has been a close race and thus far Nixon has not completely given up, apparently. I listened to a good many men who came into the dining room after me. They were all making jokes about the election. Obviously, they were Republican partisans and were making the best of the apparent loss. One of them said, "I don't know what we're going to work for today; Kennedy says we'll all be paid anyway." They turned immediately to business, however.

Smitty De France picked me up and took me to the Ames Research Center. I visited with him for a few moments and then spoke to the senior members of the staff and presented a forty-year service pin to Smitty. While at the center, I called Hugh Dryden but was too early to catch him before he went into the cabinet meeting. This was one meeting I was sorry to miss. The president had called it for 11 o'clock, presumably to discuss the transition to a new administration. Smitty took me into San Francisco, where I met the people from the Stanford Research Institute and spoke at their luncheon. Dean Taggart of the New York University School of Business spoke during the morning session and stated that we were in a slight recession, which he thought would extend for at least eight months. I gave my usual roundup of NASA business and it seemed to be very well received. There were 450 businessmen there—an excellent audience.

As I left the hotel to go back to Tom's apartment, I was handed a bag containing a bottle of liquor, a very nice memorandum pad and a beautiful alligator wallet. I reached Tom's apartment about 3:30—he was not yet home from school.[7] Martha and I had a good talk together, and when he came in we enjoyed a martini while Martha cooked up a wonderful dinner. She gave us Cornish game hens, what she called a "poor man's wild rice" and a fresh salad with green goddess dressing, finishing up with Camembert and fruit. I must say that my children seem to be able to put out really good food when they try. We had excellent conversation and then off to the airport where they left me. My plane was delayed an hour but we arrived in Baltimore the next morning less than a half hour late. I had taken two sleeping pills and managed to get a fair amount of sleep.

Thursday, November 10: [The] staff [meeting] at 8:30 was almost all given over to discussion of the results of the election. I had talked to Hugh Dryden

---

[7] He was working on a Ph.D. in economics at Stanford after graduating from Swarthmore in electrical engineering and taking a masters degree in industrial management at M.I.T.

who felt very badly about the cabinet meeting. Apparently, the president was deeply hurt by the results of the election and has stiffened in his determination to have a balanced budget, come what may. It is easy to see how the extremes of expression during the course of an election campaign can cut deeply. I do believe that Ike will be his old self in a few days, but I sympathize with his feelings today.

At 9:30, I settled with Ostrander and Hyatt the organizational changes that will bring Hyatt into my office as director of the office of program [planning] and evaluation. Then I talked to both Joe Stein and Walter Bonney to tell them what I was going to do with respect to the temporary headship of the public information office. Neither one seemed very happy. At 10 o'clock I called a meeting of the staff of the public information office and gave it a blunt and straightforward story. I was very much pleased with the general attitude and believe that, once the shock is over, the staff will perform in a very satisfactory manner. I am sure the individuals there would not have enjoyed having Shelby Thompson as their permanent boss. They do seem to respect him but he is not the same sort of easy-going good fellow that Walter has been. On the other hand, Shelby does a much better job than Walter ever did.

At 10:30 Frutkin came in to propose a cocktail party for the group of men who are working with us on post-doctoral fellowships or those from other countries. We agreed on a date of 9 December. Later in the day, I had a talk with Jim Gleason about the desirability of sending congratulatory letters to those members of our committee who had gone through the election and won. He suggested that we hold off a bit on this. At 3:15 I asked Shapley from the Bureau of the Budget to come over in an attempt to get from him some ideas of the level to which the BOB intends to cut our budget. I guessed at $1.15 billion—we now stand at $1.4 billion. Understandably I got no information. At 4 o'clock I had a long talk with Bob Nunn about his proposal to build a launch complex from which commercial satellites could be launched. He has become enamored of this idea, and it was a little difficult to make him understand just how impossible it would be to undertake such a project. He is a good man, however, and I am sure will go forward without prejudice.

Friday, November 11: This is a holiday. I had not expected it to be and thus had set up a breakfast with General Persons in the White House. Persons has been appointed by the president to be responsible for developing the liaison with the incoming administration. This was not a bad meeting. I told him what we planned to do and he approved of it. This will consist largely of providing an incoming administrator with memoranda defining our organizational philosophy and the interface relationships, both within and outside the agency. I asked Jerry to set up a meeting with Stans to talk over our budget. I've always found it is better to attack than to wait until the attack is carried to us. Returning to the office about 9 o'clock, I spent the morning preparing the speech I am to give in New York next Wednesday. At noon, I went home, worked some more on the speech, and had a long walk with Ruth. We went to the Army and Navy Club at 6 o'clock to attend a surprise party given by Captain Engleman for his wife. I had known Chris a good many years ago but had never met his wife. It seemed a good idea to have a pheasant dinner at La Salle du Bois where we had a fine meal with wine. We then moved on to the Cosmos Club to attend the Swarthmore Concert and to pick up the boy friend of one of Polly's girl friends. David Harr is a Swarthmore graduate of the class of 1960 and is presently doing his six months' turn in the army. Cay Hall, from Cleveland, one of Polly's classmates, has been going with him and he had come in from Fort Knox, Kentucky, for the weekend. We provided them with a car and I guess they had a good time.

Saturday, November 12: More speech writing after taking Dave to the appointed place so that he could take the bus back to Swarthmore with Cay. Again, Ruth and I had a good walk and I spent most of the afternoon looking at a football game.

Sunday, November 13: This was a lazy day with just enough work to keep me from becoming bored. Again, Ruth and I had a long walk and then Kent and Estelle Van Horn came in for dinner.[8] Ruth had one of her usual excellent meals— this time a couple of pheasants I had shot last year. The conversation was good and we exchanged viewpoints on education, politics, grandchildren and children.

Monday, November 14: This was a fairly full day. Several discussions with staff members in the morning were followed by an interview with Howard Simons of *Think* magazine. This is the IBM magazine, which is fairly widely distributed. It was not a bad interview. In fact, I rather enjoyed it. In the early afternoon I talked with Shelby Thompson about the release announcing his appointment and Bonney's departure. Of more importance to Shelby was the fact that he is becoming worried about money. For some reason, we have not cranked him into our budgetary operation. I just don't understand this but we certainly will have to improve in this department next year. I'll ask him to put the figures together and we will have another look.

A man by the name of Marvin Robinson came in to see me. Frutkin is interested in him. Robinson seems to be a sensible person but has very little apparent drive. Evidently, he has had a good bit of experience in working with industry in other countries, and it may well be that he will turn out to be a good second man [in the office of international programs].[9] Then came Louis Kraar of the *Wall Street Journal* and with him John Spivak who turned out to be the son of [Lawrence] Spivak on "Meet the Press." This, too, was a rather pleasant interview. It will be interesting to see what comes of it. At 4:15, John Johnson came in and we talked about the impending strike at Canaveral. As a matter of fact, it appears that the

---

[8] Kent Van Horn had a Ph.D. from Yale (1929) and was a research metallurgist with the Aluminum Company of America, having been its director of research since 1952. He and his wife were obviously friends of the Glennans.

[9] In fact, Robinson appears in the March 1961 Headquarters telephone directory as a member of that office, headed by Frutkin. He resigned his post as deputy director of the office in August 1963 to serve as scientific secretary to the Committee on the Peaceful Uses of Outer Space of the U.N. ("Marvin W. Robinson," miscellaneous NASA biographical files and telephone directory, NASA Historical Reference Collection.)

electricians have decided not to work since we have gone ahead with our plan to have government employees handle the installation in the Saturn block house. We will be prepared to seek an injunction without delay. While we were in conference, word came that the strike is on. At 4:30, Bonney came in to say good-bye and I wished him well and do hope that he finds satisfaction in his new job.

One or two more meetings and it was time for me to go to get dressed for dinner with John and Mrs. McCone. We found that we were joining General and Mrs. Schriever and Cardinal [James Francis Aloysius] McIntyre. The Cardinal is a pleasant fellow and we had quite a good evening. However, I was somewhat shocked by the complete lack of understanding on the part of Mrs. McCone of the real problems in our provision of economic aid to other countries.

Tuesday, November 15: This morning, Hugh and I spent three hours with the President's Science Advisory Committee. We listened first to a report on Nike-Zeus in which Herb York participated. The story was somewhat different from the one given by General Schomburg in Huntsville a few days ago. Clearly this is not an easy matter to deal with and most significantly so when so much emotion is involved. The second part of the session was given over to the report of the ad hoc man-in-space panel, which had heard the presentation given by our people two or three weeks ago. The members of the panel were very flattering in their remarks about that presentation, and the discussion was brisk and pertinent. Obviously, these scientists are not interested in devoting large amounts of the national treasure and the national manpower to putting a man on the moon. In this, they are not alone. I am convinced that this will be done one day, but I am not at all certain that it is a matter of prime importance. When one starts to talk about the prestige of the United States resting on the question, "When do we get a man on the moon?," it seems clear that all sense of perspective has gone out the window. Clearly, with the probability that at least ten years must elapse before we can accomplish the feat of putting a man on the moon, the leadership and stature of the United States will no longer be in question. Either we will be the leader or we will not!

Hugh and I went to the White House mess for lunch and then I came back to have further talk with [Robert] Nunn about the communications satellite business. A meeting is being arranged for later this week so a group of us can talk over the proposed paper for the cabinet and other matters that are pertinent to this project. At 1:30 Mr. S. Gacki of "Radio Free Europe," who seems to be the senior Polish editor for that organization and a Mr. Smialowski came in to record an interview for broadcast over their network. The entire process took only fifteen minutes. Either I am getting used to these interviews or they are becoming more routine and easier for me. At 2:00, Abe Silverstein and I took off for the Goddard Space Flight Center. This was my first visit to the center since the buildings have been occupied.[10] They are really quite crowded but when the entire center is

---

[10] Construction at Goddard began on 24 April 1959, and by September 1960, building 1 had been fully occupied and other buildings were well on their way to completion, although personnel for the center were still widely scattered from Anacostia in the District of Columbia to Silver Spring, Maryland. (Rosenthal, *Venture into Space*, p. 31.)

completed, they are going to have a fine research laboratory. It was good to go into the many laboratories and see the hardware being prepared for flight sometime during 1961. There is a good crew at Goddard—eager and well qualified if I am a good judge in these matters. As we drove back, Abe spoke again of his concern over the way he had "blown his stack" at JPL a couple of weeks ago. We took counsel as to how he might regain some of the ground he thinks he has lost. Once again, I am impressed with the sincerity and obvious energy of Abe Silverstein.

Wednesday, November 16: We had staff meeting early this morning because I will not be in tomorrow at the regular hour. At 9:45, Hugh Dryden and I met with General Persons and Maury Stans of the Budget Bureau. Two or three of Stans' assistants were present. We discussed generally the problems of the NASA budget and the president's determination that a balanced budget will be presented to Congress. Obviously, we are going to have to come down somewhat, and there will be an attempt to avoid a supplemental. It is pleasant, however, to find that we can talk together on matters of this kind without rancor or unnecessary strain.

At 10:45, [Assistant Director] J. D. McKenney of JPL came in to voice his concern over the possibility that a new administration would throw sand in the gears. He was concerned that a new administrator might not understand the operating mechanisms that have been established and the importance of continuing with the project manager setup we have established. I reassured him. I don't know quite what his real problem might have been.

At 11 o'clock, Hugh and I met with the directors of the various offices and went over the discussion we had had earlier with Stans. I asked the boys to come back on Friday afternoon with a cut of at least $150 million and with an indication of what they could do beyond that figure. While their faces were a bit long, they took the news in good spirit and indicated that they had been through this kind of an exercise before. At 1:45, I left for New York on Northeast Airlines Flight #118. At 6 o'clock I spoke to the International Conference on Magnetism and Magnetic Materials at the Hotel New Yorker. I think my paper was a good one for the conference although I am sure that it will make no headlines. It was a strange group—a very highly trained and intense group of research and development people—in that I seemed to know only two or three people among the three or four hundred present. I guess I just don't travel in a magnetic path.

Thursday, November 17: Up at 6:15 to catch a 7:25 plane back to Washington. I managed to get a little work done on the plane and then moved into a meeting at 9:30 immediately upon my return. This was a discussion of the legislative program we will present to Congress in January. There wasn't too much argument about it but I did enjoy resisting the pressure of certain members of the staff who were desirous of introducing legislation to give the administrator authority to pay salaries that are competitive with industry. When one considers that we are the best paid agency in government, and when one recognizes that congressmen are paid only $22,500, it would seem wise to let well enough alone. The human being is an acquisitive animal, however, and I guess I shouldn't be too critical of these fine members of my staff. We are going to introduce the same legislative package we presented last year—much of it taken directly from the bill that passed the House. It will be remembered that the Senate took no action on that bill, so we have to go through the whole fight all over again. I hope we can do as well this time as we did last time on the patent clauses.

I took lunch with Admiral Arleigh Burke and Admiral Hayward at the Pentagon. It seemed to be a purely social affair. We did a lot of guessing about the appointments yet to be made by Kennedy. Admiral Burke offered to bet me that I would remain in Washington for at least another six months. That will be the day! The rest of the afternoon was taken up with a variety of appointments, none of which seems sufficiently important to chronicle here.

Friday, November 18: The morning was a rather leisurely one—it seems to be the calm before the storm—but I managed to get some work done. At 11 o'clock Mrs. Ruth Brod who is known on [ABC-TV's] "College News Conference" as Miss Ruth Hagy came in to talk over the TV program on which I am to appear on Sunday. Apparently, the program is being taken off the air and I am to take part in the last of this nine-and-a-half-year-old series. She is a very pleasant person and we agreed on several questions to be asked. It is a pleasure to have someone want to know what should be discussed. At 12:15, I walked over to the White House mess and had a light lunch.

Back at the office, we had the meeting on the communications satellite paper for the cabinet. This turned out to be a brisk discussion and one in which I think I finally made the boys understand my attitude toward the development of this program. I am determined to get it on track before I leave NASA. There seems to be great fear—perhaps well founded—that we will be accused of avoiding competition. I think we can set up a program where competition will exist but where those who want to take the risk—in this instance, AT&T—will be given a real chance to move forward. At 3 o'clock, we started on the budget session and it lasted until six. The boys came in with a cut of $150 million, and I think we will be prepared to talk turkey with Stans. To go beyond this cut will mean taking some significant projects entirely out of the program. This can be done but I doubt the wisdom of such a move. On the other hand, when the boss is determined, it is clear to me that he will give us strict orders to comply. If this turns out to be the case, I will have my say and then will carry out his instructions.

Saturday, November 19: There is nothing much to record about the happenings of this day. I read a bit in the morning, walked with Ruth, and then watched the Yale-Harvard football game. Yale won going away. In the midst of it Kent Smith called to say that he would not be able to make the Case Club dinner tonight. I am very sorry about this—not because of the Case Club dinner but because Kent and I were going to talk at some length with John Hrones on the morrow. The dinner went very well indeed, there being about seventy in attendance. I gave a most inadequate talk—I don't know what was wrong with me—and the club gave Ruth and myself a beautiful Steuben vase. This was a great surprise to me and I suspect that Frank Gregory, the president of the club and a negro graduate of Case in the class of 1928, had much to do with this action. He is assistant superintendent of schools in the city of Washington and is a truly fine American. All-in-all, it was an interesting and pleasant evening.

Sunday, November 20: John Hrones came out for breakfast at 9 o'clock and we worked until 12:45. We made a good bit of progress in dealing with organizational and other matters of immediate concern at Case. It is too bad that Kent wasn't here for we could have dealt with some problems that must be discussed when the three of us are together. Perhaps this can be done next Tuesday. At 1 o'clock Ruth and I went to the studio and prepared for the "College News Conference." The five students were wonderful and I thoroughly enjoyed the entire operation. Ruth seemed to think I did well. It is always difficult for the participant to know exactly what the reaction of a television audience will be, but I do believe that I was responsive to the questions and I hope I did violence to no particular person or organization in government.

Back to the apartment and John and I spent another hour in going over matters requiring attention. Among these, a Ford Foundation proposal and the question of advancements in rank for several persons on the campus seemed most important. It is a pleasure to work with John. He is aggressive, demanding and full of energy. He is quite willing, however, to take suggestions and I know that our association when I return to the campus is going to be a very productive one. We took John to the airport and then Ruth and I went to the Carriage House for dinner. This was the 353rd monthly anniversary of our wedding and we celebrated with champagne cocktails and a very good dinner. Back to the house and early to bed although we spent—that is, I spent—about two hours watching television shows. What a waste of time!

Monday, November 21: Let me say a bit about the situation in Washington as I see it today. There is a kind of a hush over the whole scene. Much speculation goes on about the appointments Mr. Kennedy will make. There seems to be sort of an apprehension on the part of many of the members of the administration— I suppose this is natural. As for NASA, we are preparing the necessary briefings to give authorized people a good picture of our operation. I find that many of my staff are coming in to urge me to reconsider and stay on. I think they are sincere although it is wholly possible that they are convinced that it is better to stay with a man they know than to chance an operation with a person whose convictions and methods are unknown to them. Mr. Eisenhower has given every appearance that he is going down to the end of the wire with a firm hand on the throttle. This week he has ordered heavy restrictions on spending in foreign countries, including a drastic reduction in the number of dependents of servicemen who will be maintained abroad. Our good friend Brucker immediately opened his mouth and spoke in seeming opposition to the president's order. What a ham! Ike also ordered the Atlantic fleet to patrol the waters between Cuba and Central America. The purpose is to prevent the transport of invasion forces or arms to "rebels" in

Guatemala and Nicaragua. All of these moves, drastic as they may be, seem to meet with approval in the press and in the circles in which I move. Government is a complex mechanism and the man who sits in the White House carries a really heavy load.

This appears to be the beginning of a very tough week in spite of the fact that Thanksgiving will be a holiday. We are approaching the "moment of truth" with respect to our budget. The labor problem at Cape Canaveral is not yet settled. Nothing is known of the intentions of the new administration with respect to our agency and program. We have two shots coming up this week. Mercury-Redstone I was fired this morning without success. Something happened in the Redstone launch vehicle, which triggered off the escape power while the rocket engine was shut down and the capsule remained intact. On the bright side is the fact that the capsule can be used again, and it is probable that the Redstone can as well. Naturally, the press will beat us over the head and forecast failure for the entire program. Siepert and Rosenthal came in for a brief discussion of the budget and of a change in their organizational alignments. At 10 o'clock, I talked with Dr. Kistiakowsky about our budget. I found him resigned to the inevitable—that President Eisenhower is going to balance the budget, come hell or high water. George seems to feel that our program will not be supported as generously as we would like, but he does feel that there is a high probability that the incoming administration may increase the funding with the usual blasts at "false economy" on the part of the outgoing administration.

While with Dr. Kistiakowsky I had a chance to talk with [Jerome] Wiesner of MIT, who has been science advisor to Kennedy and is spoken of as the probable successor to Dr. Kistiakowsky.[11] I probed as much as I could with respect to the appointment calendar for new men in jobs such as mine. I was asked whether or not I would stay; the answer was no. Later in the day Dr. Wiesner called me to ask whether I would accept any other job in the administration, and again I said no. I did express my willingness to be available at a somewhat later date should the need arise. I expressed it in the sense that I felt strongly about individuals accepting responsibility in Washington under appropriate circumstances. Having done this on two occasions, I am sure they know what appropriate circumstances mean.

Frutkin came in to report about his trip to Europe. Apparently, he received excellent assurances from the British, the French and the Spanish with respect to their providing assistance in the Mercury recovery operation. We talked about a deputy for Frutkin, raising some questions in his mind about the capabilities of the man he has in mind. Seamans was gone all last week so he and I had lunch to discuss the results of his visit to a variety of industrial plants on the West Coast. Later in the day, Mr. Shapley of the Bureau of the Budget came over to review with me the progress we had made in meeting the desires of the bureau. He is a very friendly

---

[11] Wiesner was in fact special assistant to the president for science and technology from 1961-1964.

and capable person and is mindful of the necessity for avoiding promises or guesses as to the outcome of budgetary negotiations. We have been having a continuing problem with respect to the determination of the size of operation we want in the area of technical information and educational assistance. The budget has fluctuated back and forth until it is time to get down to business with it. Accordingly, Shelby Thompson, Mel Day, Seamans, Dryden, Siepert and Rosenthal came in at 5 o'clock for a discussion. It is quite apparent that we cannot afford the program Shelby would like to undertake. On the other hand, we ought to be building toward that goal and I asked that they come in with a program that will approximate the one Shelby now wants but with accomplishment dates set over two or three years rather than in the present budget year.

  Tuesday, November 22: More discussions during the morning about budget but nothing much of interest occurred. At 11:30, I left for Cleveland where I expected to spend some time on the Case campus. I was met by John Hrones and Kent Smith and we immediately started the discussions, which carried on until 4 o'clock. Decisions were taken with respect to the Ford Foundation proposal and several other matters of importance. As the time approaches when I will be back on the campus, I find it necessary to take a more definitive role in the decision-making that appears to be necessary.

  Wednesday, November 23: Up very early and off to Washington on the 8 o'clock flight. I had set up a meeting with Dryden, Seamans, Silverstein, Hyatt and Cortright to discuss the budget situation and some major problem areas. These are the areas where decisions must be made or where the deferral of a decision to the next administration is made to appear rational and sound. We talked about Project Apollo—the follow-on to Project Mercury—the C-2 version of Saturn, the Rover program, the ultimate assignment of a home for the manned space flight program, etc.[12] I found the comments of the group very helpful and achieved a degree of unanimity on my proposals that was gratifying. A Mr. William Neumeyer came in at 11:30 for a visit at my request. He seems to be a good man and I have referred him to Arnold Frutkin as a possible candidate for the job I spoke of last Monday. Lunch with Mr. [T. Roland] Berner, president of the Curtiss-Wright Corporation, and Garry Norton [president of the Institute for Defense Analysis] at the Mayflower was a rather dull affair. It appeared that Berner had asked us to come to lunch so that he might invite us to become members of the board of the National Planning Association. This was dealt with without very much loss of time and I dashed back to the office to meet with a Negro—by name, Dr. Ambrose Calever—who is the president of the Adult Education Association of the United States. Shelby Thompson picked up the ball for me and discussed our possible involvement with

---

[12] C-2 was one of three interim configurations of Saturn between the C-1 (renamed Saturn I) and the C-5 (redesignated Saturn V). C-2 actually consisted of two sub-configurations, one with four stages and the other with three, but it, the C-3, and the C-4 were each cancelled in turn for a more powerful launch vehicle. (Brooks, Grimwood, Swenson, *Chariots for Apollo*, p. 47.)

the work of the Adult Education Association. Certainly, we ought to be cooperative and will probably have to provide materials just as we are for other levels of education. One cannot have an enlightened electorate without this sort of educational activity.

At long last, Shelby Thompson and Bob Seamans got together with me on the Columbia Broadcasting TV series we have been discussing for several weeks. It appears to be difficult to make decisions that stick on matters of this kind. Shelby is very much afraid that he cannot withstand the pressure of the NBC group in seeking out opportunities for documentary programs that will conflict with the work we would be doing with CBS. I finally gave the problem back to Shelby and asked him to come up with a proper answer. I managed to keep very busy until time to go home to see Sally, Polly and Fred Watts who were to be with us over the Thanksgiving holiday. We had a pleasant evening with a good dinner followed by a walk. I think we all needed the walk in order to get in shape for the morrow.

Thursday, November 24: Thanksgiving Day! I neglected to remark yesterday on the fact that we are in trouble on our labor negotiations at Cape Canaveral. Our own Civil Service employees have gone back to work and the craft and industrial unions are beginning to walk off the job. This would be all right if it were only our own job but they are spreading the walk-out to other construction and installation jobs being done for the Air Force. Naturally, this puts pressure on us because the Air Force does not want to become the victim of a secondary strike of this kind. At 9 o'clock, Seamans, Johnson, Sohier and Siepert came to the apartment for an hour's discussion of the problem. It appeared that there was a possible solution available to us. This would require us to employ union men directly on the government payroll, brigading them with our own employees who would then simply supervise the union men. The Civil Service Commission may have something to say about this and we finally decided that we had better call the people from Huntsville to see us tomorrow morning.

During this conference, the kids were getting up and by-passing the living room by using the hallway between the kitchen and main apartment doors. Various noises were issuing from the kitchen as they rather loudly demanded their rights with respect to breakfast. Finally, the gang left and we did finish up the breakfast at 10:30. Immediately, I set to the task of making a gallon of whiskey sours and by 11 o'clock we were on the way to Harper's Ferry and the Blue Ridge Rod and Gun Club where the Adams clan will gather today for the Thanksgiving dinner. Fifty persons, thirty adults and twenty children, sat down to dinner. It was fun and everyone seemed to be enjoying themselves. We arrived home about 6:30 without incident. I tried to get a little bit of work done before the Friday sessions but was not too successful.

Friday, November 25: What a day! At 8:15, the Mercury Project group came in to discuss the posture to be taken in the press conference we are holding to explain the failure of the Mercury-Redstone shot on Monday last. They did an excellent job in the press conference and we did actually seem to make some impression on the newspaper people. Naturally, the headlines were very much smaller as the explanation of a failure—great headlines are only good when one can say there has been a failure. I forgot to mention on Wednesday that Tiros II was launched on schedule and seems to be performing well except for the wide-angle camera. Some electrical interference is present—it may clear up later but we are a bit crushed by this failure. Actually, all of the other elements in the satellite are working very well and will give us a great deal of information. The shot should be termed about 95% successful.[13]

At 9 o'clock, we began a discussion of the long-range plan. This took the combined efforts of all concerned through 1 o'clock. The plan is in very much better shape than it was a year ago and I think we will have a useful document. At 11 o'clock, I broke away from the group with Bob Seamans and we sat down with Al Siepert and Kurt Debus to discuss the labor problem. After an hour's discussion, it appeared to me necessary that we go forward with the injunction, thus reversing the decision we had made on Thanksgiving Day. Apparently, discussions with the Civil Service Commission had been fruitful and we can count on real help there. The introduction of the Labor Department into the picture with Secretary Mitchell taking a hand has brought new elements into the controversy. We did say that we would go forward with the request for an injunction through the National Labor Relations Board and so advised the mediator, Mr. Finegan, and Secretary of Labor Mitchell.

At 2:30, Dryden and I met with Stans at the Bureau of the Budget. He was shocked at our unwillingness to reduce our figure to the extent he had asked. He was willing to listen, however, and so was I. We ended up by agreeing to go back and take another hard look. There is now about $170 million dollars between us and I hope that I can buy peace with a reduction of perhaps half of that. We have set a meeting for Saturday morning to deal with this problem. I dashed directly from the Budget Bureau meeting to the Department of Labor where I found the boys in session with Jim Mitchell. He wound up the meeting by suggesting that he talk with the international presidents of the plumbers and electrical workers' unions on Saturday morning. This seemed a good solution because Jim was agreeable to our going ahead with the injunction if he did not obtain a useful result from his discussion with these union representatives.[14]

---

13 The Tiros II weather satellite had two television cameras, one with a wide-angle and the other with a narrow-angle lens. Because of defocusing of the wide-angle lens, the pictures were not of as high a quality as had been those of Tiros I, but the satellite collected much useful data through its infrared experiment and the cameras transmitted many useful cloud pictures despite the defocusing. (*NASA Historical Data Book*, Vol. II, p. 353; *NASA Space Missions since 1958*, pp. 36-37.)

14 According to the *Washington Daily News*, 28 Nov. 1960, the negotiations were at least temporarily successful. In line with what Glennan states below in the diary, it announced that construction work on launch facilities for the Saturn "missile" would resume that day following partial settlement of a labor dispute. The International Brotherhood of Electrical Workers and the United Association of Journeymen and Apprentices of the Plumbing and Pipe Fitting Industry agreed to have Secretary Mitchell name a committee to recommend a permanent settlement of the dispute. (Reprinted in NASA *Current News*, 29 Nov. 1960.)

Right away back to the office where I met with Dr. Reichelderfer of the Weather Bureau and General Yates of the Department of Defense. We were in a controversy over the best method of organizing to determine the characteristics to be built into a weather satellite beyond the present Tiros and Nimbus series. Hugh Dryden had been talking with them and they had developed a compromise solution that I bought without delay. Thus ends another day—or at least, part of a day. At home, I found Ruth about ready for dinner with friends of Polly and Fred. It was a pleasant dinner party and the kids went off to the movies about 9 o'clock.

Saturday, November 26: Down to the office at 9:30 for the budget meeting. We carried this on through lunch and managed to strip out about $75 million. I believe the budget examiners agree that we have done a sensible job and I am at the point where I will do no more without an audience with the president. I hope we can avoid this and that Stans will agree to the proposal we now have, which would bring us a budget of about $1.16 billion for FY 1962.[15] In the midst of the budget sessions, Jim Mitchell called to say that he had worked out a compromise whereby the men would all go back to work at the Cape on Monday and a fact-finding committee would be appointed to look into the basic issues in the dispute. I bought this immediately as a reasonable solution and will proceed with setting up our representation on that committee. A visit to the American Airlines ticket office relieved me of $1,311 for aircraft transportation for the family to the West Coast over the Christmas holidays. The girl said this was the largest single check she had ever taken in there. I don't wonder. Following this bit of business, I met the children and Ruth at the theater where we saw *Little Moon over Alban*. It was played beautifully by Julie Harris and company. Both Sally and I thought it could have ended a little bit earlier but all agreed that it was good theater.

Sunday, November 27: Ruth and I took a long walk this morning and are now in the process of building a luncheon for the kids before they return to Swarthmore. Sally and I will take off this afternoon for Cleveland where I am to spend some time on the Case campus tomorrow morning and to go hunting with Ab Higley in the afternoon. During the course of the day, many telephone calls resulted in our getting set on the labor negotiation problem for tomorrow. There is no rest for the wicked in these jobs!

Monday, November 28: This was another very full day but in a slightly different way. Having taken Sally to Hathaway-Brown at 8 o'clock, I went directly down to Case to await the arrival of Kent. [Describes various pieces of Case business.] Returning to the house, I heated up a bowl of soup and waited for Ab Higley to pick me up. At 1:15, we were on our way, [describes hunting trip on which the party bagged eight pheasants plus two that the dogs caught].

---

[15] In fact, the January 1961 Eisenhower budget submission for FY 1962 was roughly $1.11 billion. With amendments, supplemental appropriations, etc. by the Kennedy administration, the final adjusted appropriations for FY 1962 came to roughly $1.82 billion. (*NASA Historical Data Book*, Vol. I, p. 138.)

Tuesday, November 29: Up at 7:30, and off to the hunt with Ab and Doc about 9 o'clock. We shot quail, chukkers partridge and pheasant until noon. It was a windy and slightly chilly day but everything went well. I managed to hit a few birds although this was my first attempt at anything other than pheasant. I must say that I enjoy watching the dogs work as much as shooting—perhaps a little more. After a quick lunch, Ab, Birkett and I drove back to the city for the annual meeting of the Case Board of Trustees. Birkett went on down to his place of business, of course. The trustees' meeting was an exceedingly good one. [Describes it.] Kent took me to the airport and we managed to talk over three or four more items before time for my departure. We were about an hour late and Ruth met me in Washington.

Wednesday, November 30: This was another very full day. At 7:15, I picked up Bob Nunn and we walked over to the White House for breakfast with General Persons. Andy Goodpaster joined us as we discussed the communications satellite business. Persons is anxious that we get on with this job so that the president can mention it in his State of the Union message. He is no more anxious than I! But there are many problems facing us and we need to be reasonably certain of our ground before we step off too rapidly. Back at the office, Frutkin came in at 9:15 to talk over the results of his trip to Europe and to inform me of his decision with respect to the employment of a deputy in his office. At 10:45, Shapley of the Bureau of the Budget came together with Seamans and Rosenthal to review the results of our budgetary activities over the weekend. He seemed quite satisfied that we had done a reasonable job and agreed to so state to Mr. Stans. At 11:30, Homer Joe Stewart came in to say that he planned to leave a week from now. This is just as well; he will have finished the rewriting of the 10-year report and Abe Hyatt is already on the job.

Lunch with Seamans and Silverstein gave us an opportunity to talk about the budget, organizational and other matters. Back at the office for a briefing on zero G research by Irv Pinkel of Lewis.[16] This phenomenon is little understood and is important to us in our launch vehicle program. It appears that effective work is being done, but I am not sure that we are pushing it as hard as we should. Later in the day, Abe Hyatt came in to discuss his thinking about the operation of the program planning and evaluation office. He is an excellent choice for this job and is well thought of throughout the nation. I think he is going to do a good job. Later Commander Jenkins of the Navy Department came over to talk about the public information director's job. He seems a pleasant and able person but there is the constant problem of placing another uniformed man in our top structure. I had him

---

16 "G," of course, is the symbol used to denote gravity or its effects, in particular the acceleration it produces. Zero G denotes the state of weightlessness that occurs in space outside the gravitational influence of a planet or other large body.

talk to Shelby Thompson. At 3:15, Bob Nunn and Jim Gleason came in to counsel with me about the communications satellite program from the standpoint of congressional relations. At 5 o'clock, Reichelderfer of the Weather Bureau and Ed Cortright came in to talk about the developing picture in meteorological satellites. I think we have this one on track at long last.

A little later, Seamans, Ulmer and Rosenthal came in to review the budgetary preparations for my meeting with Stans to be held on Friday morning. I then stopped by the Mayflower for the annual party given by J. Edgar Hoover and the FBI boys. I did not stay long. I was so tired when I got home that I went to bed immediately and Ruth gave me dinner in bed.

CHAPTER THIRTEEN

DECEMBER 1960

Thursday, December 1: Only 50 more days in Washington! I was up early and down to the White House for a meeting with the National Security Council. George Kistiakowsky was presenting a paper on international science. The president agreed that we should make greater efforts to develop cooperative programs throughout the scientific area with scientists of other nations. Staff meeting was delayed this morning because of my attendance at the National Security Council. We had a good meeting, talking over the various facets of the communications satellite program, the congressional picture, the present difficulties caused by the trade press as it attempts to drive a wedge between the Department of Defense and ourselves. This was a good meeting and one has to have pride in the type of staff we now have working with us. Trevor Gardner [a former assistant secretary of the Air Force for research and development] is chairing a committee for General Schriever in an attempt to define the space program of the Air Force. He called me to ask that we assign a man to be party to a study group that is struggling with this problem. The group is to be located at Los Alamos for the next two months. After taking advice from members of my staff, I think that [Addison] Rothrock is the proper man to send. He is the senior man of the organization and well grounded in all our activities.

At 12:30, I went over to the National Press Club with Arnold Frutkin to hear Henry Wriston [president emeritus of Brown University] and Frank Pace talk about the report of the President's Commission on National Goals [of which Wriston had been the chairman and Pace the vice chairman]. Wriston was absolutely marvelous—he far outshone Pace. There is not very much glamour in the report, but I think it is a solid piece of work—at least as described by Henry Wriston. The Russians launched another space zoo early this morning. Apparently they got it into orbit all right, but they are not as yet talking about bringing it back. Two dogs, several other animals and bits of plant and biological specimens are aboard. The announcement was greeted in an almost routine fashion by the press.[1]

We have been having trouble with John McCone again. I cannot seem to get him into the same office with me to talk about our budgets for the Rover project. He is so damned busy doing other things that this one particular project gets scant attention. I fear that we will get caught in a budgetary snarl again as we did last year.

---

[1] Referred to as Sputnik VI, this spacecraft burned up on unprogrammed reentry, 2 December 1960. (Emme, *Aeronautics and Astronautics . . . 1915-1960*, p. 151.)

At 4 o'clock John Johson and Bob Nunn came in to talk about the visit we are making next week to New York to talk with the AT&T people. For an hour and a half we discussed the various facets of the communications satellite program, industry participation, patent problems, etc. I think we will be in reasonable shape for the discussions in New York. Ruth and I went out to Andrews Air Force Base to have dinner with General and Mrs. Schriever. It was a pleasant occasion with the men getting together to debate various problems in the space business. Except for George Kistiakowsky, all the rest were military people. They are sharp and able and they stay on the job—something civilians don't do. Back home at 11:30, seemingly none the worse for wear.

Friday, December 2: We worked together on the budgetary presentation and then Bob Seamans and I talked with Stans at 10:45. We had brought our budget down within $19 million of the figure desired by Stans. He kept pressing for his figure and I agreed to look at it again. Our total budget will be somewhat in excess of $1.15 billion, so it seems strange that $19 million would loom so large in these discussions. Later in the day, I talked with Jerry Persons about this problem. He will take up the cudgels for us if I want him to but I have decided that I'll fight it out myself. I did—and lost. I may have another go at it in an attempt to retrieve $10 million on Monday next. At 2 o'clock Charlie Spahr of Standard Oil of Ohio came in to seek advice about a merger problem that will involve the Justice Department. John Johnson helped with this discussion.

Saturday, December 3: This was just another day of work at home. All meals were taken rather casually and at odd hours. I am not sure how much I accomplished but at least I spent a good bit of time trying.

Sunday, December 4: Another day at home, and I managed to get some work done in preparation for two speeches next week and for the visit with the AT&T people on Wednesday morning. For dinner, the Frutkins, the Courtland Perkins and George Kistiakowsky came in.[2] The dinner was very pleasant. There was good and lively discussion about a variety of matters—mostly centering around the change in administration.

During the course of this week, Mr. Kennedy has announced several appointments. The first one—Governor [G. Mennen] Williams of Michigan as assistant secretary of state for African affairs—drew mixed comment. It is a strange appointment to make as the first selection and coming as it does, in advance of the appointment of the new secretary of state, can only be taken as a desire to show the importance placed by Mr. Kennedy on relationships with the new African nations or as a means of getting rid of G. Mennen Williams early in the game. Governor [Abraham A.] Ribicoff of Connecticut was appointed secretary of health, education and welfare, and this was well received. David Bell, a Harvard economics professor, was appointed director of the Bureau of the Budget and again, this action

---

[2] Courtland Perkins was chairman of the department of aeronautical engineering at Princeton and was at this time serving as assistant secretary of the Air Force for research and development.

was applauded. Yesterday, Kennedy appointed Governor [Luther] Hodges of North Carolina as secretary of commerce. It appears that Kennedy is working up to the tough ones—I hope all of his appointments are received as well as these have been.

As for myself, I still find the state of suspended animation a little difficult to cope with. We are attempting to put together the material that will be useful to a new administrator, but it would be much easier if we could do this having some notion of the person who will get the job. Actually, the entire Washington scene is confusing and seems to be running without too much sense of direction. I guess this is as it must be under our scheme of operation. Ruth's dinner was a masterpiece and she won many well deserved compliments. Off to bed at 11:30, just a bit worn out.

Monday, December 5: This was another tough day. I believe I stated that I had capitulated last Friday to Maury Stans of the Budget Bureau with respect to our FY 1962 budget. He had beaten me down to the point where I was a little bit uncertain about going in to see the president and I agreed, therefore, to accept a final cut of about $20 million. I had second thoughts about the matter and tried to get Stans on Saturday without luck. He called me at 8:45 this morning and I told him that I was most certain that I had been unwise and asked that he review the matter and let me have $10 million back. He expressed himself as distressed but said that he would look into it. I immediately called his assistant and gave him the story and believe that there is a reasonable chance of success in this gambit. At 9 o'clock, Dr. Howard Engstrom of Remington Rand came in to ask whether or not his company could put two men at the Marshall Space Flight Center to work in the computing center there without salary for a period of one year. This was a most unusual offer. Naturally, if we were to accept this sort of an offer from one organization, we would have to do the same for others. The purpose is a straightforward one—Remington Rand wants to know what kind of activities we engage in, which would make it a better competitor for our computing business. I will take this matter under advisement.

I got a start on the preparation of a long paper for the man who will follow me whoever he may be. I didn't get too far before I was interrupted. Hugh Dryden is back from his trip to South America and we had a little time together to review the events of the past week, particularly the budgetary problems.[3] At 11:30 I went to the White House where I joined a large group in the cabinet room for a presentation of the Collier's Award by the president to the Air Force, Space Technology Laboratories, and Convair. The award was presented this year in recognition of the activities of these three participants in the development of the Atlas intercontinental-ballistic missile. Fred Crawford [chair of the executive committee of Thompson-Ramo-Wooldridge, Inc. of which Space Technology Laboratories was a component] was on hand and I joined him at lunch as he told me

---

[3] Dryden was in Argentina 26 November-2 December for a Symposium on Space Research sponsored by the National Commission for Space Research of Argentina. (Appointment Calendar, 1960, Dryden subsection, NASA Historical Reference Collection.)

something about his visit in Minneapolis a week ago. There can be no question about his interest in and loyalty to Case [whose board of trustees he also chaired]. What a guy!

During the cocktail hour preceding lunch, [Lieutenant General] Roscoe Wilson, deputy chief of staff of the Air Force for research and development, asked me my purpose in seeing the heads of the Air Force tomorrow morning. Apparently, word had gotten around that I was going to complain about something. I told him that I was complaining about nothing but that I was interested in knowing whether or not the top men in the Air Force were concerned about the articles being printed in some newspapers and in many of the trade magazines these days indicating a rift between NASA and the Air Force. I want to be as certain as I can that our appearances before Congress this year are shoulder-to-shoulder with no possibility of congressional wedges being forced between us. Later in the day, General Wilson called to say that General White would be happy to discuss these matters with me. Further, he sent word that he hoped I was going to stay on as head of NASA. If this was not to be the case, General White stated that this was the only job in government that he desired for himself. Since he had hinted at this before in discussions with me, it was interesting now to get confirmation in such a straightforward manner.

Back at the office at 2 o'clock to meet Wayne Barrett of Midland, Michigan, who was asking me, in a very left-handed way, to speak to a group in Midland next spring. He was so earnest it was very difficult to refuse him and I have indicated that I will study the matter if he will make the date late enough in the spring. I might be able to combine it with a trip to the Dow Chemical Company. At 4 o'clock, the senior vice president of Westinghouse, E. V. Huggins, came in to seek information about the reasons for its being excluded from the airborne part of the orbiting astronomical observatory. Apparently someone has taken my name in vain in this matter. Since Westinghouse was a partner with Grumman in this proposal, I was at a loss to understand the situation. Later, it developed that the Goddard people had made a shift in the negotiations without securing agreement for doing so from headquarters. This will be looked into further, you may bet. I have no particular interest in Westinghouse but I cannot stand idly by and see a company treated in this manner.[4]

At 4:30, Dryden, Hagen, Johnson, and Frutkin came in to talk about the proposed space conference and exhibition to be held in Geneva at some point in the future if the Russians can ever make up their minds to behave like human beings. The question before us was, "Shall we attempt to go ahead with this conference even though the Russians do not indicate that they will participate?" After some discussion, my decision was yes! In 1955 they refused to go along with the Atomic Energy Conference held in Geneva but finally came in and participated in a very

[4] See entry below under 9 December. Despite what is said there, Westinghouse did win one of four subsystem development contracts from Grumman for the ground operating equipment. (*NASA Historical Data Book*, Vol. II, p. 261.)

acceptable manner. Later, meeting with Abe Silverstein and Seamans, we took another look at the budget picture, especially with respect to communications satellites. It appears that we are going to need more money even than has been set aside thus far. The most recent cut I accepted from Stans is not going to help in this matter and I must now get that money back.

Tuesday, December 6: This should be an interesting day. I will have plenty of activity both in Washington and in New York. At 8:45, Hugh Dryden, Bob Seamans, and I visited Secretary of the Air Force Dudley Sharp, General White and Joe Charyk. My purpose was to try to find out whether or not there was anything seriously wrong between NASA and the Air Force. The publication of stories of strife, vying for position, stealing each other's projects, etc. have been very frequent these last two or three weeks. It was a pleasant discussion with much agreement on both sides. Certainly at the top of our organizations there is no real difference or need for concern. I am sure, however, at the "colonel" level, there is a good deal of envy and flexing of elbows. We came away feeling that the visit was worthwhile. At 11:45, I took off for New York and was able to keep my appointment with Paul Ylvisaker of the Ford Foundation at 2:15. It was a lovely day and I walked down from the Westbury Hotel to the Ford Foundation offices. This was an interesting discussion—my opinion was requested on the desirability of providing to Cleveland through some mechanism or other an amount equivalent to $250,000 per year for each of ten years for use in civic development. It was obvious that the foundation was thinking about Case as a center for this activity. It sounds like an interesting idea—one must be careful that it is not just a simple way of buying out of a situation at less than the cost that might otherwise be incurred. I promised to think about it further. Certainly, the Ford Foundation has a difficult situation in Cleveland where so many different agencies want money from it.

At 5 o'clock, I walked on down to the Carnegie Corporation where Jim Perkins and John Gardner [the president of Carnegie Corporation] were waiting for me. We had about fifteen minutes of discussion about my interest in the Latin American scene. I hope I am not getting in over my head in this matter. Certainly, John and Jim are very much interested in my taking a part in the educational activities in Latin America and it is clear that some arrangement could be worked out that would make it an interesting and useful task as well as possible of accomplishment within the scope of my activities at Case. In any event, I will get a chance to look at the problems this spring and summer, and we can talk it over later in the fall when a decision as to the future must be made. All of this, of course, contemplates working in such a program in consonance with my work at Case.

At the Coffee House discussion group I talked to 18 of the 20 members—the largest attendance they have had in several years. It was an interesting group of lawyers, writers, professors, business and professional men. Among the group were the following: Jim Perkins of the Carnegie Corporation; Harlan Cleveland of Syracuse University; Don Price of Harvard University; Chuck Dollard, formerly president of the Carnegie Corporation; and Jack Fisher, editor of *Harpers*. The entire group apparently gets together whenever an interesting evening is in prospect. After a drink or two, we had a very excellent dinner and then I held forth on the space program. I described very briefly what we were doing and then raised four or five questions for discussion. We talked about manned space flight, international cooperation, communications satellites, etc. It was a really good evening, and I think the participants were pleased to have been involved. After a beer at the Century Club, we took a taxi to the Westbury where I spent the night.

Wednesday, December 7: Up at 6:30 for breakfast with Bob Nunn and off by taxi to 195 Broadway for a meeting with [Fred] Kappel, president of AT&T. Included in the meeting were Johnny Johnson of our office in Washington and Paul Gorman, George Best, and Jim Fisk of AT&T and Bell Laboratories. It was quite a meeting—we went over each point many times. We are trying to convince AT&T that it is not in the company's best interests to appear as a very large organization attempting to monopolize the communications satellite field. Actually, I think the whole deal will fall in its lap in time, anyway. I believe we made some headway. I suggested that AT&T might well provide the ground stations, both in this country and in Europe, at no cost to the government. This would parallel what the firm had done in Project Echo. We could pay a small sum for carrying out particular experiments for us. Having the ground terminals would mean that it had to have knowledge of the satellite circuitry and would be intimately identified with the whole project. After three hours I believe we convinced the four executives of some of our points of view and then we left for Washington. I expect we will get a letter in another day or two from Kappel.

Our plane was somewhat late so I missed a presentation by a scientist from the University of California. We did have a meeting with Dryden, Newell, Silverstein and Jastrow about the desirability of moving our theoretical group to a college campus. Jastrow feels that he can no longer attract really good people to Washington. The graduate programs in this area are just not of the caliber to make the community a lively one. Harry Goett, Jastrow's boss, is very unhappy about the prospect of having the theoretical group move away from Washington. My own feeling is that the lack of good graduate students is a really important matter to a group of this kind. We agreed to take it under advisement and to make a decision later in the week.[5]

I left immediately after this meeting to pick up Ruth and to dress for dinner. We stopped in at a party given by [Edward] Perkins McGuire, assistant secretary of defense for supplies and logistics, at the Sulgrave Club. Everybody in town was there. I moved on to the Mayflower for the dinner of the business advisory council of the Department of Commerce. Ruth went on home. This is really a wonderful

---

[5] This is a discussion about the theoretical division of the Goddard Space Flight Center. The result of the concerns Jastrow raised here was the establishment in May 1961 of the Goddard Institute for Space Studies in New York City in May 1961. For further details, see Rosenthal, *Venture Into Space*, pp. 59-60.

group of businessmen throughout the country who come together at intervals to advise the secretary of commerce and to learn something of his problems. For some reason, I was seated at a table with the president and had a chance to ask him about his visit with Kennedy and whether or not it included any concern for the space agency. Mr. Eisenhower said, "Kennedy was polite, interested and attractive and attentive. He is a sharp young man, but I found no way of learning exactly what he was thinking."

The president spoke to the group after dinner—a quiet little discussion of the importance of integrity and belief in the basic premises on which this country was founded. He said that he was not going to be a male Mrs. Roosevelt and write a column after he had left the White House. Most of the table conversation was concerned with the favorite Washington game these days—guessing about cabinet appointments and the plans of Mr. Kennedy. Never have I seen a business machine—a great organization—operating in such a vacuum. The best description is that we are in a state of "suspended animation."

Thursday, December 8: We had a staff meeting, which was not particularly exciting. Then Ben McKelway, editor of the *Washington Evening Star*, and [James] Russell Wiggins, editor of the *Washington Post*, came in to counsel with us about our public information problems. I have been concerned about the possibility —really, the probability—that the papers would make a circus of our Project Mercury as we approach the time when a man will enter the capsule to undertake a flight. Both men were quite responsive to our questions; they thought that we had been remiss in not providing to the working press interpretive materials. They thought it well for us to bring together the president of the science writers association and the science writers for the UPI and the Associated Press. It was a useful session and I think we will take their advice.

Fred Robinson, a classmate of mine whom I did not know at college, now president of the National Aviation Corporation, came in to ask me to be a director of his organization. This is quite impossible but it is always flattering to be asked. At 11 o'clock I went over to the Commerce Department to speak to the business advisory council. I gave it a run-down on our operation and answered questions. I believe it was a useful discussion. When asked, "What can the business advisory council do to help NASA?" I answered, "Just stop trying to drive a wedge between the Defense Department and NASA. We have developed effective means of cooperation and coordination. Industry ought to keep out of the act." I then attended the business advisory council luncheon at the Willard Hotel where Henry Wriston spoke. He is a wonderful person and a very able historian. It will be remembered that he is chairman of the President's Committee on National Goals.

At 2:15, we had a meeting on communications satellites with a variety of people including Abe Silverstein. Abe believes that private industry should not have a free hand in the communications satellite business. It is interesting to see the extent to which those people who have spent all of their life in government are negative in their attitude toward industry. I finally had to tell Abe that I was delighted to have his technical judgments but that he would have to leave some of the policy matters to me. I was a little bit tired today, came home rather early and went immediately to bed.

Friday, December 9: At 8:30 I met with Bob Nunn and Hugh Dryden to talk over certain aspects of the communications satellite business. Believe me, this is an involved affair. If we are successful, we will have made history. Bob Bell came in at 10 o'clock to talk about a[nother individual's] psychiatric problem involving possible perversion. We are never free of these situations. One can only have pity for a person thus afflicted. I have decided, in this case, to refrain from taking any action in the belief that the incident happened so long ago as to give reasonable indication that the man has his problem under control. A visit with Undersecretary Jim Douglas of the Department of Defense was a most useful one. I have been very much concerned about the very apparent campaign of the Air Force to stake out its claim to a larger share of the space business. Really, I think it is fighting the battle within the Pentagon more than with us but we do not escape "fringe benefits" from an action of this kind. Jim Douglas saw the point and decided that he would take action with General White.

At 1:30, several of us got together to talk over the complaint of Westinghouse that it has been dealt with unfairly in connection with its bid on the orbiting astronomical observatory. It was teamed with Grumman. It seems that the Westinghouse portion of the proposal was less than satisfactory and Grumman decided not to continue with the firm on a substantial fraction of the job. Our people seemed to have been completely in the clear although they did not keep the head office informed of the actions they were taking. I will support their position without question.

At 2:30, over to the White House to meet with General Persons. He is very much interested in the communications satellite business and wants to have the president announce his support of ownership and operation by a private organization—probably the telephone company. A discussion of our budget followed and I indicated that $10 million between Mr. Stans and myself was taking us to the president for a decision. Jerry said he thought he could handle this himself. By the time I had reached my office, Stans had called and asked me to get in touch with him. He said that he had been talking with Jerry Persons and that he felt that the $10 million was not worth stirring up any controversy. Our budget was approved! Thus do things get done in Washington. I went home shortly thereafter to dress and pick up Ruth for the reception we are giving tonight for the resident research associates who have come to us from other countries for training. There are nine of them. The reception was a pleasant one.

I took Ruth to Jack Hunt's for dinner and then we went on to the Lisner auditorium to hear C. P. Snow and his wife read from their respective writings.[6] Naturally, I fell asleep. Somehow or other, I have never been able to stay awake

---

[6] Both Snow (1905-1980) and his wife, Pamela Hansford Johnson (1912-1981), were novelists. In addition, he was a molecular physicist, university administrator, and, during World War II, scientific advisor to the British government. He is best known for his book, *The Two Cultures and the Scientific Revolution* (1959), where he argued that the sciences and the humanities constituted distinct cultures between which communication was difficult. His wife's novels often dealt in a light vein with moral concerns, but *The Unspeakable Skipton* (1959) was a piece of satire.

when anyone reads to me. Following the program, we dropped in at Bill Bundy's to meet Snow and his wife. He seems a pleasant person although it is almost impossible to get much of an impression in a throng such as that present on an occasion of this kind.

Saturday, December 10: Not much happened today. I had lunch with Dr. Mann of the National Bureau of Standards, an old friend from England. I was to see General Schriever at 4:30 in the afternoon but he found he could not make it. All in all, a quiet day!

Sunday, December 11: This morning it started to snow. I called General Schriever at 9 o'clock to suggest that he not come in from Andrews Field because of the snow. He said he would call me back in an hour when he had had a chance to survey the situation. He did just this. The upshot—we talked over the telephone for a half an hour. I gave him a full story on my concern about the actions being indulged in by the Air Force. He protested that this was all very innocent. Finally, we agreed that we would get together a week from now when he returns from the West Coast. I think I have all of these people concerned about what action I may take next. It continued to snow all day. We were going out to dinner with the Seamanses and stopped in at the Richmonds on the way for cocktails. Fortunately, they live only three blocks apart so we were able to park our car in the Richmonds' garage and thus avoid being snowed in. Already, six inches have fallen and the city is pretty well tied up. Schools will be closed tomorrow and persons driving in Washington are subject to arrest if they block traffic and are found to have neither chains nor snow tires. It was a pleasant evening—we are really very lucky to have the Seamanses with us.

Monday, December 12: I got in on time to find that the government offices had all been closed for the day. Just a few of our people came in, so we managed to accomplish quite a little bit. I was to have flown to Williamsburg to speak to a training conference we are starting at that location today.[7] Because of the snow, the airport was closed down for most of the morning. We finally took off about 3:30 in the afternoon and I managed to give my talk after dinner. I arrived back in Washington about midnight. Quite a heavy day!

Tuesday, December 13: We had an excellent presentation on the orbiting geophysical laboratory by members of the Goddard Space Flight Center staff. This is a very sophisticated satellite, interesting in concept and well presented. It looks as though Thompson-Ramo-Wooldridge would win the competition.[8] Over to the

---

[7] On this training program, see Rosholt, *Administrative History of NASA*, pp. 145-146.

[8] On 21 December 1960, NASA issued a letter contract to Space Technology Laboratories, a division of Thompson-Ramo-Wooldridge, to do preliminary analytical and design studies for three orbiting geophysical observatories. NASA ultimately contracted with TRW for six such observatories, the first of which was launched on 5 September 1964. Although only partly successful, it did return useful data about the earth's atmosphere and magnetosphere. It and the other five orbiting geophysical observatories, which went into orbit between 1965 and 1969, provided the first really automated orbiting laboratories designed to make a variety of observations; they returned a vast quantity of data on earth-sun relationships. (*NASA Historical Data Book*, Vol. II, pp. 264-270, and Vol. III, pp. 172-173.)

Pentagon at 12 o'clock to attend a presentation of an award to Dick Morse by Secretary Brucker. The latter can certainly lay it on thick! Back to the Metropolitan Club for lunch with Admiral Strauss. We had a pleasant time—I think Lewis is mellowing a bit. We discussed a good many elements of the political scene and wondered, in concert, about some of the appointments yet to be made. At 3 o'clock, we had a discussion of the NASA-AEC contract. It does seem as though we should be able to work in closer harmony than apparently is the case. I am not quite sure that we are without fault but I am sure that the AEC is a loosely run operation these days. Home to help Ruth a bit—she is really busy these days getting ready for Christmas.

Wednesday, December 14: This morning I gave a talk at the Industrial College of the Armed Forces at Ft. McNair. It went quite well; the questions were interesting and I think the audience genuinely appreciative of the story I gave them. Not much of importance happened today. Lunch with John Rubel [deputy director of defense research and engineering] was without incident or importance, and a briefing by the Defense Department on the very large radio astronomy antennae being built in West Virginia and in Puerto Rico was interesting but without too much point for us, I am afraid. Later in the afternoon, we talked over some of the problems faced by Abe Hyatt as he takes on the job of directing our office of program planning and evaluation. I think he is going to do a very much better job than did Homer Joe Stewart. At least, his approach is very different and one I can understand better than I did Stewart's operations. Home to help Ruth pack up and get the station wagon loaded for an early takeoff tomorrow. A call to the automobile club indicates that the roads ought to be reasonably clear although one may expect ice in places in the early morning. We managed to get things pretty well put away and I hope Ruth's trip [to Cleveland] tomorrow will not be too difficult.

Thursday, December 15: I started out about 6:30 this morning to get Ruth finally underway. This is going to be a long day. Down to the White House at 7:30 for breakfast with General Persons and General Goodpaster to discuss the communications satellite problem again. We talked over the cabinet paper I am to give next week. Hopefully, we will get this one under the wire before I leave for the West Coast on the 20th. Back to the office at 8:30 for a staff meeting, which was neither important nor very long this morning. At 10:50 we met in the office of the postmaster general to take part in the ceremony placing the Echo commemorative stamp on sale for the first time. It was quite an impressive show and each of us taking part was given an album containing fifty stamps autographed by the postmaster general. This is somewhat of a triumph for me—I have tried to get a commemorative stamp issued several times and have finally made it. Also, I contributed the slogan on the stamp, "Communications for Peace." There are some satisfactions in this job, after all.

I had lunch with Admiral Bennett today. He is still worrying about his future. He retires on 1 January and seems to be unable to make up his mind about any of the jobs that are being offered to him. I advised him to take one with the

United Aircraft and get it over with. Otherwise, I will bet that the Navy will convince him that he ought to stay on board. The afternoon was pretty well given over to further discussions of the communications satellite business. Three gentlemen came in from the French telephone organization to talk to us about participating in experiments in the satellite communications field. At 2:30, Jim Fisk of Bell Labs and George Best of the [American] telephone company came in to debate further with us the program they want to undertake. Earlier in the day I had received a letter from Fred Kappel setting forth a rather good program that AT&T wanted to undertake. It continues to want to be the first in the field and the only one in the field. We continue to believe that this is unwise and we must have some sort of competition so that there can be no attack by Congress on the basis of monopoly operations. I believe that we made some progress in this discussion. I pointed out there were only 35 days left and that I doubted the next administration would be quite as concerned with the support of individual and private enterprise as is the present one. At 4 o'clock, Dryden, Seamans, Thompson and Silverstein came together to talk over the public information problems with Mercury. We have decided to follow the advice given us by Messrs. Wiggins and McKelway and will do a good bit more interpretive writing that will be released for the press one to two weeks in advance of a firing. At least, we are going to try it this way.

Friday, December 16: At 8:30, Bob Nunn came in for another session on the cabinet paper on communications satellites. I believe we will get this satisfactorily underway yet. At 9 o'clock, we listened to a presentation by the Huntsville people of a study program involving the expenditure of $3 million. We were convinced and the expenditure has been approved. A little later in the morning, Clark Randt came in to discuss some of his problems in working with the military in the life sciences field. He had prepared an excellent presentation, and we were able to give him some useful advice. At 10:45, Col. Glosser came in from Andrews Field to answer our questions about the plans of the Air Force for development of passive communications satellites. He stated emphatically that the Air Force had study programs going—no hardware. He apologized for a paper given at the American Rocket Society last week saying that it had gotten through the net without proper screening. I read to him that portion of the cabinet paper defining the responsibilities of the Air Force, the Department of Defense and NASA. He said the Air Force agreed completely with what I had written. As I have said before, I think we have the animals stirred up.

At 12:15, I had lunch with Clay Blair of the *Saturday Evening Post*. I was proposing an article, descriptive in nature, that would set before the public the NASA program in its proper perspective. No one has really told the story of the difficulties of performing research and development under the bright lights of public interest. I think this could be a useful article and was glad to have Blair's agreement. I also talked to him about the possibility of doing an article on Case—at least on engineering education built around the program we are developing at Case. He promised to take a look at this and said that he felt that it could be a useful story.

Later in the afternoon, I talked with Elmer Staats at the Budget Bureau and with Secretary Douglas at the Defense Department about the cabinet paper on communications satellites. I secured their approval so that we ought to be able to go through the meeting now set for Tuesday without too much argument. Back home to a quiet house and a dull evening. Lots of work to do but without Ruth here it doesn't seem to get done very well. She, incidentally, had a fairly good trip to Cleveland yesterday.

Saturday, December 17: I spent almost all of the day working at home. Most of it was taken up with writing my Christmas letters. I then went to the vice president's for a reception—a very pleasant affair. I had a chance to talk with Dick Nixon for a few moments and he asked me to come see him after my return from the West Coast. I think I will do this. I then moved on to the Wright Memorial Dinner, which was a badly managed affair. Fred Crawford was given the Wright Memorial Trophy and gave a fine acceptance speech except that he was unable to get on his feet until 11 o'clock. The toastmaster—Walt Bonney, late of NASA—became enamored of his own words and simply would not bring the program to a close. What an evening!

Sunday, December 18: Early this morning as I returned from the dinner, I called Ruth to find that Kitty and Frank had flown to Los Angeles without incident. Today is the day on which Ruth and Sally fly out. I have insisted that Ruth call me when she arrives there. I will now record that this happened about 2 o'clock on Monday morning. I spent the entire day writing Christmas letters and then went over to George Kistiakowsky's for cocktails with the members of the President's Science Advisory Committee. This is a real powerhouse of scientists and I was glad to see many of them again. Jim Fisk was there and we had dinner together. Once again, I think I made a little progress with him in bringing him around to our point of view on some of these sticky problems. It is amazing how much time it takes to get a job of this sort done.

Monday, December 19: A meeting with the president at 10:30 gave me an opportunity to introduce Bob Seamans to him. We presented him with the cabinet paper and I discussed it in some detail. He stated his unequivocal belief in private enterprise and discussed the paper rather thoroughly so that we could be able to answer any questions tomorrow in cabinet meeting. I asked the president about my resignation. I told him that I was going to leave on 20 January and he indicated that I should then send my resignation to him to be effective on that day. We returned to the office just in time to make it to the control room as the Redstone-Mercury lifted off the pad at Cape Canaveral. This was a real thrilling bit of business. Everything went exactly on schedule; the capsule was picked up out of the water and deposited on the deck of the carrier, the *Valley Forge*, 48 minutes after lift-off. I called the president and gave him the good news. He seemed quite excited about it.

The balance of the day was rather dull after that excitement except for the fact that I left the office early to pick up Polly at the station. We had a good visit and then went on to Jack Hunt's for a lobster dinner. The gal really seems to like them. Back home to do the washing, write a few more Christmas letters, pack and get ready for the trip tomorrow. It is now 11:45 p.m. and I must be up at 6:30.[9] Goodnight!

---

[9] As the entries in the last couple of days for other members of his family would suggest, Glennan was joining the rest of his children, their spouses, and his grandson for Christmas in California.

Tuesday, December 20: This was to be quite a day. Meetings with the cabinet and the National Security Council were scheduled. At each of these, Dryden and I were to present papers and discuss elements of our program. At 8:15, Doug Lord of Kistiakowsky's office came over to meet with Dryden and myself in an attempt to reconcile some of the figures to be used by Kistiakowsky in the National Security Council meeting this afternoon. We were able to accomplish this reasonably well. At 9 o'clock, we met with the cabinet and the president to discuss the communications satellite program. I had been working very hard for a month in an attempt to develop a statement of public policy that could be enunciated by the president before he leaves office. Naturally, I wanted this statement to reflect my conviction—I am sure it is the president's as well, as demonstrated in our meeting yesterday—that private enterprise should undertake the ultimate development and operation of any non-military communications satellite system. The presentation went off without difficulty. Secretary Herter asked that the interests of the State Department in foreign negotiations be more clearly stated. Secretary Gates of the Defense Department expressed approval of the total program but wanted the Defense Department to be kept fully informed on negotiations, program, etc. The president expressed his firm conviction about the importance of avoiding governmental development and operation of the system, and we finally won from him an approval for the policy statement as included in the cabinet paper. I consider this to be a significant step forward.

At lunch with Seamans and Ostrander we reviewed the bidding on organizational changes that may take Colonel Heaton from Seamans' staff to Ostrander's operation. This seems a sensible move and it is to be accomplished by 15 January.[10] At 2:30, we met with the National Security Council. I presented our budget and 10-year plan as revised while Hugh described in more detail the activities to be undertaken under the 1962 budget, which is now set at $1.16 billion approximately. Of that amount, $50 million will be requested in a supplemental for the current fiscal year. After we had completed our story showing the NASA budget increasing to more than $2.5 billion annually by the end of the decade, Kistiakowsky talked about the manned space flight program beyond Project Mercury. Most of his information had been derived from presentations given by our people to a committee of the President's Science Advisory Committee. The total dollars estimated to be required for landing a man on the moon and returning him to earth are really quite staggering. One can support a figure anywhere from $10 billion to $35 billion and even then, not know whether or not he is in the right ballpark. The president was prompt in his response: he couldn't care less whether a man ever reached the moon. There was desultory comment by others in the meeting who were concerned over the increasing cost of space research. I pointed out that our presently planned program did not contemplate the tremendous expenditures mentioned by Kistiakowsky—that some of these decisions must be made by a later administration

---

[10] This change did occur. The December 1960 headquarters telephone directory (p. 8) shows Heaton as assistant administrator for resources under Seamans, while the directory for March 1961 (p. 13) shows him as assistant director for vehicles in Ostrander's office of launch vehicle programs. (Both directories in NASA Historical Reference Collection.)

following more significant results from research now in progress. Finally, I stood up and addressed the president saying that my toughest problem in the face of congressional, public and other pressures—some of them from within the administration—had been to develop a sound program in this area. Facts cannot be changed—this is a difficult, complex and costly business. I stated my belief that we had succeeded in avoiding the clamor for "spectacular accomplishments" that had no basic scientific interest. In some ways, the meeting was discouraging. However, I think that feeling might be considered a natural one under the circumstances.

At 3:30, I started looking for Polly. It turned out that she was to be picked up at 3:30 and thus we started for the airport at 4 o'clock, well endowed with bags and carrying flowers for Ruth's anniversary. We were able to get seats in the tourist section of the plane and, after inquiring of the counter clerk, I bought a couple of miniature Manhattans to be consumed on the plane. We boarded on time and after an appropriate interval, I asked the stewardess for a glass and some ice. She informed me that we could not have a drink in the tourist section. After a slight bit of argument, I gave in. The dinner was very ordinary and Polly seemed rather completely under the weather. There was no particular explanation for this except that I suspect she was worried about the flight. After all, only a few days ago, a jet and a Constellation had come together over Staten Island with the loss of 135 lives. The flight was without incident and we arrived just a little bit ahead of schedule. I noted, as we came in to land, that we were a bit high "over the fence!" We came to a stop on the runway and after a moment the pilot announced, "The reason we are sitting here is that we can't steer this wagon. We lost all of our hydraulic fluid in the last few moments." About thirty minutes later, we arrived at the unloading dock having been hauled in unceremoniously by a tractor.

Ruth, Tom, Kitty, Sally and Martha were there to meet us and it was a really happy reunion. We repaired immediately to the Knox establishment for dessert— some delicious pumpkin pie with ice cream. After a reasonably short interval we drove down the hill to our motel—the Travelodge—and were settled for the night. It was now about 2:30 in the morning, Washington time. What a day!

Wednesday, 21 December-Monday, 2 January 1961: This was the period of the big trek to the West Coast to see the family, get acquainted with Keith III and generally enjoy the company of good friends and relatives. I am not going to attempt to give a day-by-day account of activities. Rather, I will indicate the principal movements of members of the family and say, without hesitation, that it was a really fine holiday for all concerned. One of the first visits we made was to Arcadia where Ruth, Tommy, Martha, the baby and Sally saw Grandmother, Jessie and John. John is still shy and reticent. Mother seemed even more frail but in quite good spirits and certainly in possession of all of her faculties except for her inability to walk.[11] She does get around well with the walker, however. This was her first sight of Keith III and immediately, she retired to her bedroom to bring back a picture we had given her many years ago of Tommy when he was about six to eight weeks

---

[11] She had never fully recovered from a broken hip suffered almost two years previous to this visit.

old. The resemblance between him at that age and the baby was very marked. She gave the picture to Martha.

This day was marred by the fact that Kent Smith called me from Cleveland asking if I would come back on the 30th to appear before the city council on a hearing regarding [university matters]. After some consideration, I decided that I would leave San Francisco on Thursday rather than Saturday and thus have two days in Cleveland for [this and other business]. On Thursday, I took Polly and Sally on a trip up to Santa Barbara. On Friday, Sally and Frank accompanied me to Pomona— I should put it that we took Sally to Pomona. Because our time was short we saw only a little bit of the campus. We rushed back to Los Angeles so that Sally could join the rest of the ladies for lunch at Bullock's. This evening, all the Glennans had dinner at the Sportsman's Tavern and thus ended quite a busy day. On the 24th, we dropped down to Gordon's to visit with Mother and Jessie who had come over to see the family since there didn't seem to be much chance that they would be getting together on Christmas day. Mother had stood the trip quite well, it seemed. This evening—Christmas Eve—we had a pick-up supper that was really quite a feast. Some time after seven o'clock, a good many people came in to visit for a few minutes. Gordon and Cora came as did [others].

After all the guests had left, the stockings were hung on the fireplace and the ladies of the establishment busied themselves with preparing for the next day. During the course of the week, there had been much debate about whether or not Kitty was getting adequate rest. I think she was enjoying the debate although it was obvious that there was some cause for her discomfort. In any event, it was finally decided that she and Frank would stay at the Knox home and would not, as had been suggested, interchange with Sally and Polly at the motel. On Christmas day, we got up relatively early and joined the Knox household for the business of opening presents, drinking coffee and eating doughnuts and sweet rolls. It took us until noon to finish the job and I am sure everybody enjoyed it very much indeed. The children had put together stockings for the older folks and we enjoyed this thoughtful act on their parts. There was a good bit of picture taking, and later in the afternoon, Tom, Frank and I walked up to the Griffith Observatory. After a very good turkey dinner, we were quite happy to get back to bed.

On Monday, Kitty and Frank, Polly and Sally, and Ruth and I went to Disneyland. This was quite an experience and I think everybody had a good time. The kids went on all the rides they could crowd into the five hours we were there; then we returned to the house so that we could have dinner and pack for the drive to San Francisco on the morrow. Tom and Martha have been particularly fine throughout this week. We arose quite early and took the family to the Knox home, from which they all took off for Palo Alto about 8 o'clock. Kitty, Martha, Keith III and I took a noon plane, which was a little bit delayed. We arrived in Palo Alto without incident, having been met at the airport by one of the NASA people. We were pleased to find the automobile party already in residence. We ate a good supper consisting of tuna fish and peas at the request of Kitty. We were glad to get away to the motel for a good night's sleep.

On the morning of Wednesday, the 28th, I took Frank and Tommy over to the Ames Research Center, where they were given a good tour of the operation while I talked to Washington several times about the statement requested by the White House covering our activities for the past two years. This statement was to be accompanied by my resignation. I finally approved the documents and signed off.[12] In the afternoon, we were driven over to the Stanford campus by Tom and then we had a good roast beef dinner at the apartment. The baby was in fine spirits—in fact, he was a model of deportment throughout the entire time of our visit. Tommy took me to the airport on Thursday morning where I left him, the girls and Frank and Kitty, who were going on into San Francisco for some sightseeing. My plane was canceled but I was able to get on to the TWA jet to Chicago. I made a 10-minute transfer to the United Airlines plane to Cleveland and managed to get into that city on time. Taking a taxi home, I was able to catch up on a little reading and get to bed at a fairly early hour.

On Friday, the 30th, I spent the morning on Case [business]. I spent all of the afternoon at the city council hearing but found it unnecessary to make any comments. It was good to see [old friends]. They seemed genuine in their pleasure at the announcement that I was coming back to Cleveland. On Saturday, I spent most of the day on Case business. [Then I met the family at the airport.] They all came in in good spirits and on time. We were able to have a fine New Year's Eve dinner at home. We were in bed before midnight and probably asleep at the witching hour.

---

[12] These documents are available to researchers as part of the Glennan subsection of the NASA Historical Reference Collection.

CHAPTER FOURTEEN

JANUARY 1961

Sunday, January 1: New Year's Day was a quiet one. [Mostly] we were left pretty much to ourselves.

Monday, January 2: We drove back to Washington without incident. The roads were in fairly good condition and we managed to make it to the apartment by about 4 o'clock.

Tuesday, January 3: This is the first working day of the new year. Before going on with this chronicle, I think I want to say a few things about the events of the past few weeks. Never in my life have I seemed so frustrated in attempting an important job to bring to a conclusion. We have been in a state of suspended animation since the election. To my surprise, not one single word or hint of action has been forthcoming from the Kennedy administration. I have tried to put my house in order and to prepare materials for the new administrator whenever he may be appointed. It does seem strange that no action has been taken. After all, we are a large agency—the seventh largest in government in point of budget, and we are spending more than $4 million a working day at the present time. During the course of the holidays, and probably following my resignation, Congressman Thomas wrote to Kennedy strongly recommending the appointment of Hugh Dryden as administrator. Hugh learned from Thomas's assistant that Kennedy had responded, thanking him for the letter and recommendation, saying that he was leaving the space business entirely to Lyndon Johnson. All we know is that Lyndon has been designated by Kennedy to chair the Space Council. It is to be remembered that we have recommended the Space Council be abolished and that, in fact, it has not met for one year. Actually, the law would have to be changed if Lyndon were to be chairman of the council. As it presently reads, the president is chairman and the attorney general ruled last year that we could not have a regular meeting of the Space Council unless the president were in the chair.

It has been most interesting to note the manner in which this transition is taking place. The entire operation of government seems to slow down as the new cabinet appointees—good ones for the most part—attempt to put together their organizations. With a governmental operation totaling more than $80 billion a year and employing approximately 5 million people including the armed services, it would seem that there ought to be some regularized procedure for the transfer of powers. Naturally, as the population grows our involvements increase in world affairs and the problem becomes continuously more difficult.

Now I will get on with the chronicle. At 9 o'clock, several of our people came in to talk about the establishment of the theoretical division at some site near Columbia University in New York. I have approved this in principle and am now awaiting budgetary and logistic proposals. It seems a good idea to have this group, which is largely academic in character, associated with an academic community. At 9:45, Dryden and I visited Kistiakowsky to attempt to get a change made in the budget message of the president, which is to be delivered on 18 January. Apparently, following the National Security Council meeting on the 20th, a statement was prepared for inclusion in the message that would, in my opinion, be unwise. The president proposes to say that there is no scientific or defense need for man in space beyond Mercury. It is much better, if I am any judge of the political realities of the situation, to say that we need much more research and development before a definite decision can be made in this matter. Actually, such a statement would be in complete agreement with the facts as they will be presented in the budget message. After much telephoning, we were able to get this statement changed.

Lunch with Dryden gave Bob Seamans and me, who have both been out of the city, an opportunity to catch up on the events of the past week. At 2:30, Courtney Sheldon of the *Christian Science Monitor* came in for an interview. He has been a good friend and has reported honestly on our activities. At 3 o'clock, the Rand Corporation made a presentation on the nuclear rocket and the Rover program. John McCone, Bob Wilson and one other commissioner were present. It was a discouraging presentation although I think it quite factual. John McCone came to my office after the presentation saying that he was discouraged by the whole matter and suggesting that the AEC be allowed to proceed with the development of the reactor while NASA dropped out of the picture until the reactor was completed. I questioned this as a procedure. At 4:30, I visited Eugene Zuckert, the new secretary of the Air Force, to talk with him briefly about our relationships with that organization in the space business. Gene and I served on the Atomic Energy Commission together. He is an able fellow although I questioned this appointment as being the best that could have been arranged.

Wednesday, January 4: During the morning we had a project status review that went off in fairly good order. Some troubles have appeared in the program but generally speaking, we are in good shape and I was able to approve the new launch schedule without argument. In the afternoon, Frank Pace came in. He is the chairman of the board of General Dynamics. He talked with me for about an hour, stating that he and Earl Johnson, president of General Dynamics, wanted me to have some association, as yet undefined, with the company. Frank seems to believe that industry has a very large social responsibility that it is not discharging, and he wants to undertake some activities in the educational field. He asked that I be receptive to a proposal that he wanted to discuss with me when I come back from my vacation. I suppose that this would take the form of a consultant arrangement or a directorship. I doubt that I can do anything about it. On this same day, Jim

Hagerty, who is to be vice president of ABC, asked me to consider a position as a consultant with that television organization. He said it would be very lucrative and that he would want me to be helpful on matters of science and space. I said that I doubted that I could be involved but agreed to talk with him at some later date. On this same day, Earl Blaik of Avco wrote to me and suggested that I was to be considered for a directorship in one of the large Cleveland companies. My goodness, plums are falling from all the trees and I am quite sure that I cannot avail myself of any of them.

Thursday, January 5: We had a short staff meeting this morning and went right in to the meeting of the space exploration program council.[1] The full day was given over to a discussion of the manned flight program beyond Project Mercury. The space task group and the Marshall Space Flight Center presented alternative programs. They have no hesitancy in their planning—all they need is a vast deal of money. I suppose one has to accept that sort of philosophy but, by George, it is a little difficult for me to be very enthusiastic about it when I know that these men know how very difficult the problems are that must be solved before anything of this kind can be undertaken. It is not in consonance with our budgetary program for fiscal year 1962, yet Congress will be beating us about the ears if we do not have a long-range goal of landing a man on the moon. Government has its problems.

In the afternoon, Rich came in with [Arthur Plumb] Clow of the Chesapeake and Potomac Telephone Company to pick up my White House telephone, which they are going to present to me as a memento of my stay in Washington. This particular phone is connected directly with the White House and I have used it very heavily. We had dinner with Bob Wilson of the AEC and his wife. It was a pleasant occasion and we left immediately after dinner. This made it all the better.

Friday, January 6: We started out the day with a meeting on the Project Mercury information plan. We are trying desperately to avoid a "circus" atmosphere in connection with the first manned flights. It is clear that we must take good care of the press and give them every opportunity to have access to the facts. This will mean camera crews and reporters on the recovery ships, picture taking at the time when the astronaut enters the capsule and remote pictures of the launch. I hope we are successful. The space exploration program council continued today. It was discussing operating problems and I think a great deal of progress was made. I was called out of the meeting by John McCone, who wanted to propose that the reactor part of Rover be approved and handled entirely by AEC without interference from NASA. This is such a drastic step that I was a bit upset. During the course of the

---

[1] NASA established this body early in 1960 to "provide a mechanism for the timely and direct resolution of technical and managerial problems that are common to all Centers engaged in the space flight program." It met quarterly in the office of the associate administrator and consisted of the directors of Goddard, Marshall, and JPL; the directors of headquarters program offices except life sciences; plus the associate administrator and some of his assistants. Sometimes, as this entry indicates, Glennan and Dryden also attended. The SEPC itself did not continue after Glennan's departure from NASA, but the concept of a "super-council" did reappear as a tool for managing the human spaceflight program. (Rosholt, *Administrative History of NASA*, pp. 152-153.)

day, I talked with others of our staff and we finally decided to take it to Stans of the Bureau of the Budget and try to get some resolution of this problem before we leave Washington.

At noon I had lunch with the top brass of the Navy to talk over some of their plans for entry into the space business. They cannot be left out of it so long as the Air Force is taking a leading role. I think I was able to give them some sound advice—at least they said so. I had dinner with Secretary Jim Douglas at the Pentagon. He was entertaining in honor of Sir George Thomson of Cambridge University [Nobel Prize winner in physics]. It was a pleasant evening although we all had a little bit more to drink than was necessary. These dinners with several kinds of wine become rather heavy during the course of the night.

Saturday, January 7: We spent a full half day at the office reviewing the program of the research centers and of the life science center. The people who took part in the discussion of the manned space flight program beyond Mercury should have been here to hear the discussion of research that must be undertaken before we can embark sensibly on such a program. I took Ruth out to dinner after a long walk and we retired early. Lots of letters are coming in expressing regret at my leaving the NASA post. It is nice to have these and I think most of them are reasonably genuine.

Sunday, January 8: A quiet day without too much work. We had a long walk and then the Harknesses came in for dinner. It was a pleasant affair although Ruth was a little bit up against it for food. We thought we had two pheasants in the ice box but it turned out that we had three quail, one duck and one pheasant. She managed a delicious dinner and the evening was very worthwhile.

Monday, January 9: This was a relatively uneventful day. Some few days ago, Jim Killian had called to ask me about future commitments—how long and in what manner had I tied myself for the future. My answer, of course, was that I had undertaken the return to Case in good faith—that I had committed myself for no particular length of time, though one did not undertake a job like this for a few months. He could not tell me of his interest but said I might hear from a friend soon. Today, I was visited by that friend who asked me to consider heading his operation. Suffice it to say that I repeated my comments about commitments but agreed to consider the pros and cons of a future association. In all good conscience, it appears to me—I have not discussed this with the trustees—that John Hrones deserves a term at the top job while he is still young and vigorous. If he acquires the right kind of support and a leavening of his drive, I think he could be terrific as a leader, knowledgeable and respected for his integrity, imagination and concern for the highest standards of quality.[2] After I return from the trip, I may discuss further the matter to which my friend referred.

---

[2] Glennan's successor as president of Case and then as first president of Case Western Reserve University was not Hrones but Robert Warren Morse. Hrones remained as vice president for academic affairs until 1964, when the title changed to provost; he held that position from 1964-1967 and then became provost of science and technology for the combined university from 1967-1976. Thereafter, he was provost emeritus. (Cramer, *Case Institute of Technology*, pp. 221-222, 266.)

After lunch, we reviewed the so-called Stever report, which suggested names for the directorship of Lewis Research Center. Not too much imagination was shown in this listing. Later in the day, Mr. Patterson of General Dynamics came in to talk about long range planning for that company. I discussed this matter in the context of NASA as an R&D outfit with limited production requirements. Surely, [there will be some in] the meteorological [field]; communications and navigation satellites will be made in multiple copies and the military devices in reconnaissance and early warning will be similar; but the bulk of NASA's expenditures will be for R&D only. Therein lies the secret of future success in this business, in my opinion. Concentrate in a few areas rather than spread across the board!

In spite of admonitions to the contrary, I called Lyndon Johnson today to say that I felt a heavy responsibility in the matter of turning over my job to my successor—that I was prepared to help in any way desired; that we had prepared much information in the form of reports and briefings but that discussion was desirable; that I knew he was having trouble securing a man; what could I do [to help]? This, mind you, is the first contact with the new administration; and [it was] initiated by me. Lyndon was lavish in praise of my attitude and then said, "As soon as I have something to tell you or discuss with you, I will call you!" Transition!! No word from the Kennedy administration!

Tuesday, January 10: The morning started with General White at the Pentagon. Hugh Dryden and I took over the revised copy of the statement he proposed to make to his senior commanders' conference and to publish in the monthly policy newsletter of the Air Force. The meeting went well, with Tommy offering to say anything we wanted to have said. He is a fine person, dedicated to the development of the Air Force but cognizant of problems such as the one that has been plaguing us both in the newspapers and other media.

The rest of the day was taken up with normal business. George Low came in to talk of Mercury and the delay in launch to 31 January.[3] Then came the evening with the Hoover Medal Award Dinner at the Statler where we were the guests of Walker Cisler [president of the Detroit Edison Company]. A small party preceded it—cocktails, etc.—and we found ourselves in pleasant company. Ike got the medal after a statistically-loquacious ex-historian of the Army gave what turned out to be an interesting story of the engineering achievements of Ike's forces in Europe after D-Day in 1944. Ike was very good—informal but pertinent in his remarks—bringing in enough of the human interest stories to keep the engineering societies a bit humble. It was this group that gave the medal, of course. No news from the Kennedy administration!

---

[3] On this date, the Mercury-Redstone-2 flight from the Atlantic Missile Range sent a Mercury capsule with a chimpanzee named Ham aboard to an altitude of 157 miles. Some 418 miles down range, a recovery team found Ham in good health. (U.S. House of Representatives, Committee on Science and Astronautics, Report of NASA, *Aeronautical and Astronautical Events of 1961* [Washington, D.C.: Government Printing Office, 1962], p. 4.)

Wednesday, January 11: Off to Langley early by Convair 250 to open a press briefing on Mercury. This is part of our new policy of exposition in the hopes that greater understanding on the part of the newsmen—and thus, of the public—will result. It went quite well with a good attendance. Back to Washington by noon—lunch at the White House mess where surprise was expressed over the ignoring of NASA in the appointments lists. The AEC is in the same boat as well as the FAA. At 2:30 with Hugh and George Kistiakowsky we met with Elmer Staats and Willis Shapley. I was concerned with our inability to agree with the AEC—particularly with John McCone—on contracting matters in connection with Project Rover. Why he is so difficult—and I swear, in this case, it is he and not I or NASA—I will never fully understand. Surely, the J[oint] C[ommittee on] A[tomic] E[nergy] is in the picture and I suspect John has sold a portion of his soul to Clinton Anderson. Probably the management, at a distance, of Los Alamos with its traditions of independence and its access, wittingly or unwittingly, to the JCAE staff and members, gives him troubles. But where are the business ethics and the organizational integrity of the AEC these days? It takes months to get a decision, and then it doesn't stick!

The discussion with the BOB people resulted in an agreement that no solution could be expected in the few days remaining. My memo to my successor calls Rover a continuing and expensive problem on which decisions will need to be made. The pressures of the JCAE and of our own staff will bring action before too long. Both John McCone and I will be gone; his staff will agree with ours for the most part, and the matter will be resolved, I hope. Back to the office to meet with Jim Fisk of B[ell] T[elephone] L[aboratories]. I had Hugh in to listen to the discussion of ground stations. Jim said AT&T would provide, at its expense, the U.S. ground terminal equipment and make available time—either under contract or at no cost—for experimental work by NASA.[4] As to the European terminal, he gave assurance that AT&T preferred to work this out and was certain it would be able to provide the European terminal [either] on its own or by agreement. This is exactly what I would like to see. Recently, IT&T has made similar noises. I have stated that we should accept both—the more the merrier. The real point is that the "carriers" ought to provide and operate the ground terminal equipment so as to gain pertinent operating experience.

Later in the day, I spoke to Abe Silverstein, assuring him about the terminal facilities. His people, and even Abe, worry about this in a genuine belief that we should control and handle this sort of operation in the early stages and that competition will be lessened if AT&T is allowed to supply and operate these stations. There is validity in their concern but, practically speaking, why do things the hard way? Surely, we will not soon see a new AT&T or IT&T in the long

---

[4] The outcome of these discussions that Glennan has been outlining in the diary for some months was an announcement by NASA in July 1961, after Glennan had left NASA, to launch and track two Bell Telephone Laboratories-designed satellites, later officially designated Telstar, for AT&T on a reimbursable basis. On 10 July 1962 Telstar I was launched successfully, with demonstrations of television transmissions beginning shortly after launch. (*NASA Historical Data Book*, Vol. II, pp. 373-374.)

distance communications business. Then came Jim McDonnell [of McDonnell Aircraft] to say goodbye—and, just incidentally of course, to ask whether or not Mac's Mercury business will prevent it from winning other contracts. The answer was an emphatic no—the question, obviously, was directed at the forthcoming award of Surveyor for which Mac is one of four finalists.[5] No word from the Kennedy administration!

Thursday, January 12: The meeting with the staff was brief. I took the floor with a strong statement on the necessity for pounding away at keeping the dates for launches once they are set following designation of a contractor and subsequent negotiations as to price and schedule. The usual objections by Abe Silverstein and others gained from me a certain sympathy but a reminder that spending $1 billion or more per year promised the prospect of searching investigations; moreover, we had to be aware of the need for integrity, even in R&D. It is probable that we are seeking out the last two or three percent of assurance when sound judgment may suggest this as having little to do with the success of the flight. Such considerations are not new in R&D although it is clear that the dollars involved give them new importance. Perhaps I made some impression—at least I have beat upon this drum continuously for the past several weeks. It is probable that time is lost in the early weeks of each project, time that can never be regained. Delays in decision-making at this point always seem less important but really give the most serious trouble later on.

At 10:00, we met with Abe Hyatt, director of program planning and evaluation, and the senior staff to discuss the handling of the NASA study program in the future. We are committing as much as $20 million a year for a variety of studies in all fields. [Since the program is] uncoordinated at the moment, it is just good sense to pull together all of this activity. Some controls must be exerted and I believe Hyatt will do this. Coordination with the Air Force advanced study program will be accomplished by direct relationship between Hyatt and USAF planners. This meeting went well with still another demonstration of the increasing maturity of the organization and the development of mutual confidence among staff members. Lunch—a final one—with Rawson Bennett at Jack Hunt's was pleasant. Rawson does not yet know where he will go when he retires on 1 February but seems unconcerned.

At 1:30, more discussions of the communications satellite program. We have reached the point of proceeding with the request for bids for the satellite itself.[6]

---

[5] Of the four, on 19 January 1961 NASA selected Hughes Aircraft to build seven Surveyor landers. These spacecraft, under a project managed by the Jet Propulsion Laboratory, provided data on the lunar surface and environment in support of the Apollo program. Of the seven missions between 30 May 1966 and 9 January 1968, five were successful. (*NASA Historical Data Book*, Vol.II, pp. 325-331.)

[6] This was an apparent reference to the Relay program to launch a low-altitude active communications satellite. On 25 January NASA briefed industry on the requirements for Project Relay, and on 18 May 1961 it awarded a contract to Radio Corporation of America for the development of three Relay satellites, only the first two of which were launched on—13 December 1962 and 21 January 1964. Both were successful. Relay I showed that a satellite could function as a microwave repeater, and Relay II provided television, teletype facsimile, and digital data transmission with satisfactory results. (*NASA Historical Data Book*, Vol. II, pp. 376-379.)

Competition is the watchword, and once again, patiently, I went over my strong beliefs in this matter. Jaffe and Silverstein seem determined that anything short of having someone other than AT&T win the competition will be tantamount to following a "chosen instrument" policy. Pointing out that any company might choose to bid $0 or $1 million [and that under those circumstances] a ceiling with the company bearing costs over the $1 million seemed to me well within competitive rules unless otherwise specified in the request for proposals, I gained agreement that we would have to consider such proposals as fair—so long, of course, as the subject company provided a fully documented cost estimate for the total job, etc. Then we moved to the matter of ground stations. Agreement on the U.S. station being supplied at no cost or on a participating basis by one or more companies was easy. On the foreign station the usual desire for control was expressed. Finally, I agreed to a joint exploration with industrial participation believing that nothing was to be lost by such a procedure, with the probability that the industrial organizations would demonstrate their abilities to do the foreign job as well as and probably better than we could. My real interest here is to have the ground terminals built by operating companies who themselves will generate the traffic. Thus they will gain valuable operating experience while reserving all necessary time for experimentation by our people.[7] Finally, I approved the preliminary development plan and effectively, I guess, washed my hands of the program since nothing will happen before I depart.

At 4 o'clock, Walter Burke of McDonnell came in to say nice words and bid me goodbye. Then at 5 o'clock, John Finney came in for an hour's talk—the first such meeting with him since our bout on "Meet the Press." Asking many of the usual questions, he then concentrated on what advice, if any, I might have for my successor. Before answering that question, I asked why the press corps was so quiet about Kennedy's failure to move ahead on the appointment of my successor. It had not spared the outgoing administration in any way but seemed content to see NASA become an orphan on 20 January. No comment from Finney! Going back to his question, I pointed out the Rover, manned space flight, and very large rocket projects as real headaches for the future. Each involves long term commitments of billions of dollars. None will come to fruition short of ten years, and thus they will lose their luster as elements in the cold war. All are concerned with manned flight about which we know so little with certainty. My own program, as set out in the FY 1962 budget, delays these decisions until more definite information can be acquired by smaller, unmanned satellites. To my surprise—maybe it was a typical newsman's gambit—John seemed to be of the opinion that a constantly increasing rate of expenditures would not be palatable to the public. I pointed out that the program we had underway could fit a budget of $1 to $2 billion with little strain—that less than $1 billion would seriously cripple the scientific effort since costs of $500 millions are now and, for the next few years, will be required annually for large rocket development. Further, the development of simulators for ground environ-

---

[7] In this connection, for Telstar Bell constructed a large ground antenna in Maine, while communications agencies in England, France, and Germany constructed ground stations to operate with both Telstar and Relay. NASA stations for Relay in Italy, Brazil, Japan, and elsewhere could also work with Telstar. (*NASA Historical Data Book*, Vol. II, pp. 373, 376.)

mental tests and the problems of gaining reliability without large numbers of launches would be with us for a long and expensive period. A better than average discussion with a reporter! Home and a bit of packing as we plan a trip to Chicago on the morrow! No word from the Kennedy administration!

Friday, January 13: The last flight in the NASA Convair will take us to Chicago, Cleveland and back to Washington. It is good that we are not superstitious about dates! We took off at 7 o'clock with Dick Mittauer as a passenger and with breakfast served pleasantly. No incidents marred the flight and we were met by the director of the Museum of Science and Industry (formerly the Rosenwald Museum) where I was to participate in the opening of a NASA exhibit, attend a luncheon and make a speech. A press conference was thrown in, which took more than an hour including much picture taking, television taping, etc. Attempts to get me to discuss the Wiesner report met with complete failure, which reminds me that I have not commented on said report—so here goes! After this long and masterful silence, Johnson and Kennedy released the so-called Wiesner report. He was the chairman of a committee that included Ed Purcell, Ed Land, Bruno Rossi, Don Hornig and several other scientists.[8] Of the membership, only Land has had any administrative experience or operating responsibility. The report seemed to be a series of contradictions—we are ahead of the USSR in space science by a significant margin, but we must go faster; we have a sound program, but we must do more; and above all, we need more imaginative, technically competent management. Spelling out a rough organizational plan that was almost completely a duplicate of the pattern we are following, the report listed the administrator, deputy administrator and the directors of space science, propulsion, etc., as the positions requiring technically competent leadership. What a slap in the face at the people on whom they must rely for leadership in the future—for I doubt that they can find better people who will combine knowledge, ability, and the willingness to work in government. My own situation has no importance here. To a considerable extent, they may be right about the administrator—but I won't be here anyway.

Back to the day in Chicago. The press conference was not too bad and the luncheon went pretty well. Kitty and Frank attended with Ruth. After lunch we departed for the apartment and Frank took me down to the University Club to see John MacLean [president of MacLean-Fogg]—a pleasant hour. Then a drink and

---

[8] The report, a copy of which is available in the NASA Historical Reference Collection under "Jerome Wiesner" in the biographical file, was a "Report to the President-Elect of the Ad Hoc Committee on Space." It was dated 10 January 1961 but not released to the media until 12 January. Wiesner, who became President Kennedy's special assistant on science and technology, had been a professor of electrical engineering at MIT since 1950 and was also the director of the Research Lab of Electronics at MIT. Hornig was chair of the Princeton chemistry department; Land was president of Polaroid Corporation; Purcell was a professor of physics at Harvard; and Rossi was a professor of physics at MIT. Among other things, the report stated (p. 3): "Our review of the United States' space program has disclosed a number of organizational and management deficiencies as well as problems of staffing and direction which should receive prompt attention from the new administration. These include serious problems within NASA, within the military establishment, and at the executive and other policy-making levels of government."

some of the best sukiyaki I have tasted was prepared and served by Kitty and Frank. Time was running out as we started for the airport. Looking for the Convair took some time but we reached Cleveland about 10 o'clock to find a NASA car waiting to take us home. No word from the Kennedy administration!

Saturday, January 14: Sally had been staying at Helen Hamilton's until our arrival last night. On coming home, she seemed a bit flushed and felt a bit indisposed. She was no better this morning. From nine until after one, we (the top staff at Case) were steeped in a discussion and rehearsal for the trustees' meeting, which will be held on Monday, 23 January. Dr. Spitler had diagnosed Sally's indisposition as a "strep" throat and had given her a penicillin shot, so we took her, however reluctantly, back to the dormitory infirmary. Boarding the plane at 3:30, we arrived in Washington at 5:15. Cocktails at [Deputy Director of Defense Research and Engineering] John Rubel's in honor of the Yorks preceded a dinner with Dick and Daz Harkness. Dick, as I have probably said earlier, is the top TV newsman on the Washington scene for NBC. This seemed to be an annual party having association with the political scene. Among the guests were Marquis Childs, Walter Lippman, Milo Perkins, Senator Albert Gore, Senator Mike Monroney, Supreme Court Justice Bill Brennan, ex-Governor Leroy Collins [of Florida], Bill Foster [of Olin Mathieson Chemical Corporation?] and a few others—all with wives. Good chatter throughout the evening revealed Albert Gore as a keen analyst of the political scene. Mike Monroney asserted that FDR's New Deal had failed to quicken the economy although it had halted the panic and kept people from starving. Only as the war brought new requirements did the upturn really come. What an admission from a leading Democrat! And Milo Perkins, an early and ardent New Dealer agreed! The evening wound up with the opening of predictions made by most members of the group four years earlier as to the complexion of the political slates and the winning side in the 1960 presidential race. Most of those present were Democrats, so it was not surprising that predictions were fairly good though Stevenson had been a leading contender. Nixon had been the almost unanimous choice for the Republican nomination, and two or three had predicted his election. Choices for the slates in 1964 had Kennedy unanimously but only one or two listed Johnson as vice-presidential candidate. I proposed a Kennedy-Kennedy slate as the solution. Nixon, Rockefeller and "dark horses" dominated the choices for Republican candidates with only two predictions of a Republican victory in 1964. And so to bed—with no word yet from the Kennedy administration!

Sunday, January 15: Plans for the day called for packing, writing, packing, etc. But—early on this grey and somewhat moist day, a call from Dr. Spitler revealed that Sally was worse—might have the measles or scarlet fever—and was to be taken immediately to the hospital. Well-l-l! Plans immediately were made for Ruth to return to Cleveland at 4:45 p.m. Flights were in danger of being canceled since the weather was worsening. I decided to call the Convair 250 up and did so. Then—called Dr. Spitler who said Sally would stay at the dormitory until Ruth came home. Finally, the Convair was reported as leaving Norfolk so we started for the airport. Delay followed delay and the fog was closing in. The Convair finally came in at 4 o'clock and Ruth got off. Back home and some packing but not too much. Ruth reported in and Sally was home at last. At 8 o'clock, the Adams clan—Don, Eleanor, the children, Buck and Dorothy came in for dessert and coffee. Ruth had made preparations ahead of time, so it was not too much of a task. They are fine, warm-hearted and generous people. What a day! And still no word from Kennedy!

Monday, January 16: An early session with Seamans, Nunn, Dryden and Silverstein was interrupted by a call from Ruth. Sally has scarlet fever! A call to Dick Huffman resulted in reassurance as to the possibility of contamination, contagion and the probability of my carrying the infection to others. A few pills and no worry! This means Ruth cannot come back for the party on Wednesday night, which the staff is giving for us. But—back to business! We agreed on certain aspects of the communications satellite program. I keep hacking away at the prejudice against competitive enterprise and Abe continues to worry about our ability to justify turning over to one company the responsibility for significant parts of the system. Later in the morning, Jim Gleason and I repaired to the Hill to visit with Congressman Overton Brooks. He was cordial and sympathetic—said that Lyndon Johnson has given no indication of plans but that, sooner or later, the Senate would realize that the House of Representatives was still in existence. He seemed to have no solid views on the revival of the Space Council as proposed by Kennedy. Overton continues to feel that more money should be requested. Calling on other congressmen—Gerry Ford, Jim Fulton, Olin Teague, Harold Ostertag—we were pleasantly received but acquired no spiritual lift.

Lunch with Bob Seamans and Hugh to review a few last problems. I think we are in reasonable agreement as to organization and program although it is clear that a new administrator will have to make up his own mind—and probably will—as to the future of manned flight, Rover, the bigger rockets, etc. Dinner at the Brookings Institution with the president—Dr. [Robert D.] Calkins—and others of the staff and with Pete Quesada, Sumner Whittier of Veterans' Administration, and Fritz Mueller, [under] secretary of commerce. The purpose was to determine whether or not outgoing top administrators would be willing to devote a few days to the subject of governmental administration, transition between administrations, etc. All agreed to do so at the earliest possible time. No word from Kennedy yet!

Tuesday, January 17: The operation is slowing down. A lame duck has little authority or believability to an on-going staff, no matter how responsive the ties have been over the past months. Discussions with staff groups involve subtle but recognizable stiffening in resistance to suggestions by me or expressions of strongly held beliefs by me. No one should resent this—it is a natural and understandable fact of life. At 12 o'clock, we went to the Cosmos Club where Hugh Dryden hosted a luncheon for me with perhaps 30 of the top staff in attendance. Hugh spoke briefly and gave me a thoughtfully worded scroll while Bob Seamans, for the group, presented me with a beautiful American flag and my own agency flag.

My response was emotional—I have never been able to overcome the emotional problem. But it was a sincere effort on the part of all concerned to express gratitude to each other. Dinner with Rich and Polly and back to the apartment to pack some more. They have been real friends to us in and out of Washington.

No word yet from Kennedy! Finally, today, I called the White House—General Persons and then Dave Kendall—to say that no one in NASA had been asked to stay, at any level. Hugh had expressed his willingness to serve on an interim basis but has not been asked. Dave called back in an hour saying that he had talked with Clark Clifford. Through this channel the word was passed by me to Hugh that the new administration wanted him to stay on for the time being! Ruth called to say she has decided to fly down tomorrow afternoon for the party being given for us by the staff. She will fly back on Thursday morning with Alice taking over at home with Sally for the single night.[9]

Wednesday, January 18: This was a quite full day. Discussions with various people took up the morning hours. There has been some pressure on me to take action in raising the grade for several secretaries, etc.—an action I have refused to take. It seems clear to me that one must continue to be available and in charge right down to the wire. But to crowd in questionable last minute personnel actions is highly improper. I think I convinced the boys of the justification for my position. A last conference with Abe Hyatt resulted in my approving the outlines for action in his planning operations although I urged—and he agreed—that these plans be explained thoroughly to the new administrator without delay. I had read a column by Roscoe Drummond recently that exhibited sound analysis with respect to our operations and position vis-a-vis the U.S.S.R. This morning I decided to ask Roscoe to think over with me the position in which the new administration has placed NASA. Why, I asked, were the newspapers so loath to point out the damaging delays involving inevitable morale degradation when only a few days or weeks earlier they were reporting our problems, decrying our lack of progress, trying to build up arguments between NASA and the DOD, and patting Kennedy and Johnson on the back for their concern and brave statements about the "new look" to be given to space as a matter of urgency. Unfortunately, I had only 15 minutes with Roscoe because of my final appointment with the president. I gave him a copy of the Wiesner report and a rundown of the preparations we had made for the transition operation. I pointed out the reactions of our people to the present situation; then I had to run. I hope Roscoe got the point!

At 2:30, I saw Ike for the last time on official business. He seemed a bit preoccupied but pleasant and cordial. His appointments, most of them like mine, are being interspersed with requests for last-minute signatures, etc. Ike said, "Why we can't leave a few things for the next President to sign; I don't understand." We

---

[9] In response to a written query, on 9 July 1993 Glennan identified Alice as "our respected maid."

spoke briefly about our program and I complimented him on his courage in speaking out—in his farewell message to the nation—about the military-industry-science combination whose hold on government and, through governmental spending, on the economy, must be closely watched. This sparked him into a strong reiteration of his concern for governmental domination of the lives of the American people. Ike has been consistent in his position on the place of government in the development of the nation—a bit too conservative at times, in my opinion, but espousing a course of action much more to my liking than the Kennedy proposals as publicly discussed thus far. Ike remembered that I was going back to Case and wished me well. He expressed gratitude for my service and admiration for the work done by NASA in the short period we have been operating. He repeated his concern that man on the moon is a matter of questionable merit when compared to the many other demands on our resources. It was a pleasant 20 minutes and I shall be forever grateful for the opportunity of serving under Ike—a human being whose faults lay in his humanism, whose concern for people probably caused him to agonize over decisions affecting good people, whose belief in the separation of the legislative and executive branches caused him to refrain from putting pressure on the Hill as sometimes he might have done with profit, whose belief in democracy and the high goals of peace with justice in freedom are as bright and positively held today as they were eight years ago when he assumed office. In view of his age, illnesses and frustrations, he has performed remarkably well and leaves office with the affectionate regard of the nation— indeed, few there are who believe that Ike could be defeated if he were to run again.[10] Ike's parting quip to me as I commented on the confusion of the transition was something like this: I can take most of this clutter and clatter, but Keith, I am unhappy when I think about riding down that avenue on Friday with a man for whom I do not have the greatest admiration—and on top of all that, to have to wear that damned top hat.

After the usual short debriefing session on security matters, I went home to get Ruth for the party being held at Bolling Air Force Base officers club. Gee! It turned out to be a party of perhaps 300 with representatives from all the labs except Ames. After cocktails and a good dinner, Abe Silverstein chaired the meeting and called on representatives of each element of the organization for a few words and a presentation. Dick Horner had come in from the West Coast and sat with us at the head table. The gifts were many and interesting. Models of Echo, Mercury, the X-15, Pioneer IV and Saturn were accompanied by photo albums, a desk set with the clock timer from Mercury-Redstone 1, a Steuben bowl for Ruth, silver cigarette box and ash trays, and a quite wonderful picture made up from Tiros sequences. I had viewed with alarm the thought of a party, much preferring to fold my tent and steal silently away. Ruth spoke briefly and well and I spoke for perhaps ten minutes—this time

---

[10] Of course, Ike was prohibited from running again by Amendment XXII of the Constitution, passed in 1951 before he assumed office. Besides his heart attack in September 1955, already mentioned, he had a bout with ileitis in June 1956 and a slight stroke in November 1957.

without undue emotion and with stress on the need for maintaining integrity of purpose and action. It was a good party, warmly appreciated because of the sincerity of those present. We left the presents to be boxed and shipped by Goddard. No word from Kennedy!

Thursday, January 19: Up early with Ruth so that I could load the station wagon and get her off to the airport. I should have said that the movers were prompt. Yesterday, after I had gone to the office, the driver from Cleveland arrived to pack the dishes, books, lamps, etc. So this morning, Bill Detrich took Ruth to the airport while I waited for the movers to start loading. Bill returned and everybody pitched in for a while so that the wagon was loaded and I could get off for the office by 9:30. Bill, by the way, gave me a rifle that he had rebuilt and cared for with real pride; he is a hobbyist in the gun field. It is heavy and probably will not see much service with me but I appreciate the thought behind the gift.

My day now became a tough one. I had planned to leave at noon for Cleveland but the boys had wanted me to hear a source evaluation board on Project Surveyor presented by JPL. This I did although it took a good four hours with time out for lunch. The decision was not too difficult but I did decide also that NASA must negotiate the contract with JPL in attendance—not as the negotiator. This is sure to raise problems but it is difficult to delegate to a contractor like JPL the responsibility for such a negotiation with $35 to $50 million involved. Finally, about 3 o'clock, I was ready to leave. Snow had been falling since 11:30 and it was predicted that six inches were on the way. But I was not yet free. Josephine Dibella [Hugh Dryden's secretary] had brought in some sherry and the girls insisted that a few of us have one last drink together. As a result—it was pleasant and thoughtful, of course—it was 4:30 before I reached the apartment and started for Cleveland. Well-l-l! One hour later I had covered all of two miles. I turned around and started back—a relatively easy task. All roads were choked and snow was falling fast and furiously. Sighting a building set back from Connecticut Avenue, I turned in to borrow a telephone. Fifteen minutes later, I finally was able to get a line and called the Richmonds to beg a bed. Lucky me! They had room—so I started back into town. Two hours later, I parked the car half a mile from their apartment and walked through the snow, dodging between cars jammed together in the streets, on the road crossings and on the sidewalks. Never have I seen such a tangle—and tomorrow morning the inaugural!

Polly and I had a martini and some spaghetti since Rich was tied up entertaining business guests at the Statler. Tired out, I went off to bed without pajamas, etc., at 9:30. About 12:30 a.m. I woke and began to worry about the car, which I had left on a narrow street in a "no parking" spot. I decided to attempt to move it although the traffic was but little abated at this time. Fully dressed, I found I had no way of getting back into the apartment short of leaving the door open or waking the Richmonds—so I went back to bed. At 6:30, I rose quietly and departed, leaving a note of appreciation. Miraculously, the main streets were cleared and I was able to start for Cleveland about 7:15. Almost 5 hours later, I reached the turnpike having had a few narrow escapes—the road being icy and covered with drifted snow in places. Cleveland was a welcome sight about 4:30 in the afternoon. I shaved, bathed and attended the trustees' farewell dinner for Kent Smith after having renewed my association with Sally, whose redness has somewhat lessened by now.

        Driving home alone permitted me to listen to the inaugural ceremonies on the radio. Although I missed the TV, the commentators made it a vivid experience. Fortunately, I entered a tunnel on the turnpike just as the [Richard Cardinal] Cushing prayer started. Imagine my feelings upon emerging from said tunnel to find him still talking—actually making a speech. Much of what he said seemed to me to be in poor taste, religiously and temporally. Kennedy's speech seemed to me an excellent statement. He deserves support—will need it—and will get it from me so long as I am able to agree on the ends to be attained and, at least partially, on the means to be employed. So was ended—actually at 12:15 p.m. today—some 29 months of interesting, exciting, baffling and, at times, frustrating work in Washington. A rare privilege it has been to serve the nation and President Eisenhower in yet another new venture. After I have had time to reflect on these experiences a bit I hope to add some paragraphs to this narrative in an attempt to evaluate certain aspects of the problems with which I have been concerned on this stay in the nation's capitol city. As I sign off for the present, I am mindful of the many fine, able and devoted persons who have toiled in the bureaucratic vineyards, many of them for periods much longer than mine. Their friendship and willingness to be helpful has made the job more pleasant and has increased, most certainly, my effectiveness. Many of these were people in my own organization. Many others were people scattered throughout the other agencies of government. To them I will be grateful, always. And still—no word from the Kennedy administration!

THOUGHTS ON THE NATION'S
SPACE PROGRAM
AS IT HAS DEVELOPED

SINCE JANUARY 1961

The comments I will make in this section will cover a variety of matters and events having to do with the nation's space program as it has been conducted since 21 January 1961 by President J. F. Kennedy, Vice President Lyndon Johnson and my successor, the Honorable James Webb. I hope I can succeed in avoiding any attitude of "I told you so" or "Why didn't they continue the way we started out to accomplish the objectives of this program?" At least, I will intend so to do. But his is not a small activity nor is it one in which the average citizen can be assumed to have more than an excitable interest—rather than a real understanding. With the strong support of President Eisenhower and of my staff (although they did not always agree on the pace to which I wanted to hold the agency), I was able to do what I thought right in mounting the nation's space program. Basically, it was a broadly-based program in science and technology pursued aggressively with the intent to extend the state of the art as rapidly as possible but with no intent to endanger the lives of anyone or to undertake shots purely for propagandistic purposes. It seems to me clear that the present program has a somewhat different apparent motivation, although fundamentally it must be the same. But I have refrained from talking about this matter publicly because it is being carried out by the good people who served with me and because it seems clear that those responsible for the program today should have a fair chance to manage it in the manner and for the purposes they believe to be right. Ultimately, the public will react to this program and will require that it be changed or that it be carried out as now proposed. Already, there is evidence of substantial discontent with the "race" concept. (This is being dictated on 12 November 1963.)

In the days immediately following 20 January 1961, Ruth and I wanted nothing more than to get away and to get a little rest. Through the friendship and interest of Jim Perkins of the Carnegie Corporation, I had agreed to join an activity known as the Council for Higher Education in the American Republics (CHEAR), which seemed to meet once each year for a solid week, men and wives, in some Latin American or North American city. The 1961 meeting was scheduled for mid-February in San Francisco. It was to be preceded by a meeting in Los Angeles. Together with Jim and Jean Perkins, Ruth and I traveled to Mexico City about the end of January, hoping for a few days of rest. Early in February, a call came to me from Hugh Dryden in my hotel in Mexico City. Jim Webb had been appointed to the post of administrator and had asked Dryden to speak with me about my attitude toward an association with the program. His purpose, as Dryden reported it, was to avoid any appearance of a sharp break between management philosophies to be pursued in carrying out the program we had started. I suggested to Hugh that I thought Jim Webb ought to speak to me about a matter of this kind. Almost immediately, a call came through from Webb and it turned out that what he desired was that I serve as a consultant to the Senate Space Committee under the chairmanship of Senator Kerr (Lyndon Johnson had become vice president). Again, I suggested to Jim that I thought the senator ought to make this request himself and, in a few days, a telegram came through that I responded to in the affirmative. Actually, I have never been called as a consultant by Webb, Kerr or anyone else. This is not surprising—but it does give some insight into the way in which politicians move to cover all the bases.

As a sidelight on the above situation, after an interval of 12 months, I called to the attention of Jim Webb the fact that I had not been asked to consult with the Senate committee and thought that this association should be terminated so as to avoid any embarrassing questions on the part of the press should anyone remember that such an appointment had been announced in February 1961. Jim Webb agreed and there followed a comedy that can only suffer in the telling. An hour's discussion with Webb was needed to set the stage for a lunch with Senator Kerr. I had said that I was quite willing to submit my resignation—that I had a number of other commitments taking up my time. But at lunch, Senator Kerr in his suave way, went through quite an exercise before he asked me to write such a letter in which I was to point out that I had had "many meetings" with Senator Kerr and had finally concluded that conflicting obligation required that I resign. I simply smiled—and needless to say, I have never formally resigned from the consultantship. Why is it so hard for a person in that position to make a straightforward statement in which he admits that the situation that called for the appointment no longer exists?

It is of interest to note—and purely from the point of underscoring the long lead times involved in a program of this nature—that practically every launching undertaken in the first two years of the Kennedy administration had been in hardware status before the end of the Eisenhower administration. This is in the nature of a program of this sort, as I have pointed out. Yet it does mean that the successes, if any, of the program must be credited in the minds of the public to the administration in being at the time of the flight. Some columnists were kind enough at the time of the first Mercury flights to point out that this whole program had been brought to a state of near completion under our administration and that the Kennedy administration deserved very little credit for it even though it was claiming the entire credit. I pointed out in a letter to one of the Washington correspondents that, had the program failed in any detail, the present administration would have to take the "rap" for it—therefore, it deserves the credit for its satisfactory completion.

The program has really come of age in the years 1960-1963. The unbroken record of the Tiros weather satellites, the advent of the Telstar and Relay communications satellites, the successful flight of the Mariner probe toward Venus, the suborbital and orbital flights of the astronauts in Project Mercury—all of these and the many supporting flights that have worked in a satisfactory manner are evidence of the reasonably well-managed and well-structured program.[1] I think that the

NASA FACILITIES

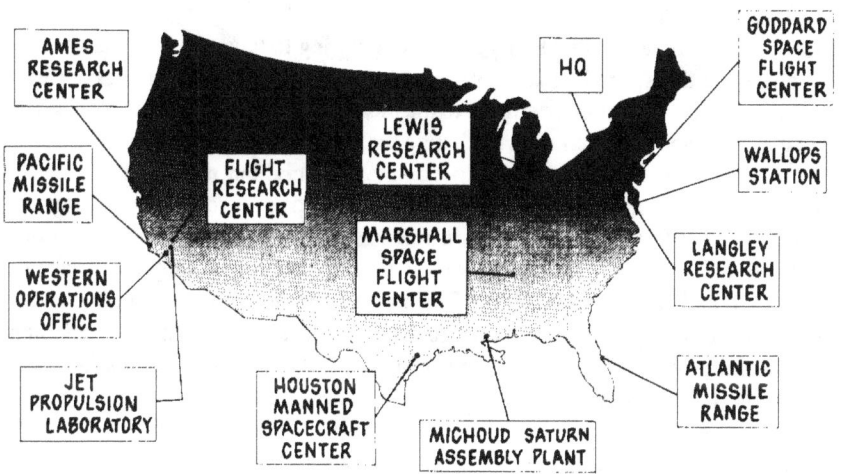

NASA facilities in late 1961 near the beginning of the tenure of James E. Webb as NASA administrator. All facilities shown except the Manned Spacecraft Center in Houston had been in existence during the tenure of T. Keith Glennan as administrator.

current record would show that about 85% of the launches are carried out successfully. During the Kennedy administration, of course, a very significant change in the program has occurred and a good many additional programs have been instituted—most of these having to do with manned space flight. Perhaps this is as good a time as any to attempt to review this period of activity and comment on it as

---

[1] All of these programs have been covered in previous notes (see index) except the Mariner probe. This was Mariner 2, launched on 27 Aug. 1962, Mariner 1 having had to be destroyed after a booster deviated from course. Both satellites were constructed by the Jet Propulsion Laboratory from components built by numerous subcontractors. Mariner 2 passed within 34,762 kilometers of Venus and became the first spacecraft to scan another planet. It surveyed the atmosphere and surface of the planet before going into heliocentric orbit, whence it made measurements of the solar wind. (*NASA Historical Data Book*, Vol. II, pp. 335-336.)

objectively as I may. Kennedy had criticized the Eisenhower administration as one lacking in leadership and neglecting the prosecution of rapid advances in science and technology. I don't recall that space, as such, came in for any great amount of individual attention in these statements, but clearly, space was very much in the national consciousness and in the mind of Congress at that time. Our last budget had been processed at a figure of $1.11 billion. I think we had begun our negotiations at $1.45 billion but had been settled back to this figure. It was lower than most of us thought it should be by at least $150 million. But NASA was now the seventh largest agency of government. Within days after taking office, Kennedy amended the budget and added another $145,000,000, as I recall it, to NASA's program. I applauded this move and subsequently, in a letter to President Eisenhower, told him so.

It will be recalled that we were attempting to have the law changed so as to eliminate the Space Council entirely—it being our purpose to replace it with a civilian advisory committee similar to the old NACA committee. Action was taken by the Kennedy administration to change the law to eliminate the necessity for the president being its chairman and to eliminate all civilian membership. Lyndon Johnson was appointed the chairman and an executive secretary, Dr. [Edward C.] Welch, was appointed.[2] So far as I am able to tell, the new arrangement has worked out better than the old. I don't think they have had nearly as many meetings of the Space Council as we had under the Eisenhower administration. But the executive secretary makes speeches from time to time and does speak with some semblance of authority about the total national program. Whether or not this is a sound situation, I don't know.

About this time—mid-April 1961—the unfortunate incident of the Cuban invasion at the Bay of Pigs took place. The nation drew back in some shock and amazement and the complete story on the reason for the invasion and the manner in which the U.S. was to support it has never been told.[3] It does seem clear that Kennedy changed his mind at the last moment and withdrew effective air cover from the landing forces, and this act is given by most reasonable people as the major reason for the failure of the "invasion." In the aftermath of that fiasco, and because

---

[2] The amendment to Section 201 of the 1958 Space Act gave the council a home in the executive office of the president, made the vice president the council chair in place of the president, eliminated the four appointed members of the council—leaving five statutory members: the vice president, the secretaries of state and defense, the heads of NASA and the AEC—and specified that the council assist as well as advise the president when he asked it to. President Kennedy signed Public Law 87-26 incorporating these changes on 25 April 1961. Meanwhile, the president had appointed Welsh, who had been an economist on Senator Symington's staff, to the position of executive secretary to the council in March 1961. The Senate confirmed him the same month. (Rosholt, *Administrative History of NASA*, p. 190n; "Chronology in Space Planning," folder, "NASC 1960-1961," in box, "NASC, 1960-1973," NASA Historical Reference Collection.)

[3] This, of course, is no longer true. As long ago as 1979 there was available Peter Wyden's *Bay of Pigs: The Untold Story* (New York: Simon & Schuster, 1979). A more recent and briefer account appears in Richard E. Neustadt and Ernest R. May, *Thinking in Time: The Uses of History for Decision Makers* (New York: The Free Press, 1986), pp. 140-155.

of the successful orbiting of astronauts by the Soviet Union, it is my opinion that Kennedy asked for a reevaluation of the nation's space program.[4] To the extent that my information is valid, a period of about ten days was spent in assessing the costs of drastically speeding up the lunar program (a program aimed at landing a man or men on the surface of the moon and retrieving them safely). I should not indulge in hearsay, but there was apparently very little more real solid thought and planning given to this problem than is indicated. In any event, Lyndon Johnson recommended it to the president over a weekend and then left on an overseas trip.[5] Kennedy made his now famous speech to Congress on 25 May in which he called for an all-out effort in a race with the Russians to get to the moon. While his words were not exactly those—mine are a not-too-liberal paraphrase of his statement—I think an unfortunate statement by Bob Seamans before a congressional committee gave the newspapers, and through them, the public, the idea that this flight was to be accomplished by late 1967.[6] Today, the objective is still "within this decade," which means by the end of 1969, I presume. In any event, this single speech and the decision made before undertaking the speech is destined to cost the people of this nation at least an extra $20 billion and probably much more to race an uncertain opponent on an uncertain course toward an uncertain goal.

I strongly hold that the United States should determine for itself the nature of the program it will undertake and the pace at which that program is to be pursued. Kennedy did call for debate in Congress—a debate that did not take place. The nature of the public's interest and congressional interest at that time was such as to insure that such a debate would not take place. The result is that only in recent months has there been any public debate of the nature and pace of the nation's program in space. As a result of that 25 May 1961 speech, the FY 1962 budget request was again increased and an appropriation of $1.825 billion was made by Congress. The 1963 appropriation totaled $3.74 billion and the requests for FY 1964—not yet acted upon by Congress in finality—totaled somewhat more

---

[4] This view is substantially the same as the one set forth in Logsdon, *Decision to Go to the Moon*, pp. 111-112. McDougall, *Heavens and the Earth*, p. 8, adds some other factors, including "the growing technocratic mentality," but also includes the Bay of Pigs and the Gagarin flight.

[5] In fact, Vice President Johnson asked Wernher von Braun, General Bernard Schriever, and Vice Admiral John T. Hayward, among others, for their views on speeding up the space program. All recommended doing so, with von Braun believing that a crash program could produce a lunar landing by 1967 or 1968. There was also a good deal of other consultation and planning before the decision was made. It is true that Johnson's overseas trip shortened the time NASA and the DOD had to make their recommendations, but the deliberations did last for a period of over two weeks before Johnson made his recommendations to President Kennedy. He then had two further weeks for deliberations before his 25 May speech on "Urgent National Needs." (Logsdon, *Decision to Go to the Moon*, esp. pp. 114-115 but also including the rest of pp. 112-127.)

[6] This refers to testimony Seamans gave before the House Science and Astronautics Committee. As the *Washington Post* (15 Apr. 1961, reproduced in NASA *Current News*, 17 Apr 1961, NASA Historical Reference Collection) summarized his comments, "Pressed as to whether this [a landing on the moon] could be advanced [from 1969-1970] to 1967, he said it would have to be 'a very major undertaking'—going 'all out on an absolutely crash basis.'"

than \$5.7 billion with the probability that Congress will appropriate somewhere between \$5 and \$5.2 billion.[7] It is my contention that such an expansion of a program in science and technology should be undertaken only when the stakes are clear and the national interest is paramount.

While others may debate this—and they are entitled to their opinion—I think events have shown that there is not a great deal of carryover value to the propaganda successes of the Russians in their spectacular shots. Whatever damage has been done to the image of the United States by the success of Sputnik I will remain. The retrieval of that image has been accomplished to a very large extent by the solid achievements of the program we had laid on, which has been carried out with significant success. To allow the Soviet Union to dictate the course to be followed by the U.S. seems to me to hold great dangers in the future. On many occasions, I have suggested that Khrushchev, if he were as smart as I think him to be, would continue a barrage of propaganda and successful space flights until he had the U.S. committed to a costly program. Having accomplished this purpose, he might then withdraw from the "race." In recent days, there seems to be some evidence that this is exactly what he has been doing although it is clearly too early, as I write this, to be certain of this assessment.

Webb has done a good job in managing the affairs of NASA with the wise counsel of Hugh Dryden and the strong support of Bob Seamans. Webb works much more closely with Congress than I ever did and I presume that his associations with the White House are, therefore, somewhat less intimate than mine were. Webb states rather frankly that with the money at his command, he is determined to improve the economic lot of various areas of the country, and I think the allocation of contracts would indicate that this is, indeed, what is taking place. There has been a great deal of talk about the "spin-off" nature of the space program but there is little substance behind such statements. Certainly, there is nothing to justify the tremendous expenditures from which some of these minor industrial applications have been derived.

The NASA staff continues in much the same configuration as when I left it. Webb has brought in a number of very good men and has managed to avoid any serious inquiries related to the management operations of the agency. Quite naturally, it takes a vastly different organization to manage a \$5 billion program than one running between \$750 million and \$1 billion. While it may be too early to make a solid evaluation, I am of the opinion that—given the directive JFK has set out—Webb, Dryden and Seamans are doing a very, very creditable job. To summarize, then, I take issue with the philosophies now guiding the activities of NASA in the sense that I cannot agree we should be engaged in a "race to the moon" with the Soviet Union. I am convinced that we will one day land a man on the moon and

---

[7] Actual budget authority for FY 1962 was exactly as reported here. For FY 1963 the actual figure totalled \$3.673 billion, and for FY 1964, \$5.1 billion. (*Aeronautics and Space Report of the President, FY 1991 Activities*, p. 180.)

retrieve him with safety. I am convinced that this should be done. I am of the opinion that a space program which might have reached the budgetary level of $2 to $2.5 billion by the end of the decade would have provided the basic information and the material resources on which such a lunar flight might have been mounted during the first five years of the decade of the 1970s. If I were to guess—and it would only be a guess—such a flight might have been accomplished under those circumstances by 1975. And I doubt very much whether Kennedy's program is going to result in a successful flight much before 1972 or 1973.[8] But these are only conjectures and what is at issue here is really only the matter of basic philosophy determining the pace and nature of the program. Certainly, the emphasis on the lunar flight is placing altogether too much of the resources of the agency on this one program. In the long run, this may work strongly to our detriment, but only time will tell!

T. Keith Glennan
Cleveland, Ohio
[12 November 1963]

---

[8] As Glennan himself notes in the Preface to this volume, "the Apollo 11 astronauts landed on the moon on 20 July 1969 . . . and [I] was as thrilled and emotionally moved as anyone could be."

BIOGRAPHICAL APPENDIX

Ira H. Abbott (1906-) began working for the Langley Aeronautical Laboratory in 1929 after graduating from MIT. He wrote many technical reports on aerodynamics and was instrumental in setting up programs in high-speed research, rising to the position of assistant chief of research at Langley in 1945. Transferring to NACA Headquarters in 1948 as assistant director of research (aerodynamics), he was promoted to the post of director, advanced research programs in NASA in 1959 and to that of director, advanced research and technology in 1961. As such, he supervised the X-15, the supersonic transport, the nuclear rocket, and the advanced reentry programs. He retired in 1962. ("Ira H. Abbott," biographical file, NASA Historical Reference Collection.)

John G. Adams (1912-) had been a Republican party official since World War II and had held a succession of government positions as an attorney since 1949. He became counselor and later general counsel to the Department of the Army in 1953 and served until 1955. He was a consultant on organization and management to the Atomic Energy Commission in 1956-1957 and served as director of the bureau of enforcement in the Civil Aeronautics Board, 1958-1965. President Lyndon B. Johnson appointed him a member of the Civil Aeronautics Board in 1965, and he served until 1971.

Sherman Adams (1899-1986) had the title of assistant to the president and served as Eisenhower's chief of staff between 1953 and 1958. Previously he had been a member of the House of Representatives (R-NH) between 1945 and 1947 and governor of New Hampshire from 1949 to 1953. Adams resigned from the Eisenhower administration in 1958 following House subcommittee revelations that he had accepted expensive gifts, including a vicuna overcoat, from a textile manufacturer seeking government favors. On his career, see Kenneth E. Shewmaker, "The Sherman Adams Papers," *Dartmouth College Library Bulletin*, 10 (April 1969): 88-92; John E. Wickman, "Partnership for Research," *Dartmouth College Library Bulletin*, 10 (April 1969): 93-97; *Historical Materials in the Dwight D. Eisenhower Library* (Abilene, KS: Dwight D. Eisenhower Library, 1989), pp. 8, 48; *New York Times*, 28 October 1986, p. D-28.

Konrad Adenauer (1876-1967) was a former mayor of Cologne, Germany, who had been twice imprisoned during the Nazi era. He was chancellor of West Germany from 1949-1963, during which time he did much to consolidate Germany's first effective democratic and republican form of government. He sponsored a Western European union and a close alliance with France and presided over a resurgence of German industry.

Bruce Alger (1918-) (R-TX) was first elected to the U.S. House of Representatives in 1954 and served for a decade.

Gordon L. Allott (1907-1989) (R-CO) was elected to the Senate in 1954 and served until 1973.

Milton B. Ames, Jr. (1913-) earned a B.S. in aeronautical engineering from Georgia Tech in 1936 and joined the Langley Aeronautical Laboratory the same year. In 1941 he transferred to NACA Headquarters where he served on the technical staff. In 1946 he became chief of the aerodynamics division. With the establishment of NASA, he became chief of the aerodynamics and flight mechanics research division. In 1960 he assumed deputy directorship of the office of advanced research programs at NASA Headquarters and then directorship of space vehicles in 1961. He retired in 1972. ("Milton B. Ames, Jr.," biographical files, NASA Historical Reference Collection.)

Clinton P. Anderson (1895-1975) (D-NM) was elected to the House of Representatives in 1940 and served through 1945, when he was appointed secretary of agriculture. He resigned that position in 1948 and was elected to the Senate, where he served until 1973.

Robert B. Anderson (1910-1989) was secretary of the treasury between 1957 and 1961. A firm believer like Eisenhower in fiscal restraint, he attempted to hold federal spending within narrow limits and to balance the budget. Previously, he had held a number of important government posts in Texas and Washington, including secretary of the Navy, 1953-1954, and deputy secretary of defense, 1954-1955. In 1961 he became a partner with Carl M. Loeb, Rhoades, and Co.

Victor L. Anfuso (1905-1966) (D-NY) served in the House of Representatives 1950-1952, 1954-1962.

John A. Barclay (1909-) became commander of the Army Ballistic Missile Agency on 31 March 1958 after serving as deputy commander since 1 May 1956 under the more famous John B. Medaris. Barclay had been promoted to the rank of brigadier general on 29 September 1955 after a career in the field artillery and ordnance that included command of Picatinny Arsenal from 1954 to 1956.

Perkins Bass (1912-) (R-NH) was first elected to the House of Representatives in 1954 and served through 1962.

Page Belcher (1899-1980) (R-OK) was elected to the House of Representatives in 1950 and served through the 92d Congress (1971-1973).

Robert L. Bell was director of the security division in NASA's office of business administration in 1960. (Headquarters Telephone Directory, May 1960, NASA Historical Reference Collection).

Rawson Bennett (1905-1968) became an ensign in the U.S. Navy in 1927, earned an M.S. in electronic engineering at the University of California in 1937, and rose through the ranks to become a rear admiral in 1956. He served as chief of naval research in Washington, D.C., from 1956 through 1961 and then became senior vice president and director of engineering for Sangamo Electric Co. in 1961.

Paul F. Bikle (1916-1991) earned a B.S. in aeronautical engineering from the University of Detroit in 1939 and was employed by Taylorcraft Aviation Corp. for a year before working for the Air Corps and Air Force as a civilian from 1940 to 1959, both at Wright Field, Ohio, and at the Flight Test Center at Edwards Air Force Base, California. At Edwards, he rose to the position of technical director of the center. In 1959 he became director of NASA's Flight Research Center, also at

Edwards, a position he held until his retirement in 1971. Both with the Air Force and with NASA, he was associated with many major aeronautical research programs from the XB-43, the first jet bomber, through the successful rocket-powered X-15, to wingless lifting bodies that lead to the Space Shuttle and reusable boosters. ("Paul F. Bikle," biographical file, NASA Historical Reference Collection.)

Henry E. Billingsley (1906-) was appointed as NASA's director of the office of international cooperation in January 1959. Previously he had served in the Navy in World War II and with the Department of State. ("Henry E. Billingsley," biographical file, NASA Historical Reference Collection).

Earl Henry ("Red") Blaik (1897-1989) had been a star end on the Army football team before he graduated in 1920. He later became head football coach there in 1941, and in 18 seasons as a coach achieved a 121-33-10 record. He was named coach of the year in 1946 and elected to the National Football Foundation's Hall of Fame in 1964. He became a vice president and director of Avco Corporation, 1959-60, and then director and chairman of its executive committee in 1960.

Clay D. Blair, Jr. (1925-) had been successively a correspondent for *Time* and *Life* magazines before becoming associate editor for the *Saturday Evening Post* from 1957-1961. He later rose through the position of editor in 1963-1964 to become executive vice president and director of Curtis Publishing Co. in the latter year. He was also the author of numerous books about a variety of subjects, including the atomic submarine and the hydrogen bomb.

Hendrik W. Bode (1905-?) was vice president of military development and systems engineering at Bell Telephone Laboratories from 1958-1967. He was a research engineer and worked for Bell from 1926 to 1967, when he became a professor at Harvard.

Charles E. ("Chip") Bohlen (1904-1974) was a career U.S. foreign service officer and diplomat who served as ambassador to the Soviet Union, 1953-1957; to the Philippines, 1957-1959; and to France, 1962-1968. Among other posts, he was special assistant to the secretary of state for Soviet affairs, 1959-1961.

Walter T. Bonney (1909-1975) was NASA's first director of the office of public information (1958-1960). From 1949 to 1958 he had worked for NASA's predecessor organization, the National Advisory Committee for Aeronautics, and before that, for Bell Aircraft Corp. as manager of public relations. From 1960 to 1971, he served as director of public relations for the Aerospace Corp. ("Walter T. Bonney," biographical file, NASA Historical Reference Collection.)

Styles Bridges (1898-1961) (R-NH) had served as governor of his state, 1935-1937, and was elected to the Senate in 1936. He was at this time the ranking Republican member of the Appropriations Committee, a member of the Armed Services Committee and its preparedness investigating subcommittee, as well as the Aeronautical and Space Sciences Committee. He was the leader of his party's conservative wing and a strong proponent of military preparedness. Bryce Harlow told Eisenhower in 1958 that he was "a walking 25 votes in the Senate, the most skilled maneuverer on the Republican side." (Quoted in Divine, *The Sputnik Challenge*, p. 140.)

Wallace R. Brode (1900-1974) was a chemist and scientific consultant who received a Ph.D. from the University of Illinois in 1925 and became a Guggenheim fellow in Europe, 1926-1928. From then until 1948 he was on the chemistry faculty at Ohio State, rising in 1939 to the rank of full professor. He worked with the Office of Scientific Research and Development during World War II and became head of the science department, U.S. Naval Ordnance Test Station, Inyokern, CA, 1945-1947. From then until 1958, he served as associate director of the National Bureau of Standards. For the next two years, he was scientific advisor to the secretary of state, following which he became a scientific consultant, also serving on numerous committees advisory boards, etc.

Luigi Broglio was chairman of the Italian National Committee on Space Research. He was also a professor at the Aeronautical Engineering School (*Scuola d'Ingegneria Aeronautica*) in Rome.

Detlev W. Bronk (1897-1975) was president of the National Academy of Sciences, 1950-1962, and a member of the National Aeronautics and Space Council. A scientist, he was president of Johns Hopkins University, 1949-1953, and Rockefeller University, 1953-1968.

Arthur B. Bronwell (1909-) was an electrical engineer who had been a professor at Northwestern University and became president of the Worcester Polytechnic Institute from 1955-1962, then the dean of engineering at the University of Connecticut at Storrs.

Overton Brooks (1897-1961) (D-LA) had been elected to represent his home state in the House to 12 successive terms since 1937. He became chairman of the House Committee on Science and Astronautics in January 1959 and was reappointed to this chairmanship in 1961.

Wilber M. Brucker (1894-1968), was secretary of the Army between 1955 and 1961. An attorney, he had also held a number of important government positions, including governor of Michigan in 1930-1932, prior to becoming secretary. Brucker had served with the Army in World War I. After leaving federal service Brucker returned to his law practice in Detroit (William Gardner Bell, *Secretaries of War and Secretaries of the Army: Portraits & Biographical Sketches* [Washington, D.C.: Center of Military History, 1982], p. 140; *New York Times*, 29 October 1968, p. 41).

Edmond C. Buckley (1904-1977) went to work for the NACA at Langley in 1930 after earning his B.S. in electrical engineering from Rensselaer Polytechnic Institute. He became chief of the instrument research division in 1943 and was responsible for instrumentation at Wallops Island and at the Flight Research Center at Edwards, California. In 1959 he became assistant director for space flight operations at NASA Headquarters. Two years later, his title changed to director for tracking and data acquisition, and from 1962 to 1968 he was associate administrator for tracking and data acquisition. He retired in 1969 as special assistant to Administrator James E. Webb. ("Edmond C. Buckley," biographical file, NASA Historical Reference Collection.)

Hugh Bullock (1898-), son of Calvin Bullock, was an investment banker. He was president and director of several investment funds including Calvin Bullock, Limited, and Bullock Fund, Ltd.

William A.M. Burden (1906-1984) was an aviation consultant of wide experience. He had advised Brown Bros., Harriman & Co., on aviation financing, 1928-1932; headed aviation research for Scudder, Stevens, & Clark, 1932-1939; and directed National Aviation Corp., 1939-1941. In 1942-1943 Burden had served as a special assistant to the secretary of commerce, with supervision of the Civil Aeronautics Authority, and between 1943 and 1947 he had been assistant secretary of commerce for air. Thereafter he was an independent consultant. He served on the National Aeronautics and Space Council from its creation until March 1959, when he resigned to serve as ambassador to Belgium.

Carter L. Burgess (1916-) was a corporate executive who served as assistant secretary of defense for manpower, personnel, and resources, 1954-1957; president and director of Trans World Airlines, Inc. in 1957; president and director of American Machine & Foundry Co., 1958; its chairman and chief executive officer thereafter.

Arleigh A. Burke (1901-) was a career naval officer who served as commander of a destroyer squadron and then chief of staff of Task Force 58 during World War II. During the Korean War he was commander of Cruiser Division 5. He was chief of naval operations, 1955-1961, and then retired to become a corporate executive.

Charles P. Cabell (1903-1971) was a career officer in the Army Air Corps and later the Air Force, rising to the rank of general in 1958. During World War II he commanded a combat wing in the European theater and later was director of operations and intelligence, Mediterranean Allied Air Forces from 1944-1945. In 1948 he was director of intelligence, Headquarters, U.S. Air Force, and from 1953-1962 he served as deputy director of the Central Intelligence Agency.

Joseph Campbell (1900-) worked as an accountant and then comptroller with a couple of private firms from 1924 to 1932, became a partner in another for two years, and then formed his own accounting firm. He became comptroller general of the U.S. in 1954 and remained in that position until 1965.

Howard W. Cannon (1912-) (D-NV) was first elected to the Senate in 1958 and served until 1983.

William Monte Canterbury graduated from West Point in 1934 and was commissioned in the Army Air Corps in that year. He held a number of positions involving research and development before becoming deputy chief of staff, research, engineering in the Air Research and Development Command in 1959-60. He subsequently served in 1960-61 as commander, Air Force Missile Development Center at Holloman Air Force Base, New Mexico, before retiring as a major general to become senior staff scientist, Lockheed Missiles and Space Co. in 1961.

Clifford P. Case (1928-1982) (R-NJ) was a member of the House of Representatives from 1945 to 1953 and was elected to the Senate the following year, serving until 1979.

Francis H. Case (1896-1962) (R-SD) was elected to the House in 1936 and served seven consecutive terms until he was elected to the Senate in 1950 and reelected in 1956.

Joseph V. Charyk (1920-) was under secretary of the Air Force at this time (1960-1963) and later returned to aerospace industry, whence he had come, serving as president of Communications Satellite Corporation after 1963.

Peter T. Chew worked in the reports section of the office of public information at NASA Headquarters.(Headquarters Telephone Directory, May 1960, pp. 2, 6, NASA Historical Reference Collection.)

Archie Trescott Colwell (1895-1979) went to work in 1922 as a sales engineer for Steel Products Co. (later Thompson Products, Inc. and then Thompson-Ramo-Wooldridge, Inc.—still later, TRW, Inc.) and rose to become vice president in charge of engineering from 1937-1960.

Arthur Holly Compton (1892-1962) earned a Ph.D. in physics from Princeton in 1916 and became physics department chair at Washington University in St. Louis in 1920. He became a professor of physics at the University of Chicago in 1923. In 1927 he and C. T. R. Wilson of England jointly won the Nobel Prize in physics for their discovery and explanation of the wavelength changes in diffused X-rays when they collide with electrons. From 1942-1945, Compton directed the metallurgical laboratory at Chicago, which developed the first self-sustaining atomic chain reaction. He became chancellor of Washington University in 1945 and was a professor of natural history there from 1953 to 1961.

Emerson W. Conlon (1905-) received an aeronautical engineering degree from MIT in 1929 and spent 12 years in private engineering before joining the aeronautical engineering department at the University of Michigan in 1937. He went on active duty with the Navy in 1942 and later directed the development of the Douglas D-558, a transonic research aircraft. He returned to Michigan as chair of the aeronautical engineering department and remained in that position until 1953, with a year's leave of absence in 1950-1951 as technical director of the Air Force's Arnold Engineering Development Center. He spent some years with Fairchild Engine Division and as general manager of Curtiss-Wright's Turbomotor Division before becoming research director of Drexel Institute of Technology in 1958. In 1959-1960 he was on a leave of absence from Drexel to serve as NASA's assistant director for power plants in the office of advanced research programs. (biography in NASA miscellaneous biographical files, NASA Historical Reference Collection.)

Silvio O. Conte (1921-) (R-MA) was first elected to the House of Representatives in 1958 and was reelected to every succeeding Congress through the 101st (1989-1990).

John J. Corson (1905-1990) had been a management consultant with McKinsey & Co., since 1951, remaining there until 1966. Glennan contracted with McKinsey & Co. for a series of studies. These included: "Organizing Headquarters Functions," 2 volumes, December 1958; "Financial Management—NASA-JPL Relationships," February 1959; "Security and Safety—NASA-JPL Relationships,"

February 1959; "Facilities Construction—NASA-JPL Relationships," February 1959; "Procurement and Subcontracting—NASA-JPL Relationships," February 1959; "NASA-JPL Relationships and the Role of the Western Coordination Office," March 1959; "Providing Supporting Services for the Development Operations Division," January 1960, on the transfer of the Army Ballistic Missile Agency to NASA; "Report of the Advisory Committee on Organization," October 1960; "An Evaluation of NASA's Contracting Politics, Organization, and Performance," October 1960. (All in T. Keith Glennan, administrator, correspondence, NASA Historical Reference Collection.)

Edgar M. Cortright (1923-) earned an M.S. in aeronautical engineering from Rensselaer Polytechnic Institute in 1949, the year after he joined the staff of Lewis Laboratory. He conducted research at Lewis on the aerodynamics of high-speed air induction systems and jet exit nozzles. In 1958 he joined a small task group to lay the foundation for a national space agency. When NASA came into being, he became chief of advanced technology in NASA Headquarters, directing the initial formulation of the agency's meteorological satellite program, including projects Tiros and Nimbus. Becoming assistant director for lunar and planetary programs in 1960, Cortright directed the planning and implementation of such projects as Mariner, Ranger, and Surveyor. He became deputy director and then deputy associate administrator for space science and applications in the next few years, then (1967) deputy associate administrator for manned space flight. In 1968 he became director of the Langley Research Center, a position he held until 1975, when he went to work for private industry, becoming president of the Lockheed-California Co. in 1979. ("Edward M. Cortright," biographical file, NASA Historical Reference Collection.)

Albert Scott Crossfield (1921-) learned to fly with the Navy during World War II. He became an aeronautical research pilot with the NACA in 1950, flying the X-1 and D558-II rocket planes and other experimental jets. From 1955 to 1961 he was the chief engineering test pilot for North American Aviation, Inc. The first man to fly at twice the speed of sound (Mach 2) in the D558-II in 1953, Crossfield reached Mach 2.11 and an altitude of 52,341 in the first powered flight of the X-15 in 1959. His last flight in the X-15 apparently occurred on 6 December 1960. ("Albert Scott Crossfield," biographical file, NASA Historical Reference Collection.)

John W. ("Gus") Crowley, Jr. (1899-1974) joined the Langley Aeronautical Laboratory in 1921 after earning his mechanical engineering degree from MIT the year before. He became head of the research department at Langley in 1943, then transferred to the NACA's Washington headquarters in 1945 to become acting director of research there. He assumed the post of associate director for research in 1945, and when NASA replaced the NACA, he became director of aeronautical and space research. He retired in 1959. ("John W. Crowley, Jr.," biographical file, NASA Historical Reference Collection.)

Robert Cutler (1895-1974) was a lawyer and banking executive. He practiced law in Boston from 1922-1942 and then became president and director of the Old Colony Trust Co., 1946-1953, and its chairman for the next several years. In 1960-1962 he served as executive director of the Inter-American Development Bank.

Emilio Quincy Daddario (1918-) (D-CT) was first elected to Congress in 1958 and served until 1971.

Melvin S. Day (1923-) earned a B.S. in chemistry from Bates College in 1943 and worked in private industry for a year before serving in the Army from 1944-1946 at Oak Ridge as a laboratory foreman. He joined the Atomic Energy Commission at Oak Ridge in 1947 and rose through various positions to become chief of the technical information service there in 1958. The same year he transferred to Washington to become assistant chief of the AEC technical information service. He became director of the office before joining NASA in 1960 as deputy director of the office of technical information and educational programs. From 1962-1967 he headed the scientific and technical information division for NASA before becoming successively deputy assistant administrator for technology utilization (1966) and acting assistant administrator for technology utilization (1969). In 1970 he left NASA to head the office of scientific information in the National Science Foundation. ("Melvin S. Day," biographical file, NASA Historical Reference Collection.)

Kurt H. Debus (1908-1983) earned a B.S. in mechanical engineering (1933), an M.S. (1935) and Ph.D. (1939) in electrical engineering, all from the Technical University of Darmstadt in Germany. He became an assistant professor at the university after receiving his degree. During the course of World War II he became an experimental engineer at the A-4 (V-2) test stand at Peenemünde (see entry for Wernher von Braun), rising to become superintendent of the test stand and test firing stand for the rocket. In 1945 he came to the United States with a group of engineers and scientists headed by von Braun. From 1945-1950 the group worked at Fort Bliss, Texas, and then moved to the Redstone Arsenal in Huntsville, Alabama. From 1952-1960 Debus was chief of the missile firing laboratory of the Army Ballistic Missile Agency. In this position, he was located at Cape Canaveral, Florida, where he supervised the launching of the first ballistic missile fired from there, an Army Redstone. When ABMA became part of NASA, Debus continued to supervise missile and space vehicle launchings, first as director of the Launch Operations Center and then of the Kennedy Space Center as it was renamed in December 1963. He retired from that position in 1974. ("Kurt H. Debus," biographical file, NASA Historical Reference Collection.)

Smith J. De France (1896-1985) was a military aviator with the Army's 139th Aero Squadron during World War I, then earned a B.S. in aeronautical engineering from the University of Michigan in 1922 before beginning a career with the NACA and NASA. He worked in the flight research section at Langley Aeronautical Laboratory and designed its 30-by-60-foot wind tunnel, the largest ever built until that time (1929-1931). He directed the research in that tunnel and designed others as well before becoming director of the new Ames Aeronautical Laboratory in 1940. He remained its director until his retirement in 1965. During that time, the center built 19 major wind tunnels and conducted extensive flight research, including the blunt-body research necessary for returning spacecraft from orbit to the earth's atmosphere without burning up. ("Smith J. De France," biographical file, NASA Historical Reference Collection; Muenger, *History of Ames*, esp. pp. 12-14, 67-68, 131-132.)

James R. Dempsey (1921-) was manager of the Astronautics Division for Convair in San Diego, California, from 1957-1958 and then became vice president of the Convair division, 1958-1961. In the latter year, his title became president, General Dynamics Astronautics. He remained in that position until 1965. ("J.R. Dempsey," biographical file, NASA Historical Reference Collection.)

C. Douglas Dillon (1909-) was under secretary of state, 1958-1959, and secretary of the treasury, 1960-1965.

Thomas J. Dodd (1907-1971) (D-CT) received his law degree from Yale in 1933 and served in the Justice Department's civil rights section, 1938-1945, then as a chief trial counsel in the prosecution of Nazi war criminals at Nuremberg in 1945-1946. He began the practice of law in Hartford in 1947 and was a member of the U.S. House of Representatives, 1953-1957. He was elected to the Senate in 1957 and served until 1971 but was defeated for a third term in 1970 after the Senate censured him in 1967 for financial irregularities.

James H. Doolittle (1896-) was a longtime aviation promoter, air racer, Air Force officer, and aerospace research and development advocate. He had served with the U.S. Army Air Corps between 1917 and 1930, and then was manager of the aviation section for Shell Oil Co. between 1930 and 1940. In World War II, Doolittle won early fame for leading the April 1942 bombing of Tokyo, and then as commander of a succession of air units in Africa, the Pacific, and Europe. He was promoted to the rank of lieutenant general in 1944. After the war he was a member of the Air Force's Scientific Advisory Board and the President's Scientific Advisory Committee. At the time of Sputnik he was chair of the National Advisory Committee for Aeronautics and the USAF Scientific Advisory Board. In 1985 the Senate approved his promotion in retirement to four-star general. (General James H. (Jimmy) Doolittle with Carroll V. Glines, *I Could Never Be So Lucky Again: An Autobiography* [New York: Bantam Books, 1991]; Carroll V. Glines, *Jimmy Doolittle: Daredevil Aviator and Scientist* [New York: Macmillan, 1972]; "James H. Doolittle," biographical file, NASA Historical Reference Collection.)

James H. Douglas, Jr. (1899-1988), was secretary of the Air Force between 1957 and 1959 and deputy secretary of defense, 1959-1961. Trained as an attorney, Douglas practiced most of his career in Chicago but served as fiscal assistant secretary of the treasury, 1932-1933, and undersecretary of the Air Force, 1953-1957, prior to becoming secretary of the Air Force. At the conclusion of the Eisenhower administration, Douglas rejoined his old law firm, Gardner, Carton, Douglas, Chilgren & Waud.

Charles Stark Draper (1901-1987) earned his Ph.D. in physics at MIT in 1938 and became a full professor there the following year. In that same year, he founded the Instrumentation Laboratory. Its first major achievement was the Mark 14 gyroscopic gunsight for Navy antiaircraft guns. Draper and the lab applied gyroscopic principles to the development of inertial guidance systems for airplanes, missiles, submarines, ships, satellites, and space vehicles, notably those used in the Apollo moon landings. (John Noble Wilford, "Charles S. Draper, Engineer: Guided Astronauts to the Moon," *New York Times*, 27 Jul. 1987, p. 2; Donald MacKenzie, *Inventing Accuracy: A Historical Sociology of Nuclear Missile Guidance* [Cambridge, MA: MIT Press, 1990], esp. pp. 64-94; C. Stark Draper, "The Evolution of Aerospace Guidance Technology at Massachusetts Institute of Technology, 1935-1951: A Memoir," *Essays on the History of Rocketry and Astronautics*, R. Cargill Hall, ed. [Washington, D.C.: NASA Conf. Pub. 2014, 1977], Vol. II, pp. 219-252.)

Roscoe Drummond was a journalist and editor with the Washington bureau of the *New York Herald Tribune* Syndicate, serving as chief of the bureau between 1953 and 1955, and as a syndicated columnist into the 1960s.

Hugh L. Dryden (1898-1965) was a career civil servant and an aerodynamicist by discipline who had also begun life as something of a child prodigy. He graduated at age 14 from high school and went on to earn an A.B. in three years from Johns Hopkins (1916). Three further years later (1919) he earned his Ph.D. in physics and mathematics from the same institution even though he had been employed full-time in the National Bureau of Standards since June 1918. His career at the Bureau of Standards, which lasted until 1947, was devoted to studying airflow, turbulence, and particularly the problems of the boundary layer—the thin layer of air next to an airfoil that causes drag. In 1920 he became chief of the aerodynamics section in the Bureau. His work in the 1920s on measuring turbulence in wind tunnels facilitated research in the NACA that produced the laminar flow wings used in the P-51 Mustang and other World War II aircraft. From the mid-1920s to 1947, his publications became essential reading for aerodynamicists around the world. During World War II, his work on a glide bomb named the Bat won him a Presidential Certificate of Merit. He capped his career at the Bureau by becoming its assistant director and then associate director during his final two years there. He then served as director of the NACA from 1947-1958, after which he became deputy director of NASA under Glennan and James Webb. (Richard K. Smith, *The Hugh L. Dryden Papers, 1898-1965* [Baltimore, MD: The Johns Hopkins University Library, 1974].)

Lee A. DuBridge (1901-), a physicist with a Ph.D. from the University of Wisconsin (1926), became director of the radiation laboratory at MIT after an academic career capped to that point by a deanship at the University of Rochester, 1938-1941. He was president of the California Institute of Technology between 1946 and 1969, when he resigned to serve as science advisor to Richard M. Nixon. He had been involved in several governmental science advisory organizations before taking up his formal White House duties in 1969 and serving in that capacity until 1970. ("Lee A. DuBridge," biographical file, NASA Historical Reference Collection).

Allen W. Dulles (1893-1969), brother of President Eisenhower's more famous secretary of state, served as director of the CIA from 1953-1961.

Louis G. Dunn (1908-1979), born in South Africa, earned a B.S. (1936), two M.S.s—in mechanical engineering (1937) and aeronautical engineering (1938)— and a Ph.D. (1940) from Caltech and then joined the faculty there. He became assistant director of the Jet Propulsion Laboratory in 1945-1946 and its director from 1947-1954, presiding over its early program in rocketry leading up to the development of the Sergeant missile. He left JPL to take over the beginning Atlas missile project for the recently-formed Ramo-Wooldridge Corporation. He remained there through 1957 as associate director and then director and vice president of the guided missile research division, before becoming executive vice president and general manager, then president, and finally chairman of the firm's Space Technology Laboratories. He left the firm in 1963 to assume various management positions for Aerojet-General Corporation. Besides the Atlas (built by General Dynamics), he played a key role in developing the Thor (McDonnell Douglas), the Titan and Minuteman missiles (Martin Marietta). (Koppes, *JPL*, pp. 31-32, 63-64; "Louis G. Dunn," industry miscellaneous biographical file, NASA Historical Reference Collection.)

Henry C. Dworshak (1894-1962) (R-ID) was elected to the U.S. House of Representatives in 1938 and served there from 1939 to 1946, when he was elected to the Senate, where he remained through 1962.

Frederick M. Eaton (1905-1984) was a lawyer and served in 1960 as chairman of the American delegation to the Disarmament Commission in Geneva.

Allen Joseph Ellender (1890-1972) (D-LA) was first elected to the Senate in 1936 and served until 1972.

Eugene M. Emme (1919-1985) became the NASA chief historian in 1959 and served until his retirement in 1979. Previously he had been a historian with the Air University of the U.S. Air Force. (Sylvia D. Fries, "Eugene M. Emme [1919-1985]," *Technology and Culture*, 27 [July 1986]: 665-67).

Maxime A. Faget (1921-), an aeronautical engineer with a B.S. from LSU (1943), joined the staff at Langley Aeronautical Laboratory in 1946 and soon became head of the performance aerodynamics branch of the pilotless aircraft research division. There, he conducted research on the heat shield of the Mercury spacecraft. In 1958 he joined the space task group in NASA, forerunner of the NASA Manned Spacecraft Center that became the Johnson Space Center, and he became its assistant director for engineering and development in 1962 and later its director. He contributed many of the original design concepts for Project Mercury's manned spacecraft and played a major role in designing virtually every U.S. crewed spacecraft since that time, including the Space Shuttle. He retired from NASA in 1981 and became an executive for Eagle Engineering, Inc. In 1982 he was one of the founders of Space Industries, Inc. and became its president and chief executive officer. ("Maxime A. Faget," biographical file, NASA Historical Reference Collection.)

Philip J. Farley (1916-) earned a Ph.D. from the University of California, Berkeley, in 1941 and was on the faculty at Corpus Christi Junior College from 1941 to 1942 before entering government work—for the Atomic Energy Commission, 1947-1954, and for the State Department, 1954-1969. From 1957 to 1961 he was a special assistant to the secretary of state for disarmament and atomic energy and from 1961 to 1962 his responsibilities shifted to atomic energy and outer space. After several years of assignment to NATO, he returned to Washington and became deputy secretary of state for political-military affairs, 1967-1969. Then from 1969 to 1973 he became deputy director of the U.S. Arms Control and Disarmament Agency.

George J. Feldman (1904-) was a lawyer and financier. He served as consultant for the House Select Committee on Science and Astronautics in 1960. He then became 1 of 14 directors of Communications Satellite Corp. from 1962-1965. (COMSAT, as it is called, is a mixed private-government entity established by legislation in 1962; it had a mandate to cooperate with other countries to set up an international communications satellite system, and it helped set up INTELSAT [International Telecommunications Satellite Organization] in 1964 for that purpose.)

Harold B. Finger (1924-) joined the NACA in 1944 as an aeronautical research scientist at the Lewis facility in Cleveland, where he worked with compressors until 1957 when, having received training in nuclear engineering, he became head of the nuclear radiation shielding group and the nuclear rocket design analysis group. In 1958 he moved to NASA Headquarters to assume duties as chief of the nuclear engine program. By 1962 he had become director of nuclear systems. From 1967 to 1969 he was NASA's associate administrator for organization and management before becoming assistant secretary for research and technology in the Department of Housing and Urban Development from 1969-1972. ("Harold B. Finger," biographical file, NASA Historical Reference Collection.)

John F. Floberg (1915-) was one of six members of the Atomic Energy Commission from 1957 to 1960. A lawyer, he also served as assistant secretary of the Navy from 1949-1953. In 1960 he became general counsel for the Firestone Tire and Rubber Co., later rising to be vice president and then director and member of its executive committee.

Gerald R. Ford (1913-) (R-MI) was elected to the House of Representatives in 1948 and served there until he became vice president in 1973 following the resignation of Spiro Agnew and president, 1974-1978, following Richard M. Nixon's resignation in the wake of the Watergate break-in.

Arnold W. Frutkin (1918-) was deputy director of the U.S. National Committee for the International Geophysical Year in the National Academy of Sciences when NASA hired him in 1959 as director of international programs, a title that changed in 1963 to assistant administrator for international affairs. In 1978 he became associate administrator for external relations, a post he relinquished in 1979 when he retired from federal service. During his career, he had been NASA's senior negotiator for almost all of the important international space agreements. ("Arnold W. Frutkin," biographical file, NASA Historical Reference Collection.)

J. William Fulbright (1905-) (D-AR) became president of the University
of Arkansas in 1939, then served as congressman from 1943 to 1944 and senator
from 1945 to 1974. In 1946 he sponsored the so-called Fulbright act providing for
exchanges of scholars with other countries. From 1959 he served as chairman of the
powerful Senate Foreign Relations Committee. He is best known for his opposition
to the U.S. war in Vietnam.

James G. Fulton (1903-1971) (R-PA) was first elected to the House of
Representatives in 1944 and served through 1971.

Clifford C. Furnas (1900-1969) earned his Ph.D. from the University of
Michigan in 1926 and served as a chemist with the U.S. Bureau of Mines from 1926-
1931; he then taught chemical engineering at Yale from 1931-1942. He became
director of research at Curtiss-Wright Airplane Division, 1943-1946 and served as
vice president for Cornell Aeronautical Lab. from 1946-1954. Becoming chancellor at the University of Buffalo from 1954 to 1962, he then became president of the
State University of N.Y. at Buffalo.

Thomas S. Gates, Jr. (1906-1983) was secretary of the Navy between
1957 and 1959, deputy secretary of defense in 1959, then secretary of defense from
1959-1961. Before that time, Gates had been undersecretary of the Navy, 1953-
1957; director of the Scott Paper Co.; and on active duty with the Navy in World
War II.

Robert R. Gilruth (1913-) was a longtime NACA engineer working at
the Langley Aeronautical Laboratory from 1937-1946, then as chief of the pilotless
aircraft research division at Wallops Island from 1946-1952, who had been
exploring the possibility of human spaceflight before the creation of NASA. He
served as assistant director at Langley from 1952-1959 and as assistant director
(manned satellites) and head of Project Mercury from 1959-1961, technically
assigned to the Goddard Spaceflight Center but physically located at Langley. In
early 1961 Glennan established an independent Space Task Group (already the
group's name as an independent subdivision of the Goddard center) under Gilruth
at Langley to supervise the Mercury program. This group moved to the Manned
Spacecraft Center, Houston, Texas, in 1962. Gilruth was then director of the
Houston operation from 1962-1972. See, Henry C. Dethloff, *"Suddenly Tomorrow
Came...": A History of the Johnson Space Center, 1957-1990* (Washington, D.C.:
NASA SP-4307, 1993); James R. Hansen, *Engineer in Charge: A History of the
Langley Aeronautical Laboratory, 1917-1958* (Washington, D.C.: NASA SP-4305,
1987), pp. 386-88.

James P. Gleason had been appointed to head the NASA office of
congressional relations in late 1958 or early 1959 and served through March 1961.
Thereafter, he practiced law in Washington, served as county executive for
Montgomery County, Maryland, and as an administrative judge with the Nuclear
Regulatory Commission. ("James P. Gleason," biographical file, NASA Historical
Reference Collection.)

Frank E. Goddard, Jr. (1915-) was a long-time employee of the Jet Propulsion Laboratory, having held positions as chief of the high-speed wind tunnel section, chief of the aerodynamics division, and chief of the aerodynamic and propellants department, before assuming duties from 1959 to 1961 as assistant director for NASA relations, with offices in NASA Headquarters. He then became director of planning back at JPL and, in 1962, assistant laboratory director for research and advanced development. ("Frank E. Goddard," biographical files, NASA miscellaneous, NASA Historical Reference Collection.)

Harry J. Goett (1910-) earned a degree in physics from Holy Cross College in 1931 and one in aeronautical engineering from NYU in 1933. After holding a number of engineering posts with private firms, he became a project engineer at Langley Aeronautical Laboratory in 1936. He later moved to Ames Aeronautical Laboratory, where he was chief of the full-scale and flight research division, 1948-1959. In the latter year he became director of the Goddard Space Flight Center, a post he held until July 1965, when he became a special assistant to NASA Administrator James E. Webb. Later that year he became director for plans and programs at Philco's Western Development Labs in California and ultimately retired from a position with Ford Aerospace and Communications. ("Harry J. Goett," biographical folder, NASA Historical Reference Collection.)

Barry M. Goldwater (1909-) (R-AZ) was a U.S. senator from 1953-1965. In 1964 he ran unsuccessfully for president of the U.S. against Lyndon Johnson. He was an outspoken conservative and became the leader and later elder statesman for the right wing of the Republican party.

Nicholas E. Golovin (1912-1969), born in Odessa, Russia, but educated in this country (Ph.D. in physics, George Washington University, 1955) worked in various capacities for the government during and after World War II, including for the Naval Research Laboratory, 1946-1948. He held several administrative positions with the National Bureau of Standards from 1949 to 1958. In 1958 he was chief scientist for the White Sands Missile Range and then worked for the Advanced Research Projects Agency in 1959 as director of technical operations. He became deputy associate administrator for NASA in 1960. He joined private industry before becoming, in 1961, the director of the NASA-DOD large launch vehicle planning group. He joined the Office of Science and Technology at the White House in 1962 as a technical advisor for aviation and space and remained there until 1968 when he took a leave of absence as a research associate at Harvard and as a fellow at the Brookings Institution. (Obituaries, *Washington Star*, 30 Apr. 1969, p. B-6, and *Washington Post*, 30 Apr. 1969, p. B14.)

Andrew Jackson Goodpaster (1915-) was a career Army officer who served as defense liaison officer and secretary of the White House staff from 1954 to 1961, being promoted to brigadier general during that period. He later was deputy commander, U.S. forces in Vietnam, 1968-1969, and commander-in-chief, U.S. forces in Europe, 1969-1974. He retired in 1974 as a four-star general but returned to active duty in 1977 and served as superintendent of the U.S. Military Academy, a post he held until his second retirement in 1981.

Albert A. Gore (1907-) (D-TN) was elected to the U.S. House of Representatives in 1938. He was reelected to each succeeding Congress until 1952 when he won election to the Senate. He served there until 1970.

Melvin N. Gough (1906-) earned a B.S. in mechanical engineering from Johns Hopkins in 1926 and joined the wind tunnel staff of the Langley Aeronautical Laboratory. Taking a leave of absence, he learned to fly with the Navy at Pensacola and became an NACA test pilot in 1929. He logged more than 6,000 hours of flying time and flew more than 300 different airplanes under test conditions. In 1943 he became director of flight research activities at Langley. He was assigned in 1958 as director of NASA activities at the Atlantic Missile Range, Cape Canaveral, Florida. In 1960 he became director of the bureau of safety at the Civil Aeronautics Board. Two years later, he joined the FAA as director of the new aircraft development service. He retired in 1963. ("Melvin N. Gough," biographical file, NASA Historical Reference Collection.)

Theodore Granik was a lawyer and the founder of "American Forum of the Air," a weekly radio program for discussion of national problems, inaugurated in 1928 and later broadcast on television as well.

Edward Z. Gray (1915-) worked for Boeing Co. from 1943-1963 as a design engineer for the Boeing jet aircraft series as well as the DynaSoar and Minuteman programs. He held a number of positions in systems engineering management, the last one being as development program manager of advanced space systems. He served on numerous committees for the government and aerospace industry, including the NASA research advisory committee on structural loads in 1958-1959, of which he was chairman. In 1963 NASA appointed him to the directorship of its advanced manned missions programs. He worked in that position through 1967, transferred to a position as assistant to the president of Grumman Aircraft Engineering Corp. from 1967-1973, and then returned to NASA as assistant administrator for industry affairs and technology utilization. By 1978 he had assumed a position as director of government/industry affairs. In 1979 he joined Bendix Corp.'s aerospace-electronics group as director of systems development. ("Edward Z. Gray," biographical files, NASA Historical Reference Collection.)

Gordon Gray (1909-1982) was a former publishing company executive and past president of the University of North Carolina who had served in various positions in the DOD and presidential administrations, including a period as secretary of the Army from 1949-1950. He served as special assistant to the president for national security affairs from 1958 to 1961. (Bell, *Secretaries of War and Secretaries of the Army*, p. 134; *New York Times*, 28 November 1982, p. 44.)

Crawford H. Greenewalt (1902-) had been president of E.I. du Pont de Nemours & Co. since 1948 and had been with the company in a series of positions since 1922. The Greenewalt committee, appointed by Glennan to advise him on the goals of the NASA space program and consisting of ten members including Greenewalt, convened in late 1959 and continued into 1960 to explore and identify the national objectives to be served by a program of non-military space activities and in particular to examine the significance of competition with the Soviet Union in that arena. The committee deemphasized the issue of preeminence in space. ("Greenewalt Committee" and "Ad Hoc Advisory Committee" files, Glennan subseries, NASA Historical Reference Collection.)

H. R. Gross (1899-1987) (R-IA) was elected to the House of Representatives in 1948 and served through the mid-1970s.

Robert Ellsworth Gross (1897-1961) had worked for Lee Higginson Corp. (1919-1927), Stearman Aircraft Co. (1927-1928), and the Viking Flying Boat Co. (as president, 1928-1930). In 1932 he became president and chairman of the board of Lockheed Aircraft Corp., a position he held until at least 1960.

John P. Hagen (1908-1990) was a solar radio astronomer who earned an M.A. from Wesleyan in 1931 and began working for the Naval Research Laboratory in 1935. There he worked on improving radar techniques and helped develop an automatic ground speed indicator for aircraft. After World War II he headed NRL's radio physics research group, which developed the world's most precise radio telescope in 1950, a year after he earned a Ph.D. in astronomy at Georgetown. In 1955 he became director of the Vanguard earth satellite program and, when that program became part of NASA on 1 October 1958, he remained chief of the NASA Vanguard division and then (1958-1960) became assistant director of space flight development. In February 1960 he became director of NASA's office for the United Nations conference and later, assistant director of NASA's office of plans and program evaluation. In 1962 he set up a graduate program of radio astronomy at Pennsylvania State University, retiring from there as head of the astronomy department in 1975. ("John P. Hagen," biographical file, NASA Historical Reference Collection.)

James C. Hagerty (1909-1981) had been on the staff of the *New York Times* from 1934 to 1942, the last four years as legislative correspondent in the paper's Albany bureau. He served as executive assistant to New York Governor Thomas Dewey from 1943 to 1950 and then as Dewey's secretary for the next two years before becoming press secretary for President Eisenhower from 1953 to 1961.

Leonard W. Hall (1900-1979) (R-NY) was a congressman from 1939-1952 and then served as chairman of the Republican national committee from 1953-1957. A lawyer, he then became the senior partner in the Long Island law firm of Hall, Casey, Dickler, & Brady.

Richard Harkness (1907-1977) was a radio and television news commentator who became NBC's Washington correspondent in 1943 and stayed in that position through 1970.

Bryce N. Harlow (1916-1987) was deputy assistant to the president for congressional affairs, a position he had held since 1959. He had held other positions on the White House staff since 1953. From 1938 to 1951 he was on the congressional staff, rising to be chief clerk, 1950-1951. In 1951-1952 he was vice president of Harlow Pub. Corp in Oklahoma City. He became director of governmental relations for Proctor and Gamble Manufacturing Company from 1961 to 1969 and then rejoined the White House as assistant to the president for legislative and congressional Affairs, becoming counselor to the president, 1969-1970. He served as vice president of Proctor and Gamble, 1970-1973, then returned as counselor to the president at the height of the Watergate scandal, remaining until April 1974, when he resigned and returned to private life.

Karl G. Harr, Jr. (1922-) was special assistant to the president and vice chair of the Operations Coordinating Board (OCB) between 1958 and 1961. Before that he had been a special intelligence officer with the U.S. Army in World War II, attended Yale Law School, and practiced law until 1954 when he began work with the Department of State. In 1963 he assumed the presidency of the Aerospace Industries Association and served until 1988. In 1989 he was named a senior fellow with the Eisenhower Institute for World Affairs. ("Karl G. Harr, Jr.," biographical file, NASA Historical Reference Collection.)

Rupert Vance Hartke (1919-) (D-IN) was first elected to the Senate in 1958 and served until 1977.

Carl T. Hayden (1877-1972) (D-AZ) served the new state of Arizona in Congress from 1912 to 1927. Elected to the Senate in 1926, he remained a senator until 1969 and was president pro tempore of the Senate from 1957-1969.

John Tucker Hayward (1910-) was a career naval officer and naval aviator whose assignments had included the Manhattan Project and the Armed Forces Special Weapons Base, Sandia, NM. He was serving as deputy commanding officer for research and development, Navy Department, from 1957 to 1963. He became a vice admiral in 1959, retired in 1968, and became vice president of General Dynamics Corporation in that year.

Donald H. Heaton was an Air Force officer who from 1951 to 1957 as a lieutenant colonel and colonel had served on various subcommittees of the NACA committee on power plants for aircraft as well as on the committee itself. Available information does not indicate just when he joined NASA Headquarters, but the August 1959 telephone directory shows him working in the office of the assistant director of propulsion within the office of space flight development. He served in a variety of positions connected with launch vehicles, and in June 1961 Associate Administrator Robert Seamans appointed him chairman of an ad hoc task group to formulate plans and determine the resources necessary to carry out a manned lunar landing. His group submitted its summary report in August 1961. He appears to have left NASA Headquarters sometime between June and October 1963. ("Donald H. Heaton," biographical file, NASA Historical Reference Collection and headquarters telephone directories for the period; on his committee's report, see especially Brooks, Grimwood, and Swenson, *Chariots for Apollo*, pp. 45, 70-72.)

Harlow J. Heneman (1906-1983), besides being a general partner of Cresap, McCormick and Paget, was a management consultant. He held a Ph.D. from the University of London and was on the political science faculty of the University of Michigan from 1933-1945. Thereafter, he held a number of positions in and outside of government, including that of management analyst with the Bureau of the Budget, 1944-1945.

Christian A. Herter (1895-1966) was undersecretary of state, 1957-1959, and then succeeded John Foster Dulles as secretary of state from 1959-1961. He never achieved the level of mutual understanding with President Eisenhower that Dulles had enjoyed, however, and thus failed to have the sort of influence in developing the administration's foreign policy that his predecessor had achieved. (Pach and Richardson, *Presidency of Eisenhower*, p. 204.)

Bourke B. Hickenlooper (1896-1971) (R-IA), a former governor of Iowa (1943-1944), was first elected to the U.S. Senate in 1944 and served until 1969.

Daniel C. Hickson (1906-) had been the vice president of Bankers Trust Co. since 1947.

Lister Hill (1894-1984) (D-AL) was elected to fill a vacant position in the House of Representatives in 1923 and served until 1939. The previous year, he had been elected to fill an unexpired term in the Senate, where he served through 1969.

John H. Hinrichs (1904-) was a career army officer and at this time was deputy chief of ordnance, U.S. Army Field Services Division. He had been promoted to major general in 1954.

Wesley L. Hjornevik (1926-) began federal service in 1949 as a budget examiner and program analyst. In 1957 he became assistant to the under secretary in the Department of Health, Education, and Welfare, and in October of the next year he moved to NASA as assistant to T. Keith Glennan. He became deputy director of business administration on 15 December 1959. In 1961 he became associate director for manned space flight in Houston, Texas, serving until October 1969 when he became the deputy director of the Office of Economic Opportunity. In 1974 he became the director of public administration for the state of Texas. ("Wesley L. Hjornevik," biographical file, NASA Historical Reference Collection.)

Alfred S. Hodgson was NASA's director of management analysis from 1958-1960 and then became director of business administration in the headquarters. By 1962 he was assistant to the director of administration, and thereafter he became the director of the headquarters administration office, a position from which he appears to have retired during 1968. (Miscellaneous biographical file and headquarters telephone directories, Sept. and Dec. 1968, NASA Historical Reference Collection.)

William M. Holaday (1901-) was special assistant to the secretary of defense for guided missiles between 1957 and 1958, then DOD director of guided missiles in 1958 and chairman of the civilian-military liaison committee, 1958-1960. Previously Holaday had been associated with a variety of research and development activities, notably as director of research for the Socony-Mobil Oil Co., 1937-1944. ("William M. Holaday," biographical file, NASA Historical Reference Collection.)

Chester E. Holifield (1903-) (D-CA) was first elected to the House of Representatives in 1942 and served until the mid-1970s.

Spessard L. Holland (1892-1971) (D-FL) was first appointed and then elected to the Senate in 1946 and served there until 1970.

George W. Hoover was an early space enthusiast who had entered the Navy in 1944 and become a pilot. He moved to the Office of Naval Research to conduct a program in all-weather flight instrumentation. Later he helped originate the idea of high-altitude balloons that were used in a variety of projects like Skyhook, which supported cosmic-ray research and served as a research vehicle for obtaining environmental data relevant to supersonic flight, among other uses. In 1954 he was project officer in the field of high-speed, high-altitude flight, with involvement in the Douglas D558 project leading to the X-15. Hoover was also instrumental in establishing Project Orbiter with von Braun and others, resulting in the launch of Explorer I, the first American satellite. ("George W. Hoover," biographical file, NASA Historical Reference Collection.)

Richard E. Horner (1917-) was associated with aerospace activities throughout his career. He served as a pilot in the U.S. Army Air Forces during World War II and on active duty between 1945 and 1949 as director of flight test engineering at Wright Field, Ohio (1944-1945 and 1947-1949). He was promoted to colonel in 1948. Between 1950 and 1955 he was first technical director and then senior engineer for the Air Force Flight Test Center at Muroc, California. In May 1955, Horner became deputy for requirements in the office of the assistant secretary of the Air Force, R&D, and in 1957 he became assistant secretary for research & development. In June 1959 he left the USAF to become NASA associate administrator. He resigned from NASA in July 1960 and became senior vice president of the Northrop Corp. In 1970 he joined the E.F. Johnson Co. as president and chief executive officer. ("Richard E. Horner," biographical file, NASA Historical Reference Collection.)

Donald F. Hornig (1920-), a chemist, was a research associate at the Woods Hole Oceanographic Lab, 1943-1944, and a scientist and group leader at the Los Alamos Scientific Laboratory, 1944-1946. He taught chemistry at Brown University starting in 1946, rising to the directorship of Metcalf Research Lab, 1949-1957, and also serving as associate dean and acting dean of the graduate school from 1952-1954. He was Donner Professor of Science at Princeton from 1957-1964 as well as chairman of the chemistry department from 1958-1964. He was a special assistant to the president of the U.S. on science and technology from 1964-1969 and president of Brown University from 1970-1976.

Roman Lee Hruska (1904-) (R-NB) was elected to the House of Representatives in 1952 and to the Senate in 1954. He remained in the Senate until 1976.

Abraham Hyatt (1910-) earned a B.S. in aeronautical engineering from Georgia Tech in 1933. After working for the U.S. Geodetic Survey and private industry, in 1948 he became head of the design research branch for the Navy's Bureau of Aeronautics and advanced to chief scientist and research analysis officer there, 1956-1958. In 1959 he became assistant director for propulsion in NASA. The following year he became director of NASA's office of program planning and evaluation. He remained in that position until 1964 when he became a professor at MIT and then, in 1965, executive director for corporate planning at North American Aviation, Inc. ("Abraham Hyatt," biographical files, NASA Historical Reference Collection.)

Henry M. ("Scoop") Jackson (1912-1983) (D-WA) was first elected to the House of Representatives in 1940 and to each succeeding Congress until 1952, when he was elected to the Senate, where he served until the mid-1980s. During the Eisenhower administration, he was a leading advocate of greater attention to the development of the U.S. missile program.

Robert Jastrow (1925-) earned a Ph.D. in theoretical physics from Columbia in 1948 and pursued post-doctoral studies at Leiden, Princeton (Institute for Advanced Studies), and the University of California at Berkeley before becoming an assistant professor at Yale in 1953-1954. He then served on the staff at the Naval Research Laboratory from 1954-1958. In the latter year he was appointed chief of the theoretical division of the Goddard Spaceflight Center. He became director of the Goddard Institute of Space Studies in 1961 and stayed at its helm for 20 years before becoming professor of earth sciences at Dartmouth. He specialized in nuclear physics, plasma physics, geophysics, and the physics of the moon and terrestrial planets. ("Robert Jastrow," biographical files, NASA Historical Reference Collection.)

Jacob K. Javits (1904-1986) (R-NY) was elected to the House of Representatives in 1946 and served through 1954. After a term as attorney general of New York, he was elected to the Senate in 1956 and served until 1980.

Ben Franklin Jensen (1892-1970) (R-IA) was first elected to the House of Representatives in 1938 and served into the mid-1960s.

John A. Johnson (1915-), after completing law school at the University of Chicago in 1940, practiced in Chicago until 1943 when he entered military service with the Navy. From 1946 to 1948 he was an assistant for international security affairs in the Department of State. He joined the office of the general counsel of the Department of the Air Force in 1949 and served until 7 October 1958 (for the last six years as the general counsel) when he accepted the general counsel position in NASA. In 1963 he left NASA to become director of international arrangements at the Communications Satellite Corporation. The next year he became a vice president of COMSAT, and in 1973, senior vice president and then chief executive officer. He retired in 1980. ("John A. Johnson," biographical file, NASA Historical Reference Collection.)

Lyndon B. Johnson (1908-1973) (D-TX) was elected to the House of Representatives in 1937 and served until 1949. He was a senator from 1949-1961, vice president of the U.S. from 1960-1963, and president from then until 1969. Best known for the social legislation he passed during his presidency and for his escalation of the war in Vietnam, he was also highly instrumental in revising and passing the legislation that created NASA and in supporting the U.S. space program as chairman of the Committee on Aeronautical and Space Sciences and of the preparedness subcommittee of the Senate Armed Services Committee, then later as chairman of the National Aeronautics and Space Council when he was vice president. (On the NASA legislation, Griffeth, *National Aeronautics and Space Act*, passim; on his role in support of the space program, Robert A. Divine, "Lyndon B. Johnson and the Politics of Space," in *The Johnson Years*, vol. II: *Vietnam, the Environment, and Science*, ed. Robert A. Divine [Lawrence, KS, 1987] and Robert Dallek, "Johnson, Project Apollo, and the Politics of Space Program Planning," unpublished paper delivered at a symposium on "Presidential Leadership, Congress, and the U.S. Space Program," sponsored by NASA and American University, 25 March 1993.)

Roy W. Johnson (1906-1965) was named director of the Advanced Research Projects Agency for the Department of Defense in 1958, serving until 1961. Previously he had been with the General Electric Co. He was a strong proponent of exploiting space for national security objectives. ("Roy W. Johnson Dead; First U.S. Space Chief," *Washington Post*, 23 July 1965.)

Roger W. Jones (1908-) worked in various capacities for the Bureau of the Budget from 1939 to 1959, rising in the last two years to be deputy director. He was chairman of the Civil Service Commission from 1959 to 1961. He held various other government posts thereafter, including that of assistant director, Office of Management and Budget, 1969-1971.

Robert W. Kamm (1917-) graduated from New York University in 1939 with a bachelors degree in aeronautical engineering and joined the Langley Aeronautical Laboratory, where he investigated spin characteristics of various military aircraft in wind tunnels. In 1946 he left the NACA to become senior aerodynamicist with the Glenn L. Martin Company. In 1950 he went to work for the Air Force's Arnold Engineering Development Center, where he became chief of the plans and policy office in 1957. In 1959 he accepted an appointment as director of NASA's western operations office in Santa Monica, responsible for contract negotiations and administration, public information, technical representation, financial management, security, legal and patent administration. In 1968 he retired from that position and NASA to become assistant to the director of the Space Institute at the University of Tennessee, Tullahoma.

Arthur Kantrowitz (1913-) earned his Ph.D. in physics from Columbia in 1947 after having worked as a physicist for the NACA from 1936 to 1946. He taught at Cornell for the next decade, meanwhile founding the Avco-Everett Research Lab in Everett, Massachusetts, in 1955. He served as its director, senior executive officer, and chairman until 1978 when he became a professor at Dartmouth. From 1956 to 1978 he also served as a vice president and director of Avco Corporation.

Joseph E. Karth (1922-) (Democrat-Farmer-Labor-MN) was first elected to the House of Representatives in 1958 and served through the mid-1970s.

Kenneth B. Keating (1900-1975) (R-NY) was elected to the House of Representatives in 1946 and served there through 1958. Elected to the Senate the latter year, he served through 1965 and then became an associate justice in the New

York Court of Appeals for three years before becoming ambassador to India in 1969.

William B. Keese (1910-) was a career Air Force officer who became a major general in 1960 and was the director of developmental planning at Headquarters, U.S. Air Force from 1960-1962.

Robert F. Keller (1913-1980) had worked for the General Accounting Office from 1935-42 and 1946-69. He became general counsel in 1958. In 1969 he became controller general of the U.S.

Mervin J. Kelly (1894-1971) was a longtime research physicist with Bell Telephone Laboratories, becoming director of research in 1934, vice president in 1944, and president of the organization between 1951 and 1959. His work at the laboratories focused on radar, gunfire control, and bombsights. After his retirement from Bell, Kelly was named advisor to NASA Administrator James E. Webb in 1961. (Obituary, *New York News Herald*, 20 March 1971, p. 32.)

David W. Kendall (1903-1976) served as special counsel to the president from 1958-1961. He had previously been general counsel of the U.S. Treasury, 1954-1955 and then assistant secretary of the Treasury, 1955-57.

Robert S. Kerr (1896-1963) (D-OK) had been governor of Oklahoma from 1943-1947 and was elected to the Senate the following year. In 1961 he chaired the Aeronautical and Space Sciences Committee.

Seymour S. Kety (1915-) was a physician who worked with the National Institute of Mental Health (NIMH) throughout the 1950s. In 1951 he became associate director in charge of research for NIMH and for neurological diseases and blindness, and in 1956 he moved to the directorship of the Laboratory of Clinical Sciences. In 1959 Kety was the chair of NASA's bioscience advisory committee. In 1967 he left NIMH and became a professor of psychiatry at Harvard Medical School, where he assumed emeritus status in 1983. (See esp. "Seymour S. Kety" biographical file, NASA Historical Reference Collection.)

David Keyser (1918-) became chief congressional liaison officer for NASA in 1959. He had worked from 1951-1955 as administrative assistant to Congressman Charles J. Kersten of Wisconsin. Just before his NASA appointment, he had worked as a municipal consultant to various city governments.

Nikita S. Khrushchev (1894-1971) was premier of the USSR from 1958 to 1964 and first secretary of the Communist party from 1953 to 1964. He was noted for an astonishing speech in 1956 denouncing the crimes and blunders of Joseph Stalin and for gestures of reconciliation with the West in 1959-1960, ending with the breakdown of a Paris summit with President Eisenhower and the leaders of France and Great Britain in the wake of Khrushchev's announcement that the Soviets had shot down an American U-2 reconnaissance aircraft over the Urals on 1 May 1960. Then in 1962 Khrushchev attempted to place Soviet medium range-missiles in Cuba. This led to an intense crisis in October, following which Khrushchev agreed to remove the missiles if the U.S. promised to make no more attempts to overthrow Cuba's Communist government. Although he could be charming at times, Khrushchev was also given to bluster (extending even to shoe-pounding at the U.N.) and was a tough negotiator, although he believed, unlike his predecessors, in the possibility of Communist victory over the West without war. See his *Khrushchev Remembers: The Last Testament* (Boston: Little, Brown, 1974); Edward Crankshaw, *Khrushchev: A Career* (New York: Viking, 1966); Michael R. Beschloss, *Mayday: Eisenhower, Khrushchev and The U-2 Affair* (New York: Harper & Row, 1986); and Robert A. Divine, *Eisenhower and the Cold War* (New York: Oxford University Press, 1981) as well as Eisenhower's *Waging Peace* for further information about him.

   James R. Killian, Jr., (1904-1988) was president of the Massachusetts Institute of Technology between 1949 and 1959, on leave between November 1957 and July 1959 when he served as the first presidential science advisor. President Dwight D. Eisenhower established the President's Science Advisory Committee (PSAC), which Killian chaired, following the Sputnik crisis. After leaving the White House staff in 1959, Killian continued his work at MIT but in 1965 began working with the Corporation for Public Broadcasting to develop public television. Killian described his experiences as a presidential advisor in *Sputnik, Scientists, and Eisenhower: A Memoir of the First Special Assistant to the President for Science and Technology* (Cambridge: MIT Press, 1977). For a discussion of the PSAC see Gregg Herken, *Cardinal Choices: Science Advice to the President from Hiroshima to SDI* (New York: Oxford University Press, 1992).

   David S. King (1917-) (D-UT) was elected to the House of Representatives in 1958 and served through 1962. He was reelected in 1964 for one term and then became an ambassador to the Malagasy Republic.

   George B. Kistiakowsky (1900-1982) was a pioneering chemist at Harvard University, associated with the development of the atomic bomb, and later an advocate of banning nuclear weapons. He served as science advisor to President Eisenhower from July 1959 to the end of the administration. He later served on the advisory board to the United States Arms Control and Disarmament Agency from 1962 to 1969. (*New York Times*, 9 December 1982, p. B21; "George B. Kistiakowsky," biographical file, NASA Historical Reference Collection.)

   William F. Knowland (1908-1974) (R-CA) served in the Senate between 1945 and 1959. (*Washington Post*, 5 October 1959, p. C3; *Guide to Research Collections of Former United States Senators, 1789-1982* [Washington, D.C.: Government Printing Office, 1983], p. 291).

   Robert L. Krieger (1916-1990) began his career with the NACA and NASA at the Langley Aeronautical Laboratory in 1936 as a laboratory apprentice. Leaving the NACA for college, he earned a B.S. in mechanical engineering at Georgia Tech in 1943 and returned to Langley. From there, he was part of the group that set up the Pilotless Aircraft Research Station at Wallops Island under Robert R. Gilruth in 1945. In 1948 he became the head of the Wallops facility, which performed aerodynamic tests on instrumented models propelled at high speeds. In 1958 Wallops became an independent field center of NASA; there, Krieger led the first successful test flight of the Mercury capsule. During his career there, Wallops launched thousands of test vehicles, including 19 satellites. He retired as director in 1981. ("Robert L. Krieger," biographical file, NASA Historical Reference Collection.)

Hermann H. Kurzweg (1908-) was born in Germany and earned his Ph.D. from the University of Leipzig in 1933. During the Second World War, he was chief of the research division and deputy director of the aerodynamic laboratories at Peenemünde, where he did aerodynamic research on the V-2 rocket and the antiaircraft rocket Wasserfall as well as participated in the design of the supersonic wind tunnels there. In 1946 he came to the U.S. and worked for the Naval Ordnance Laboratory at White Oak, Maryland, doing aerodynamics and aeroballistics research and becoming associate technical director of the lab in 1956. He joined NASA Headquarters in September 1960 as assistant director for aerodynamics and flight mechanics in the office of advanced research programs. In 1961 he became director of research in the office of advanced research and technology. Nine years later, he was appointed chief scientist and chairman of the research council in the same office. He retired in 1974. ("Hermann H. Kurzweg," biographical file, NASA Historical Reference Collection.)

Robert J. Lacklen had joined the NACA in 1945 as classification and organization officer. He became head of the NACA personnel administration two years later. When NASA succeeded the NACA, he became director of the personnel division, a position he held until 1964, when he resigned to become head of a personnel research institute at the Richardson Foundation in Greensboro, North Carolina. ("Robert J. Lacklen," biographical file, NASA Historical Reference Collection.)

Thomas G. Lanphier, Jr. (1915-) was a special assistant to the secretary of the Air Force, 1949-1950, and became vice president of Consolidated Vultee Aircraft Corp., 1951-1960; then president of Fairbanks, Morse, and Co, 1960-1962; before becoming vice president for corporate planning of Raytheon in 1962.

Richard E. Lankford (1914-) (D-MD) was elected as a representative of Maryland's fifth district in Congress in 1954, a seat he retained through 1964.

Ludwig George Lederer (1911-1978) was a specialist in internal and aviation medicine. He was medical director for Capital Airlines from 1942-1960 and was simultaneously medical examiner and physician in chief of the Washington National Airport. In 1960 he became medical director for American Airlines. At about this time he was president of the aerospace medical association.

Max Lehrer was the assistant staff director of the Committee on Aeronautical and Space Sciences of the U.S. Senate at this time. (Letter, T. Keith Glennan to Max Lehrer, 30 March 1960, "Congress" file, Glennan subseries, NASA Historical Reference Collection.)

Lyman L. Lemnitzer (1899-1988) was a career army officer who served as Army vice chief of staff between 1957 and 1959; Army chief of staff, 1959-1960; chairman of the Joint Chiefs of Staff, 1960-1962; commanding general of United

States Forces, Europe, 1962-1969; and Supreme Allied Commander, Europe, 1963-1969. (Bell, *Commanding Generals and Chiefs of Staff*, p. 132; *New York Times*, 13 November 1988, p. 44).

William E. Lilly (1921-) entered federal civilian service in 1950 as a budget and program analyst with the Navy Ordnance Test Station in California and held a variety of positions with the Navy and the Bureau of Standards until 1960 when he joined NASA as chief, plans and analysis, office of launch vehicles. He served NASA for 21 years, becoming its first comptroller—a position with associate administrator status—in 1973. He retired in 1981 with 37 years of federal service including service in the Navy from 1940-1946.

Ernest K. Lindley (1899-1979) was a Rhodes scholar in 1923 and served as a reporter and political writer for the *New York World* from 1924-1931, then wrote for the *Herald Tribune* from 1931-1937. He became chief of the Washington bureau of *Newsweek* from 1937-1961 and also served as a political commentator for the *Washington Post* for part of that period.

John V. Lindsay (1921-) (R-NY) served in the House of Representatives from 1959-1965, when he became mayor of New York City.

Albert W. Lines (1914-) was a British physicist who had previously been the principal scientific officer at the U.K. Ministry of Supply. His appointment to the directorship of the Royal Aircraft Establishment at Farnsborough apparently was quite recent because the 1959 *Directory of British Aviation* listed Sir George Gardner in that position and its "Who's Who in British Aviation" did not even mention Lines.

Donald P. Ling was on the staff at Bell Telephone Laboratories. In 1954 he had co-authored a two-volume report entitled "Command Guidance for a Ballistic Missile." He continued working in this area and later became vice president of Bell Labs and, in 1970, president of Bellcom, Inc., a subsidiary incorporated in 1962. He retired in 1971. (*A History of Engineering and Science in the Bell System: National Service in War and Peace (1925-1975)*, M. D. Fagen, ed. [Bell Telephone Laboratories, 1978], pp. 396, 447, 506, 699.)

Walter Lippmann (1889-1974) was perhaps the most eminent and influential journalist of his day. He helped found and then edit *The New Republic* in 1914. He wrote editorials for and then edited the *World* from 1921 to 1931 and then began a column for the *New York Herald Tribune* that was eventually syndicated in more than 250 newspapers and won two Pulitzer Prizes (1958, 1962).

William Littlewood (1898-1967) was a vice president of American Airlines in charge of engineering from 1937 to 1963. Before that, he was regarded as the developer of the DC-3 that helped revolutionize air travel. From 1946 to 1964 he chaired the NACA committee on operating problems and its NASA successor, the committee on aircraft operating problems and was one of the country's most highly regarded aircraft engineers, known as an advocate of the government's devoting more resources to the research and development of aircraft rather than spacecraft. ("William Littlewood," biographical file, NASA Historical Reference Collection.)

Henry Cabot Lodge, Jr. (1902-1985) had been a senator from Massachusetts (1937-1944, 1947-1952) and had been active in promoting President

Eisenhower's presidential candidacy. Eisenhower appointed him permanent representative to the U.N. (1953-1960), from which position he advised the president on domestic affairs as well as U.N. issues. He was Richard M. Nixon's vice presidential running mate in 1960 and then U.S. ambassador to South Vietnam (1963-1964, 1965-1967). See his *As It Was: An Inside View of Politics and Power in the '50s and '60s* (New York: W. W. Norton, 1976).

A. C. Bernard Lovell (1913-) taught physics at the University of Manchester in England before World War II, specializing in cosmic ray investigations. During the war he worked on radar development. Upon return to Manchester, he established the Jodrell Bank station in nearby Cheshire, setting up a radio telescope with a 250-foot antenna. It was completed in 1957 and was used by NASA to receive signals from the Pioneer series of moon probes and as a sensitive receiver of signals bounced off the passive Echo satellite in the period of this diary. ("Bernard Lovell," biographical file, NASA Historical Reference Collection.)

George M. Low (1926-1984), a native of Vienna, Austria, came to the U.S. in 1940 and received an aeronautical engineering degree from Rensselaer Polytechnic Institute (RPI) in 1948 and an M.S. in the same field from that school in 1950. He joined the NACA in 1949 and at Lewis Flight Propulsion Laboratory he specialized in experimental and theoretical research in several fields. He became chief of manned space flight at NASA Headquarters in 1958. In 1960, he chaired a special committee that formulated the original plans for the Apollo lunar landings. In 1964 he became deputy director of the Manned Spacecraft Center in Houston, the forerunner of the Johnson Space Center. He became deputy administrator of NASA in 1969 and served as acting administrator in 1970-1971. He retired from NASA in 1976 to become president of RPI, a position he still held at his death. In 1990 NASA renamed its quality and excellence award after him. ("George M. Low," biographical file, NASA Historical Reference Collection.)

Alvin R. Luedecke (1910-) served in the Army Air Corps and the Air Force from 1934 to 1958, rising through the ranks to become a major general. He served as the executive secretary of the military liaison committee to the Atomic Energy Commission from 1949-1951 and was thereafter deputy chief and then chief of the Armed Forces Special Weapons Project, 1951-1957, and commander of Joint Task Force 7, 1957-1958. From 1958 to 1964 he was general manager of the AEC. Thereafter, he became deputy director of the Jet Propulsion Laboratory, 1964-1967; associate dean of engineering at Texas A&M, 1968-1970; acting president of Texas A&M, 1970; and the university's executive vice president beginning in 1971. ("Alvin R. Luedecke," biographical file, NASA Historical Reference Collection.)

Bruce T. Lundin (1919-) earned a B.S. in mechanical engineering from the University of California in 1942 and worked for Standard Oil of California before joining the staff at Lewis Laboratory in 1943. He investigated heat transfer and worked to improve the performance of World War II aircraft engines. Then in 1946 he became chief of the jet propulsion research section, which conducted some of America's early research on turbojet engines. He became assistant director of Lewis in 1958 and directed much of the center's efforts in space propulsion and power generation. He advanced through the positions of associate director for development (1961) at Lewis, managing the development and operation of the

Centaur and Agena launch vehicles, and of deputy associate administrator for advanced research and technology at NASA Headquarters (1968), before becoming acting associate administrator for advanced research and technology there (1969). Later that year, he received the appointment as director of the Lewis Research Center, where he remained until his retirement in 1977. ("Bruce T. Lundin," biographical file, NASA Historical Reference Collection.)

Warren G. Magnuson (1905-1989) (D-WA) was elected to the House of Representatives in 1936 and served until 1944 when he was appointed to fill an unexpired term in the Senate. He was subsequently elected to the Senate later that year and remained a senator until 1981.

George C. Marshall (1880-1959) was a career Army officer who served as general of the army and U.S. Army chief of staff during World War II. He became secretary of state (1947-1949) and of defense (1950-1951) and was the author of the European recovery program known to the world as the Marshall Plan; it played a critical role in reconstructing a Europe ravaged by the war that Marshall had done so much to direct to a victorious end. In recognition of the effects of the Marshall Plan and his contributions to world peace, he received the Nobel Prize for Peace in 1953. It was fitting that a NASA center should be named after the only professional soldier to receive the prize, given NASA's charter to devote itself to the peaceful uses of outer space and yet to cooperate with the military services. (The standard source on Marshall is the magisterial, multivolume biography by Forrest C. Pogue, *George C. Marshall* [New York: Viking, 1963-1966], but there are several recent one-volume studies, including Mark A. Stoler, *George C. Marshall: Soldier-Statesman of the American Century* [Boston: Twayne, 1989].)

Joseph W. Martin, Jr. (1884-1968) (R-MA) had been a member of Congress since 1924 and until 1958, minority leader of the House of Representatives every session since 1939 except for the years 1947-1949 and 1953-1955, when he was speaker of the House.

Paul Logan Martin (1912-1978) worked for a variety of newspapers and the Press Association, Inc. before becoming the political and legal correspondent for Gannett Newspapers from 1947-1950. From 1950 to 1966, he was chief of Gannett's bureau before becoming an editor for *U.S. News and World Report*.

Edward A. McCabe (1917-) was part of Eisenhower's congressional liaison staff. His formal titles were associate counsel to the president, 1956-1958, and administrative assistant, 1958-1961. After the end of the Eisenhower administration, McCabe became a partner in the law firm Hamel, Park, McCabe & Saunders.

John A. McCone (1902-1991) began his career as a construction engineer in 1929 and became executive vice president and director of the Consolidated Steel Corporation, 1933-1937. He was organizer and president of Bechtel-McCone Corporation (an engineering firm), 1937-1945, and subsequently served as a business executive with several other firms. He worked in the Defense Department from 1948 to 1951 and as chairman of the U.S. Atomic Energy Commission, 1958-1960. He was appointed director of the Central Intelligence Agency in 1961 and remained in that position until 1965.

John W. McCormack (1891-1980) (D-MA) was a member of the House of Representatives serving the district in which Boston was located. He first entered the House in 1929 to fill the unexpired term of the late James A. Gallivan and served until his retirement in 1970. He was House majority leader from 1955 to 1962 and speaker of the House between 1962 and 1970. (*Official Congressional Directory for the Use of the United States Congress* [Washington, D.C.: Government Printing Office, 1970], p. 81).

James S. McDonnell, Jr. (1899-1980) graduated from MIT with an M.S. in aeronautical engineering in 1925 and worked as an engineer and pilot with a variety of aircraft companies before he founded McDonnell Aircraft Corp. in St. Louis in 1939. He served as its president until 1962. The FH-1 Phantom, which first flew in 1946, was the first in a line of fighter aircraft his company produced, including the F-4 Phantom 2, the F-15 Eagle, and the F-18 Hornet. In 1959 the company became the contractor for the Mercury spacecraft; almost three years later it also became contractor for the Gemini spacecraft. In 1967, McDonnell Aircraft merged with Douglas Aircraft Co. to form McDonnell Douglass Corp., with James McDonnell serving as chairman and chief executive officer until 1972 and chairman thereafter. (See esp. obituaries in *Aviation Week and Space Technology*, 1 Sept. 1980, p. 50 and the *New York Times*, 23 Aug. 1980, p. 11.)

Neil H. McElroy (1904-1972) became secretary of defense in 1957 and served through 1959. He had previously been president of Procter & Gamble and returned there in December 1959 to become chair of the board. He served in that position until October 1972, a month before his death.

H. Roemer McPhee (1925-) served at this time as associate special counsel to the president in the White House, where he began work in 1957. A lawyer educated at Princeton and Harvard, at the end of the Eisenhower administration he became a partner in the law firm of Hamel, Morgan, Park and Saunders.

John M. McSweeney (1916-1979) was a career foreign service officer. From 1959 to 1961 he was deputy director and then director of the office of Soviet Union affairs in the State Department. He became the U.S. ambassador to Bulgaria from 1967 to 1970.

George Meader (1907-) (R-MI) began serving in the House of Representatives in 1950 and served until 1964.

John B. Medaris (1902-1990) was a major general commanding the Army Ballistic Missile Agency when Glennan tried to incorporate it into NASA. He attempted to retain the organization as part of the Army, but with a series of DOD agreements the Air Force obtained primacy in space activities and Medaris could not succeed in his effort. Medaris also worked with Wernher von Braun to launch *Explorer I* in early 1958. He retired from the Army in 1969 and became an Episcopal priest, later joining an even more conservative Anglican-Catholic church. ("John Bruce Medaris," biographical file, NASA Historical Reference Collection; John B. Medaris with Arthur Gordon, *Countdown for Decision* [New York: Putnam, 1960].)

John T. Mengel (1918-) taught physics at Lafayette College from 1939-1940, worked for General Electric from 1940-1942, and then developed and evaluated special detection devices at the Bureau of Ships from 1942-1946. He joined the Naval Research Laboratory in 1946, becoming head of the electronic instrument section in 1947. In 1955 he became head of the tracking and guidance branch for Project Vanguard. He joined NASA in 1958 in the same position. From 1959-1973 he was director for tracking and data systems at Goddard Space Flight Center. ("John T. Mengel," biographical file, NASA Historical Reference Collection.)

Livingston T. Merchant (1903-1976) was under secretary of state for political affairs during the period of the Glennan diary. He served in the Department of State from 1942 to 1962.

Robert E. Merriam (1918-1988) was deputy assistant to the president for interdepartmental affairs between 1958 and 1961. Previously he had been an urban planner and housing administrator in Chicago, 1946-1955, and deputy director of the Bureau of the Budget, 1955-1958. At the conclusion of the Eisenhower administration, Merriam became president successively of Spaceonics, Inc., 1961-1964 and Universal Patents, Inc., 1964-1971; then chair of the board of MGA Tech., Inc., 1971-1988.

Elliott Mitchell earned a B.S. in chemistry from William and Mary in 1941 and served from 1942 to 1950 as a physical chemist and chemical engineer in the Department of the Navy. From then until 1958 he was physical sciences administrator and then chief of propulsion research and development in the Navy's Bureau of Ordnance. In 1958 he joined NASA as chief of the solid rocket development program. When he left NASA in 1961, he was assistant director of manned space flight programs for propulsion. Thereafter, he became a consultant. ("Elliott Mitchell," biographical files, NASA Historical Reference Collection.)

James P. Mitchell (1900-1964) had served as director of personnel and industrial relations for R. H. Macy & Co. from 1945-1947 and became vice president of Bloomingdale's in the latter year, with responsibility for labor relations. He became secretary of labor in 1953 after his predecessor had resigned over the failure of the Eisenhower administration to amend the Taft-Hartley law to abolish its right-to-work provision in favor of organized labor. Mitchell served as secretary of labor until 1961, making some efforts to recommend amendments to Taft-Hartley but without either success or the fervor his predecessor had exhibited in the interests of organized labor. (See Ambrose, *Eisenhower, the President*, pp. 116-118.)

Richard T. Mittauer (1927-1973) had worked as a news editor for radio station WOW in his native Omaha, Nebraska, and as a newswriter for ABC in Chicago before coming to Washington, D.C., in 1954 as a congressional intern at the American Political Science Association. From 1955-1959 he was press secretary for Senator Roman L. Hruska before joining NASA's office of public information the latter year. He became director of that office in 1972.

James J. Modarelli had headed the research reports division at Lewis Laboratory when NACA Executive Secretary John Victory requested suggestions for a NASA seal. Members of the illustration section in Modarelli's division sent in some designs, one of which (referred to as the "meatball" to distinguish it from a later insignia called the "worm") was selected and approved. Modarelli is generally credited as the designer. By 1959, Modarelli had moved to NASA Headquarters as head of the exhibits branch of the office of public information. ("James J. Modarelli," biographical file and headquarters telephone directories, 1959-1960, NASA Historical Reference Collection.)

Jack Pendleton Monroe (1904-) was a career naval officer who became a rear admiral in 1956. He served as commander of the Pacific Missile Range from 1957-1961 before becoming the director of astronautics for the chief of naval operations from 1961-1963.

A. S. Mike Monroney (1902-1980) (D-OK) was first elected to the House of Representatives in 1938 and served there through 1950 when he won election to the Senate, where he served through 1969.

Gerald D. Morgan (1908-1976) served in a variety of capacities in the Eisenhower White House—special assistant, 1953; administrative assistant to the president, 1953-1955; special counsel to the president, 1955-1958; and deputy assistant, 1958-1961. Previously he had been a partner with the Washington law firms of Morgan and Calhoun (1946-1950) and Hamel, Park, and Saunders (1950-1953), and assistant legislative counsel with the U.S. House of Representatives (1935-1945).

Robert S. Morison (1906-1986) was at this time director of the natural and medical sciences for the Rockefeller Foundation and a member of the Kety committee (see entry under Kety). He had worked for the Rockefeller Foundation in various capacities since 1944.

Delmar M. Morris (1913-1961) was deputy director for administration at the Marshall Space Flight Center. Until his untimely death from a heart attack, he had worked since March 1960 helping Wernher von Braun transfer his organization from the Army Ballistic Missile Agency to NASA. At his death, Morris had almost 25 years of government service with a variety of agencies, most recently the Atomic Energy Commission and NASA. ("Delmar M. Morris," biographical file, NASA Historical Reference Collection.)

Richard S. Morse (1911-1988) was at this time director of research and development for the Army (1959-1961). He had previously served as director of the National Research Corporation from 1940-1959.

Frank E. "Ted" Moss (1906-) (D-UT) was first elected to the Senate in 1958 and served until 1977.

Karl E. Mundt (1900-1974) (R-SD) was a member of Congress, 1939-1949, and was elected to the Senate in 1948, being reelected thereafter until 1973.

Jason John Nassau (1893-1965) earned a Ph.D. from Syracuse in 1920 and became an assistant professor of astronomy at Case Institute of Technology in 1921. He continued to teach there, serving as chairman of the graduate division from 1936-1940, and became the director of the Warner and Swasey Observatory from 1924-1959. Thereafter he was a professor emeritus at Case.

Homer E. Newell (1915-1983) earned his Ph.D. in mathematics at the University of Wisconsin in 1940 and served as a theoretical physicist and mathematician at the Naval Research Laboratory from 1944-1958. During part of that period, he was science program coordinator for Project Vanguard and was acting superintendent of the atmosphere and astrophysics division. In 1958 he transferred to NASA to assume responsibility for planning and development of the new agency's space science program. He soon became deputy director of space flight programs. In 1961 he assumed directorship of the office of space sciences; in 1963, he became associate administrator for space science and applications. Over the course of his career, he became an internationally known authority in the field of atmospheric and space sciences as well as the author of numerous scientific articles and seven books, including *Beyond the Atmosphere: Early Years of Space Science* (Washington, D.C.: NASA SP-4211, 1980). He retired from NASA at the end of 1973. ("Homer E. Newell," biographical file, NASA Historical Reference Collection.)

Frank Clarke Newlon (1905-) had been city editor and then managing editor for the *Dallas Dispatch-Journal* and then managing editor of the National Education Association Service in Cleveland. Following military service during World War II and a subsequent career in the Air Force, he became editor of *Missiles and Rockets Magazine* from 1958-1961. Thereafter, he became a free-lance writer.

Paul H. Nitze (1907-) had been with the investment firm of Dillon, Read, and Co., before World War II, and then entered federal service. He held a variety of posts, including director of the U.S. Strategic Bombing Survey, 1944-1946, and served with the State Department during the remainder of the Truman administration. Between 1953 and 1961 he was president of the Foreign Service Educational Foundation. He was assistant secretary of defense for international security affairs, 1961-1963; secretary of the Navy, 1963-1967; and deputy secretary of defense, 1967-1969.

Warren J. North (1922-) earned a B.S. from the University of Illinois in 1947. From then until 1955 he was an engineer and test pilot for the Lewis Laboratory. From 1956-1959 he served as assistant chief of the aerodynamics branch at Lewis. He then transferred to NASA Headquarters, where he took part in early planning for Project Mercury, including the selection and training of the seven Mercury astronauts. He moved in 1962 to the Manned Spacecraft Center (later the Johnson Space Center), where he headed the division responsible for training the astronauts for the Gemini rendezvous and docking operations and the Apollo lunar landings. He continued to work in the fields of astronaut selection and training until he retired in 1985 as special assistant to the director of flight operations in planning space shuttle crew training. ("Warren North," biographical file, NASA Historical Reference Collection.)

Robert G. Nunn, Jr. (1917-1975) earned a law degree from the University of Chicago in 1942. After four years in the Army during World War II, then private practice of law for eight years in Washington, D.C., and in his home town of Terre Haute, Indiana, he joined the office of general counsel of the Air Force in 1954. He became NASA assistant general counsel in November 1958 and then special assistant to Glennan in September 1960. He helped draft many legal and administrative regulations for NASA, then went to work for the Washington law firm of Sharp and Bogan. Later he formed the firm of Batzell and Nunn, specializing in energy legislation and administrative law.

John B. Oakes (1892-) had been a writer and then editor for the *New York Times* since 1946 and became editor of the editorial page in 1961.

Hugh Odishaw (1916-1984) became assistant to the director of the National Bureau of Standards 1946-1954, served as executive director of the U.S. National Committee for the International Geophysical Year from 1954-1965, and then became the executive secretary of the division of physical sciences in the National Academy of Sciences, 1966-1972.

Frank C. Osmers, Jr. (1907-1977) (R-NJ) was first elected to Congress in 1951 and served through the 88th Congress (1963-1965).

Harold C. Ostertag (1896-1985) (R-NY) was first elected to the House of Representatives in 1950 and served through 1964.

Don Richard Ostrander (1914-1972) was a career Air Force officer who became a major general in 1958. He was deputy commander of the Advanced Research Projects Agency in 1959 and became director of NASA's launch vehicle programs in late 1959 as NASA began taking over responsibility for the Saturn program. He left NASA in 1961 and retired from the Air Force in 1965 as vice commander of the Ballistic Systems Division, Air Force Systems Command, to become vice president for planning of the Bell Aero Systems Corporation. ("Don Richard Ostrander," biographical file, NASA Historical Reference Collection.)

Carl F. J. Overhage (1910-) earned his Ph.D. in physics at Caltech in 1937 and served as acting director of research for Technicolor Motion Picture Corp. until 1941, when he joined the staff of the radiation laboratory at MIT from 1942-1945. After a stint with Eastman Kodak from 1946-1954, he joined the Lincoln Laboratories of MIT, becoming its director from 1957-1964, after which he served as a professor of engineering.

Frank Pace, Jr. (1912-1988) was president of General Dynamics, Inc. Previously he had been Secretary of the Army, 1950-1953, and had held several other key posts in the Truman administration during the latter 1940s. (Bell, *Secretaries of War and Secretaries of the Army*, p. 136).

John F. Parsons (1908-1969) had been associate director of the Ames Research Center since 1952. He had joined the staff of the Langley Aeronautical Laboratory as a junior aeronautical engineer in 1931. He worked there with wind tunnels. He moved to Ames in the 1939-1940 period when it was being set up and worked on planning, design, and construction of the new center. He continued wind tunnel work there and was also chief of the construction division. In 1948-1949 he was assistant to the director of the center. Then from 1949-1956 he supervised the wind tunnel construction program among other duties. ("John F. Parsons," biographical file, NASA Historical Reference Collection.)

Morehead Patterson, chairman of the board of the American Machine & Foundry Co., was a member of the [Kimpton] Advisory Committee on Organization (see Chapter One, note 31).

Nathan W. Pearson, vice president of T. Mellon & Sons, was also a member of the Kimpton committee.

James A. Perkins (1911-) was vice president of the Carnegie Corp. from 1951 to 1963 and president of Cornell University, 1963-1969. He, too, served on the Kimpton committee.

Milo Randolph Perkins (1900-1972) began his career as a salesman in 1919, but by 1926 he had become a partner in the King-Perkins Bag Co. He served in a variety of capacities in the Roosevelt administration, ending as executive director of the Board of Economic Warfare in 1941. Thereafter he became a foreign investment consultant.

Wilton B. Persons (1896-1977) was a career Army officer who had entered the U.S. Army Coast Artillery in 1917 and advanced through the ranks to major general in 1944. He had served in both the A.E.F. in World War I and in Europe in World War II. He headed the office of legislative liaison for the Department of Defense between 1948 and his retirement in 1949. He was called back to active duty as a special assistant to General Dwight D. Eisenhower at Supreme Headquarters Allied Powers in Europe from 1951-1952 and was active on behalf of Eisenhower's presidential campaign in 1952. He became a deputy assistant to the president in 1953 and then was made an assistant to the president in 1958. He served throughout the Eisenhower presidency, handling congressional liaison before he replaced Sherman Adams in 1958 as, effectively, Eisenhower's chief of staff.

Franklyn W. Phillips (1917-) graduated from MIT in 1941 with a degree in mechanical engineering and went to work at the Langley Aeronautical Laboratory, later moving to Lewis where he did research on aircraft engine materials and stresses. In 1945 he became a member of the NACA director's staff and served as administrator for a variety of NACA research programs in aircraft engines and aircraft and missile structures and loads. In October 1958 he became special assistant to Glennan. He gave up that position in January 1959 to become acting secretary of the National Aeronautics and Space Council, but in February 1960 he returned to his position as Glennan's assistant. He continued in that job under James E. Webb until 1962, when he became director of NASA's new North-Eastern Office. In 1964 he became assistant director for administrative operations at the new NASA Electronics Research Center in Cambridge, Massachusetts.

Clifford P. Phoebus (1910-1984 [or early 1985]) was a naval aviator, flight surgeon and medical corps officer who rose to the rank of captain in 1953 and was commander of the U.S. Naval School of Aviation Medicine in Pensacola,

Florida, from 1960-1964. (See obituary notice, *Aviation, Space, and Environmental Medicine* [February 1985]: 192, which does not give a date of death but does state he died "recently" and gives his age at death as 74.)

William H. Pickering (1910-) obtained his bachelors and masters degrees in electrical engineering, then a Ph.D. in Physics from Caltech before becoming a professor of electrical engineering there in 1946. In 1944 he organized the electronics efforts at the Jet Propulsion Laboratory to support guided missile research and development, becoming project manager for Corporal, the first operational missile JPL developed. From 1954 to 1976 he was director of JPL, which developed the first U.S. satellite (Explorer I), the first successful U.S. cislunar space probe (Pioneer IV), the Mariner flights to Venus and Mars in the early to mid-1960s, the Ranger photographic missions to the moon in 1964-65, and the Surveyor lunar landings of 1966-67. ("William H. Pickering," biographical files, NASA Historical Reference Collection.)

Harvey F. Pierce (1909-) was an electrical engineer who worked in the 1930s for the Florida Power & Light Co. and then became a partner in the firm of Maurice H. Connell & Assoc., 1937-1946; its secretary-treasurer in 1946; and a partner in Connell, Pierce, Garland & Friedman beginning in 1956.

I. Irving Pinkel (1913-) graduated from the University of Pennsylvania, where he studied physics and mathematics, in 1934 and entered government service in 1935 as a physicist with the U.S. Bureau of Mines, with which he did research on the problem of synthesizing liquid fuels from coal. He joined the Langley Aeronautical Laboratory in 1940 and transferred to the Lewis Laboratory in 1942. There, he worked on hydraulics problems of aircraft engine lubricating systems operating at high altitudes. This effort led to the development of a method to control foaming in lubricating oil and to a new lubricant pump that met the stringent demands of high-altitude flight. In 1949 he became associate chief of the physics division at Lewis with responsibility for studying aircraft operating problems. Among other results of this work was a means of reducing the incidence of fire after airplane crashes. Pinkel became chief of the flight problems division in 1956 and of the aerospace safety research and data institute in 1968. ("I. Irving Pinkel," biographical file, NASA Historical Reference Collection.)

Allen E. Puckett (1919-) earned his Ph.D. at Caltech in 1949 and went to work for Hughes Aircraft Co. that year, becoming its executive vice president from 1965-1977 and its president thereafter.

Edward M. Purcell (1912-) was a professor of physics at Harvard University during this period and also served on the president's Scientific Advisory Committee from 1957 to 1960 and 1962 to 1965. He had been co-winner of the Nobel prize in physics in 1952 (with Felix Bloch) for the discovery of nuclear magnetic resonance in solids.

Donald A. Quarles (1894-1959) was a deputy secretary of defense between 1957 and 1959. Just after World War II he had been a vice president first at Western Electric Co. and later at Sandia National Laboratories, but in 1953 he accepted the position of assistant secretary of defense (research and development). He was also secretary of the Air Force between 1955 and 1957.

Elwood R. ("Pete") Quesada (1904-1993) was a career Army aviation and Air Force officer who rose through the ranks from private to lieutenant general. He commanded fighter forces in Africa and Europe during World War II and served as the first commander of Tactical Air Command from 1946-1948. A supporter of air-ground cooperation with the Army, he retired in 1951, perhaps because he believed the new Air Force's treatment of tactical air power as less important than the strategic air arm violated a promise to General Eisenhower that there would always be a tactical force to support the Army. He subsequently served as manager at Olin Industries, organizer and director of Lockheed's Missile Systems Division, the controversial but highly successful first administrator of the Federal Aviation Agency, and president of L'Enfant Properties, among other duties. (John Schlight, "Elwood R. Quesada: TAC Air Comes of Age," *Makers of the United States Air Force*, pp. 177-204; Rochester, *Takeoff at Mid-Century*, esp. pp. 288-289.)

James M. Quigley (1918-) was a member of Congress from Pennsylvania in 1955-1957 and 1959-1961, then served as assistant secretary of the Department of Health, Education, and Welfare from 1961 to 1966.

Clark T. Randt (1917-) worked throughout the 1950s as a professor of neurology at Western Reserve University, moving to NASA in 1959 as director of life sciences. In 1961 he left NASA to accept a professorship in neurology at the New York University School of Medicine. In 1970 he became chair of the department of neurology at NYU. ("Clark T. Randt," biographical file, NASA Historical Reference Collection.)

Eberhard F. M. Rees (1908-) was at this time deputy director for technical and scientific matters at the Marshall Space Flight Center. A graduate of the Dresden Institute of Technology, he began his career in rocketry in 1940 when he became technical plant manager of the German rocket center at Peenemünde. He came to the U.S. in 1945 with von Braun's rocket team and worked with von Braun at Fort Bliss, Texas, moving to Huntsville in 1950 when the Army transferred its rocket activities to the Redstone Arsenal. He served as deputy director of development operations at the Army Ballistic Missile Agency from 1956 to 1960. In 1970 he succeeded von Braun as director of the Marshall center. He retired in 1973. ("Eberhard Rees," biographical file, NASA Historical Reference Collection.)

Francis W. Reichelderfer (1895-1983) joined the Naval Air Service in World War I after graduating from Northwestern in science in 1917 and studying meteorology at Harvard. He became a weather forecaster despite earning his pilot's wings in 1919. While in the Navy, he earned his masters degree at the University of Bergen in Norway. In 1938 President Franklin D. Roosevelt appointed him head of the Weather Bureau. He stated that one of the most significant advances in weather forecasting during his tenure was the orbiting of the Tiros I weather satellite, although he also expanded hurricane forecasting, instituted the use of the telephone in the weather service, provided crop and marine forecasts, instituted frost warnings, and provided hourly reports for aircraft. On his retirement, President John F. Kennedy wrote him, "You presided over the evolution of meteorology . . . from an art to a science." ("Francis W. Reichelderfer," biographical file, NASA Historical Reference Collection.)

Henry J. E. Reid (1895-1968) graduated from Worcester Polytechnic Institute in 1919 with a B.S. in electrical engineering. After a brief stint in private industry, he joined Langley Aeronautical Laboratory in 1921 as one of a small group of engineers and scientists then on the professional staff there. His principal field of research was the design and improvement of basic instruments for flight research. He became director of the center in 1926 and presided over the extensive growth that accompanied its becoming a leading aeronautical and space research facility. He retired in 1961. ("Henry J. E. Reid," biographical file, NASA Historical Reference Collection.)

John T. Rettaliata (1911-) was president of the Illinois Institute of Technology between 1952 and 1973. He then became chair of the Board of Directors of the Banco di Roma, Chicago. Previously Rettaliata had been involved in business, government, and educational activities associated with scientific research and development. Among many other duties and honors, in 1959 he became a nongovernmental member of the National Aeronautics and Space Council.

Victor G. Reuther (1912-) became director of the United Automobile, Aircraft and Agricultural Implement Workers of America (AFL-CIO) in Indiana in 1937 and rose through a number of other positions to become administrative assistant to the president of the union, a position he held throughout the 1960s.

Richard V. Rhode (1904-) received a B.S. in mechanical engineering from the University of Wisconsin in 1925 and joined the NACA as an aeronautical engineer at the Langley Aeronautical Laboratory. In 1945 he became chief of the aircraft loads division. In 1949 he transferred to the NACA Headquarters in Washington, D.C., and became assistant director for research (aircraft construction and operating problems). When NASA came into existence in 1958, he became assistant director for advanced design criteria in the space vehicle technology division. There, he was responsible for advanced technology supporting the development of space vehicles. He retired in early 1967 and was awarded the NASA Medal for Exceptional Scientific Achievement. ("Richard V. Rhode," biographical file, NASA Historical Reference Collection.)

Sid Richardson was a Texas oil millionaire who had contributed money to the Democratic Congressional Campaign Committee and to Lyndon Johnson's campaigns. Robert Dallek describes him as "an exceptionally charming conversationalist and a lonely bachelor" who invited Johnson to visit his "privately owned island in the coastal Gulf." (*Lone Star Rising: Lyndon Johnson and His Times, 1908-1960* [Oxford: Oxford University Press, 1991], pp. 160, 201-2, 249, 308-9, quotations from p. 309.) Eisenhower had first met Richardson in 1941, knew he was a Johnson supporter, and through an intermediary got him to put pressure on Johnson at least once. (Greenstein, *Hidden-Hand Presidency*, p. 59.)

R. Walter Riehlman (1899-1978) (R-NY) was first elected to the House of Representatives in 1946 and served through 1964.

Charles E. Robbins (1906-) worked for the *Wall Street Journal* in various capacities from 1929-1941. He departed the job there of managing editor to become a member of the business department of the *New York Times*, 1941-1949. In 1953 he became executive manager of the Atomic Industrial Forum, Inc.

Herbert H. Rosen was deputy director of the office of public information in NASA in early 1960. He was an engineer who had previously worked for Hayden Publications Corp. as editor of *Electronic Design* and for the Bureau of Standards both as an engineer and in the areas of public relations and technical information. ("Herbert H. Rosen," biographical file, and Headquarters Telephone Directory for May 1960, NASA Historical Reference Collection.)

Milton W. Rosen (1915-), an electrical engineer by training, joined the staff of the Naval Research Laboratory in 1940, where he worked on guidance systems for missiles during World War II. From 1947 to 1955, he was in charge of Viking rocket development. He was technical director of Project Vanguard, the scientific earth satellite program, until he joined NASA in October 1958 as director of launch vehicles and propulsion in the office of manned space flight. In 1963 he became senior scientist in NASA's office of the deputy associate administrator for defense affairs. He later became deputy associate administrator for space science (engineering). In 1974 he retired from NASA to become executive secretary of the National Academy of Science's Space Science Board. ("Milton W. Rosen," biographical file, NASA Historical Reference Collection; see also his *The Viking Rocket Story* [New York: Harper, 1955].)

Aaron Rosenthal became director of financial management at NASA in February 1960, assuming oversight over the budget and fiscal offices. Before coming to NASA, he had been controller for the Veterans Administration, an agency for which he had worked since 1936. In September 1961 he transferred to the National Science Foundation. (Two announcements under his name in "biography NASA miscellaneous," NASA Historical Reference Collection.)

Walt W. Rostow (1916-) has spent his career, since completing his Ph.D. at Yale in 1940, moving between academic and government positions. He began his career in the economics department of Columbia University, but he served with the State Department in European recovery efforts following the war. Throughout the 1950s he was a professor of economic history at MIT. In 1961 he returned to Washington as an assistant to the president for national security affairs and served in that and other similar capacities throughout the Kennedy and Johnson administrations. Thereafter he accepted an academic post with the University of Texas.

Addison M. Rothrock (1903-1971) graduated from Penn State in 1925 with a B.S. in physics and began working at the Langley Aeronautical Laboratory the next year. He worked in the areas of fuel combustion and fuel rating, rising to the position of chief of the fuel injection research laboratory and writing more than 40 papers and reports. In 1942 he made the move to Lewis Laboratory, where he was chief of the fuels and lubricants division and then chief of research for the entire laboratory. In 1947 he became assistant director for research at NACA Headquarters. With the foundation of NASA, he assumed the duties of assistant director of research (power plants). Two months later, he became the scientist for propulsion in and then (1961), the associate director of, the office of program planning and evaluation. He retired in 1963 and taught for five years at George Washington University. ("Addison M. Rothrock," biographical file, NASA Historical Reference Collection.)

John Edward Roush (1920-) (D-IN) served in the House of Representatives from 1959-1964.

Leverett Saltonstall (1892-1979) (R-MA) was governor of Massachusetts from 1939-1944, when he won election to the U.S. Senate. He served in the Senate from then until 1967 and became one of its Republican leaders.

August Schomburg (1908-1972) was a career Army officer who rose to the rank of lieutenant general. From 1960 to 1962 he was commander of the Army missile ordnance command at Redstone Arsenal.

Bernard A. Schriever (1910-) earned a B.S. in architectural engineering from Texas A&M in 1931 and was commissioned in the Army Air Corps Reserve in 1933 after completing pilot training. Following broken service, he received a regular commission in 1938. He earned an M.A. in aeronautical engineering from Stanford in 1942 and then flew 63 combat missions in B-17s with the 19th Bombardment Group in the Pacific Theater during World War II. In 1954, he became commander of the Western Development Division (soon renamed the Air Force Ballistic Missile Division), and from 1959-1966 he was commander of its parent organization, the Air Research and Development Command, renamed Air Force Systems Command in 1961. As such, he presided over the development of the Atlas, Thor, and Titan missiles, which served not only as military weapon systems but also as boosters for NASA's space missions. In developing these missiles, Schriever instituted a systems approach, whereby the various components of the Atlas and succeeding missiles underwent simultaneous design and test as part of an overall "weapons system." Schriever also introduced the notion of concurrency, which has been given various interpretations but essentially allowed the components of the missiles to enter production while still in the test phase, thereby speeding up development. He retired as a general in 1966. (Jacob Neufeld, "Bernard A. Schriever: Challenging the Unknown," *Makers of the United States Air Force*, pp. 281-306; Perry, "Atlas, Thor . . .," *A History of Rocket Technology*, Eugene M. Emme, ed. (Detroit, MI: Wayne State University Press, 1964), pp. 144-160; Divine, *The Sputnik Challenge*, p. 25.)

Robert C. Seamans, Jr. (1918-) had been involved in aerospace issues since he completed his Sc.D. degree at MIT in 1951. He was on the faculty at MIT's department of aeronautical engineering between 1949 and 1955, when he joined the Radio Corporation of America as manager of the Airborne Systems Laboratory. In 1958 he became the chief engineer of the Missile Electronics and Control Division and joined NASA in 1960 as associate administrator. In December 1965 he became NASA deputy administrator. He left NASA in 1968, and in 1969 he became secretary of the Air Force, serving until 1973. Seamans was president of the National Academy of Engineering from May 1973 to December 1974, when he became the first administrator of the new Energy Research and Development Administration. He returned to MIT in 1977, becoming dean of its School of Engineering in 1978. In 1981 he was elected chair of the board of trustees of Aerospace Corp. ("Robert C. Seamans, Jr.," biographical file, NASA Historical Reference Collection.)

Willis H. Shapley (1917-), son of famous Harvard astronomer Harlow Shapley, earned a B.A. from the University of Chicago in 1938. From then until 1942, he did graduate work and performed research in political science and related fields at the latter institution. He joined the Bureau of the Budget in 1942 and became a principal examiner in 1948. From 1956-1961 he was assistant chief (Air Force) in the bureau's military division, becoming progressively deputy chief for programming (1961-1965) and deputy chief (1965) in that division. He also served as special assistant to the director for space program coordination. In 1965 he moved to NASA as associate deputy administrator, with his duties including supervision of the public affairs, congressional affairs, DOD and interagency affairs, and international affairs offices. He retired in 1975 but rejoined NASA in 1987 to help it recover from the Challenger disaster. He served as associate deputy administrator (policy) until 1988, when he again retired but continued to serve as a consultant to the administrator. ("Willis H. Shapley," biographical file, NASA Historical Reference Collection.)

Timothy E. Shea (1898-) was a manufacturing executive who had held a variety of positions. Since 1957 he had been vice president for engineering with Western Electric.

Gerald Siegel served during the 1950s in various staff positions in the Senate, including those of counsel to the Democratic Policy Committee and the preparedness investigating subcommittee of the Senate Armed Services Committee. He also served *de facto* as staff director of the Senate Special Committee on Space and Aeronautics during 1958 when it considered the Eisenhower administration's proposal for what became the Aeronautics and Space Act. Soon thereafter, Siegel left the Senate to lecture for three years at Harvard and then became vice president and chief counsel of the *Washington Post*. ("The Legislative Origins of the Space Act," [transcribed] proceedings of a videotape workshop conducted on 3 April 1992 by Professor John Logsdon of the George Washington University Space Policy Institute, copy available in NASA Historical Reference Collection; Griffith, *National Aeronautics and Space Act*, p. 28.)

Albert F. Siepert (1915-) was a longtime federal employee who entered federal service in 1937 and moved from being executive officer for the National Institutes of Health to NASA in 1958. In 1959 he was NASA's chief negotiator in the transfer of the Army Ballistic Missile Agency to the space agency from his position as director of business administration, and in 1963 he moved to the deputy director position at the Kennedy Space Center in Florida. In 1969 Siepert left NASA to become a program associate at the University of Michigan's Institute for Social Research. ("Albert F. Siepert," biographical file, NASA Historical Reference Collection.)

Abe Silverstein (1908-), who earned a B.S. in mechanical engineering (1929) and an M.E. (1934) from Rose Polytechnic Institute, was a longtime NACA manager. He had worked as an engineer at the Langley Aeronautical Laboratory between 1929 and 1943 and had moved to the Lewis Laboratory (later, Research Center) to a succession of management positions, the last (1961-1970) as director of the center. Interestingly, in 1958 Case Institute of Technology had awarded him an honorary doctorate. When Glennan arrived at NASA, Silverstein was on a rotational assignment to the Washington headquarters as director of the office of space flight development (later, space flight programs) from the position of associate director at Lewis, which he had held since 1952. During his first tour at Lewis, he had directed investigations leading to significant improvements in reciprocating and early turbojet engines. At NASA Headquarters he helped create and direct the efforts leading to the space flights of Project Mercury and to establish the technical basis for the Apollo program. As Lewis's director, he oversaw a major expansion of the center and the development of the Centaur launch vehicle. He retired from NASA in 1970 to take a position with Republic Steel Corp. On the career of Silverstein see, Dawson, *Engines and Innovation*, passim; "Abe Silverstein," biographical file, NASA Historical Reference Collection.

B[ernice] F. Sisk (1910-) (D-CA) was first elected to Congress in 1954 and served in every successive Congress through the 95th (1977-1979), although he represented three different districts over that period.

Howard K. Smith (1914-) graduated from Tulane University in 1936 and became a Rhodes scholar the next year. He was the foreign correspondent of United Press in London in 1939 and the Berlin correspondent for CBS in 1941. He served as a war correspondent in Europe in 1944 and covered the Nuremberg trials in Germany the following year. He was the chief European correspondent and European director for CBS in London from 1946-57 and then moved to the CBS Washington bureau from 1957-1961 as chief correspondent and general manager. In 1962 he became a news analyst for ABC, also in Washington, D.C. A winner of many awards for his journalism, Smith also wrote several books, including *Last Train from Berlin* (1942).

Kent H. Smith (1894-1980) served as acting president of Case Institute of Technology during Glennan's absence at NASA. A life-long Cleveland resident, he was the son of Albert W. Smith, who had been on the faculty at CIT. In 1917, as a young man, Smith had enlisted in the aviation section of the U.S. Army Signal Corps, been commissioned a second lieutenant, and sent to France. During World War I he commanded ground personnel supporting combat air operations. He also learned to fly and maintained a lifelong interest in aerospace developments.

Afterward he entered business, in several capacities, but especially as the head of the Lubrizol firm, which he helped found and which made lubrication products for automobiles and other vehicles. Kent was chairman of the board of Lubrizol from 1951-1959 and served as its director from 1928-1980. (Cramer, *Case Institute of Technology*, pp. 202-206).

Murray Snyder (1911-1969) had been assistant White House press secretary between 1953 and 1957 and then became assistant secretary of defense for public affairs (1957-1961). Prior to that time he had been political journalist for several media organizations, and after leaving public office he became vice chairman of a public relations firm.

Walter D. Sohier (1924-), a graduate of Columbia Law School, had worked for the CIA and the Air Force before joining NASA in 1958 as assistant general counsel. He became deputy general counsel in 1961 and general counsel in 1963. He left NASA in 1966 to become a partner in a New York law firm. ("Walter D. Sohier," biographical file, NASA Historical Reference Collection.)

John J. Sparkman (1899-1985) (D-AL) was elected to the House of Representatives in 1936 and served until 1946, when he was elected to the Senate. He served in the Senate into the late 1970s.

Elmer B. Staats (1914-) was deputy director at the Bureau of the Budget. A career government official, he received a Ph.D. from the University of Minnesota in 1939 and joined the BOB staff that year. He became deputy director from 1950 to 1953 and again from 1958 to 1966, serving in the interim with the National Security Council. In 1966 he became comptroller general of the U.S., a post he held until the early 1980s.

John Stack (1906-1972) graduated from MIT in 1928 and joined the Langley Aeronautical Laboratory as an aeronautical engineer. In 1939 he became director of all the high-speed wind tunnels and high-velocity airflow research at Langley. Three years later he was named chief of the compressibility research division there. He was promoted to assistant chief of research in 1947 and subsequently had that title changed to assistant director of the research center. He guided much of the research that paved the way for transonic aircraft, and in 1947 he was awarded the Collier Trophy together with the pilot of the X-1 who broke the sound barrier, (by then) Major Charles E. Yeager. He won the award again in 1952 and later won the Wright Brothers Memorial Trophy among other awards. From 1961-1962 he was director of aeronautical research at NASA Headquarters before retiring to become vice president for engineering at Republic Aircraft Corp (later part of Fairchild Industries) from which he retired in 1971. ("John Stack," biographical file, NASA Historical Reference Collection.)

Maurice H. Stans (1908-) was a longtime Republican in Washington. He served in several positions with the Eisenhower administration, notably as deputy director of the Bureau of the Budget between 1957 and 1958, then director from 1958-1961. In 1969 he was appointed as secretary of commerce for the Nixon administration and served until 1972. He was finance director of the 1972 Nixon re-election campaign and pleaded guilty in 1975 to five misdemeanor charges of violating campaign laws during the campaign. ("Maurice H. Stans," biographical file, NASA Historical Reference Collection.)

Frank Stanton (1908-) earned a Ph.D. from Ohio State University in 1935 and went on to become a business executive, serving most notably as president of CBS, Inc. from 1946 to 1971 and its vice chairman from 1971 to 1973.

Joseph A. Stein (1912-) was an aviator in the U.S. Navy during World War II and then became a reporter for Portland, Oregon, newspapers. In 1954-1955 he was an information specialist at Lewis Laboratory before becoming (1955-1958) an aeronautical information specialist for NACA Headquarters. With the creation of NASA he became chief of the news division in the office of public information, and in July 1960 he was promoted to deputy director of the office.

John C. Stennis (1901-) (D-MS) was elected to the Senate in 1947 and served until 1989. He was a member of the Appropriations, Armed Services, and Aeronautical and Space Sciences committees in the early 1960s. In 1988, NASA's National Space Technology Laboratories in Mississippi became the John C. Stennis Space Center in his honor. ("John C. Stennis," biographical file, NASA Historical Reference Collection.)

Robert Ten Broeck Stevens (1899-1983) was secretary of the Army for President Eisenhower between 1953 and 1955. He had long been associated with the textile industry, notably as president of the J.P. Stevens & Co. between 1929 and 1942 and chair of the board between 1945 and 1953. Stevens served in the Army in World Wars I and II. After leaving the Department of the Army in 1955 he returned to his business activities in New York City, serving as president of J.P. Stevens Co., 1955-1959, and chair of its executive committee, 1969-1974. (Bell, *Secretaries of War and Secretaries of the Army*, p. 138; *New York Times*, 1 February 1983, p. D23; "Stevens, Robert T[en Broeck])," *Current Biography 1953*, pp. 591-92).

Horton Guyford Stever (1916-) earned a Ph.D. in physics at Caltech in 1941 and became a member of the staff at the Radiation Lab of MIT until 1942, when he went to London as a scientific liaison officer through the end of World War II. He then returned to MIT as a member of the faculty, rising to become associate dean of engineering from 1956-1959. He remained a professor of aeronautical engineering until 1965 and then served as president of Carnegie-Mellon University until 1972. Meanwhile, he had begun an extensive and distinguished additional career of service to government. For example, he was chief scientist of the Air Force from 1955-1956 and served on its scientific advisory board from 1947-1969 (as chairman, 1962-1969). He was director of the National Science Foundation, 1972-1976, and in 1973 he became scientific and technical advisor to President Gerald R. Ford, Jr., a post he held until 1977. Along the way, he served on advisory committees to the NACA, including its special committee on space technology, and to NASA, including a stint as chairman of an independent panel of experts established by the National Research Council to advise NASA and monitor its compliance with the recommendations of the Rogers Commission that investigated the Challenger explosion in 1986. ("H. Guyford Stever," biographical file, NASA Historical Reference Collection.)

Homer J. Stewart (1915-) earned his doctorate in aeronautics from Caltech in 1940, joining the faculty there two years before that. In 1939 he participated in pioneering rocket research with other Caltech engineers and scientists, including Frank Malina, in the foothills of Pasadena. Out of their efforts, the Jet Propulsion Laboratory (JPL) arose, and Stewart maintained his interest in rocketry at that institution. He was involved in developing the first American satellite, Explorer I, in 1958. In that year, on leave from Caltech, he became director of NASA's program planning and evaluation office, returning to Caltech in 1960 to a variety of positions, including chief of the advanced studies office at JPL from 1963 to 1967 and professor of aeronautics at Caltech itself. ("Homer J. Stewart," biographical file, NASA Historical Reference Collection; Koppes, *JPL*, pp. 23, 32, 44, 47, 79-80, 82.)

Julius A. Stratton (1901-) earned an Sc.D. from the Eidgenossische Technische Hochschule (technical institute of the Swiss Confederation), Zurich, in 1928 and began as a research associate at MIT in 1924, rising to the rank of professor in 1941. He became chancellor of MIT in 1956 and president in 1959.

Samuel S. Stratton (1916-1990) (D-NY) was first elected to Congress in 1958 and served into the late 1980s. In the early 1960s he was a member of the House Armed Services Committee.

Lewis L. Strauss (1915-1974) was chairman of the Atomic Energy Commission from 1953-1958 and secretary of commerce 1958-1959.

William G. Stroud, Jr. (1923-) was chief of meteorology in the satellite applications systems division of the Goddard Space Flight Center and had been project manager for Tiros I. He later became chief of the aeronomy and meteorology division at Goddard and, still later, special assistant to the director of flight projects there. ("William G. Stroud," biographical file, NASA Historical Reference Collection.)

Ernst Stuhlinger (1913-) was a physicist who earned his Ph.D. at the University of Tübingen in 1936 and continued research into cosmic rays and nuclear physics until 1941 while serving as an assistant professor at the Berlin Institute of Technology. He then spent two years as an enlisted man in the German army on the Russian front before being assigned to the rocket development center at Peenemünde, Germany. There he worked principally on guidance and control of rockets. After World War II, he came to the United States as part of Project Paperclip and worked with Wernher von Braun at Fort Bliss, Texas, and then at the Redstone Arsenal in Huntsville, Alabama. Transferred to the Marshall Space Flight Center in 1960, he was director of its space science lab from 1960 to 1968 and then its associate director for science from 1968 to 1975, when he retired and became an adjunct professor and senior research scientist with the University of Alabama at Huntsville. He directed early planning for lunar exploration and the Apollo telescope mount, which flew on Skylab and produced a wealth of scientific information about the sun. He was also responsible for the early planning on the high energy astronomy observatory and contributed to the initial phases of the space telescope project. His work included studies of electric propulsion and of scientific payloads for the Space Shuttle. ("Ernst Stuhlinger," biographical file, NASA Historical Reference Collection.)

Stuart Symington (1901-1988) (D-MO) served in the Senate between 1953 and 1977. He entered government in 1945 when his fellow Missourian, Harry S. Truman, appointed him chair of the Surplus Property Board. He later served Truman as secretary of the Air Force and was an outspoken advocate of building a strong aerospace presence. As such, he repeatedly charged the Eisenhower administration with balancing the budget at the expense of national security and was one of its most vocal critics after the launch of Sputnik, predicting what proved to be a fallacious missile gap between the U.S. and the Soviet Union. He left the Senate in 1977. (*New York Times*, 15 December 1988, p. D26; Divine, *Sputnik Challenge*, pp. 20, 43, 125, 178-183.)

John Taber (1880-1965) (R-NY) was first elected to the House of Representatives in 1923 and served through 1962.

Olin E. Teague (1910-1981) (D-TX) was first elected to the House of Representatives in 1946 and served in each succeeding Congress through the 95th (1977-1979). He was appointed to the new Science and Astronautics Committee in the 86th Congress (1959-1961).

Edward Teller (1908-) was a naturalized American physicist born in Hungary who made important contributions to the development of both fission- and fusion-type bombs. As a member of the advisory committee of the AEC, he advocated the hydrogen bomb as a U.S. tactical weapon, arousing a great deal of controversy. He also spoke publicly about Sputnik as showing that the Soviets were beginning to gain a lead on the U.S. in the fields of science and technology. Among other works on Teller, see the view of the insider, Herbert York, *The Advisors: Oppenheimer, Teller, and the Superbomb* (San Francisco: W. H. Freeman, 1976). For one perspective on Teller's more recent and still controversial activities in the world of science and defense technology, see William J. Broad, *Teller's War: The Top-Secret Story Behind the Star Wars Deception* (New York: Simon & Schuster, 1992).

Morris Tepper (1916-) earned a Ph.D. in fluid mechanics from Johns Hopkins in 1952. Before that but after earning an M.A. in mathematics from Brooklyn College, he had joined the Air Force in 1943 and served as a meteorologist in the Pacific Theater during World War II. In 1946 he joined the Weather Bureau as a research meteorologist. After becoming chief of the severe local storms research unit there in 1951, he transferred to NASA in 1959 as a meteorologist in the office of space flight development. By 1962, he had become director of meteorological systems in the office of applications (later, office of space science and applications). In 1969 he added the title of deputy director of the earth observations programs division. He also worked through the U.N. Committee on Space Research and elsewhere to promote international cooperation in the field of meteorology in space. After serving as a special project officer at Goddard in 1978-1979, he left NASA in the latter year to become a professor of mathematical physics at Capitol College, Maryland. ("Morris Tepper," biographical file, NASA Historical Reference Collection.)

Albert Thomas (1898-1966) (D-TX), a lawyer and World War I veteran, had first been elected to the House of Representatives in 1936 and served successively until 1962. In 1960 he was chair of the independent offices subcommittee of the House Appropriations Committee and thus exercised considerable congressional power over NASA's funding.

Floyd L. Thompson (1898-1976) served in the Navy for four years after 1917 and entered the University of Michigan, earning a B.S. degree in aeronautical engineering in June 1926. He then joined the Langley Aeronautical Laboratory as part of a staff of only about 150. He worked in the flight research division, where he was author or co-author of more than 20 technical reports. He became chief of the division in 1940 and assistant chief of research for all of Langley in 1943. From 1945-1952 he served as chief of research before becoming associate director of the center in 1952 and director in 1960. He was briefly a special assistant to the NASA administrator in 1968 before retiring later that year. ("Floyd L. Thompson," biographical file, NASA Historical Reference Collection; see also Hansen, *Engineer in Charge*, passim.)

Shelby Thompson (1907-) worked for the Atomic Energy Commission from 1947-1955 as the chief of its public information service and from 1955-1960 as deputy director, division of information services. He joined NASA in June 1960 as director of the office of technical information and educational programs. In 1964 he became special assistant to the NASA assistant administrator for public affairs. He retired in 1970. ("Shelby Thompson," biographical file, NASA Historical Reference Collection.)

Arthur G. Trudeau (1902-1991) was a career army officer and in 1958 was serving as commanding general of the I United States Corps Korea. He was promoted to lieutenant general in 1956.

Nathan F. Twining (1897-1982) was a career pilot in the Army and the Air Force, commanding the 13th Air Force in the Pacific, the 15th Air Force in Europe, and then the 20th Air Force again in the Pacific during World War II. He became chief of staff of the Air Force in 1953 and chairman of the Joint Chiefs of Staff from 1957 to 1960. (Donald J. Mrozek, "Nathan F. Twining: New Dimensions, a New Look," in *Makers of the United States Air Force*, pp. 257-280.)

Ralph E. Ulmer (1917-1985) began working as an aeronautical engineer at Langley Aeronautical Laboratory in 1938. The next year he transferred to NACA Headquarters and served after 1940 as a technical assistant to the NACA senior staff. In 1950 he became NACA budget officer. From 1958-1961 he was the NASA budget officer. He then became director of facilities coordination in the NASA office of programs. He retired as a program analyst in 1973. ("Ralph E. Ulmer," biographical files, NASA Historical Reference Collection.)

John F. Victory (1893-1975) began work for the government in 1908 as a messenger for the patent office. After becoming the first employee of the NACA in 1915, he became its secretary in 1921 and its executive secretary in 1948, in general charge of its administration. When NASA came into being, he served as a special assistant to Glennan until his retirement at the end of July 1960. Over the years, he became known as "Mr. Aviation" to his friends, who ranged from Orville Wright to the builders of the fastest jet fighters. Although not an engineer or a technician, he assisted the NACA achieve working relationships with Congress, where he frequently testified, the military services, aerospace industry, and related groups engaged in government-sponsored research and development. ("John F. Victory," biographical file, NASA Historical Reference Collection.)

Wernher von Braun (1912-1977) was the leader of what has been called the "rocket team," which had developed the German V-2 ballistic missile in World War II. At the conclusion of the war, von Braun and some of his chief assistants—as part of a military operation called Project Paperclip—came to America and were installed at Fort Bliss in El Paso, Texas, to work on rocket development and use the V-2 for high altitude research. They used launch facilities at the nearby White Sands Proving Ground in New Mexico. Later, in 1950 von Braun's team moved to the Redstone Arsenal near Huntsville, Alabama, to concentrate on the development of a new missile for the Army. They built the Army's Jupiter ballistic missile, and before that the Redstone, used by NASA to launch the first Mercury capsules. The story of von Braun and the "rocket team" has been told many times. See, as examples, David H. DeVorkin, *Science With a Vengeance: How the Military Created the US Space Sciences After World War II* (New York: Springer-Verlag, 1992); Frederick I. Ordway III and Mitchell R. Sharpe, *The Rocket Team* (New York: Thomas Y. Crowell, 1979); Erik Bergaust, *Wernher von Braun* (Washington, D.C.: National Space Institute, 1976).

David Wade (1910-) was a career Air Force officer who advanced to the rank of lieutenant general in 1964. He was commander of the 1st Missile Division from 1958 to 1961 and later commanded the 16th, 2d, and 8th Air Forces.

James H. Wakelin, Jr. (1911-1990) was assistant secretary of the Navy (research and development) from 1959-1964. He had previously served in various capacities as a research director and administrator and later became president and chairman of the board of Research Analysis Corp.

Abbott McConnell Washburn (1915-) was deputy director of the U.S. Information Agency from 1954-1961.

Eugene W. Wasielewski (1913-1972) earned a B.S. degree in mechanical and aeronautical engineering and an M.S. in engineering mechanics from the University of Michigan. He worked in the private aircraft industry before going to work for Lewis Laboratory from 1947 to 1956. There he directed the design and construction of major engine-testing laboratories and supersonic wind tunnels; he also served as chief of the engine research division and as assistant director. He returned to private industry and then became associate director of Goddard Space

Flight Center in October 1960. As such, he was the principal institutional manager under the director of the center. ("Eugene W. Wasielewski," biographical file, NASA Historical Reference Collection.)

Alan T. Waterman (1892-1967) was the first director of the National Science Foundation (NSF) from its founding in 1951 until 1963. Waterman received his Ph.D. in physics from Princeton University in 1916, served with the Army's Science and Research Division in World War I, on the faculty of Yale University in the interwar years, with the War Department's Office of Scientific Research and Development in World War II, and then with the Office of Naval Research between 1946 and 1951. He and NASA leaders contended over control of the scientific projects to be undertaken by the space agency, with Waterman's NSF being used as an advisory body in the selection of space experiments. See, "Waterman, First NSF Head, Dies at 75," *Science*, 158 (8 December 1967): 1293; Norriss S. Hetherington, "Winning the Initiative: NASA and the U.S. Space Science Program," *Prologue: The Journal of the National Archives*, 7 (Summer 1975): 99-108; John E. Naugle, *First Among Equals: The Selection of NASA Space Science Experiments* (Washington, D.C.: NASA SP-4215, 1991).

Mark S. Watson (1887-1966) was a longtime journalist with the *Baltimore Sun*, and had been the military correspondent since 1941. (*New York Times*, 26 March 1966, p. 29.)

Thomas J. Watson, Jr. (1914-) started with International Business Machines Corp. in 1937, with a break for service in the Air Corps from 1940-1945, and became president of the firm in 1952. In 1961 he became chairman of IBM.

James E. Webb (1906-1992) was NASA administrator between 1961 and 1968. Previously he had been an aide to a congressman in New Deal Washington, an aide to Washington lawyer Max O. Gardner, and a business executive with the Sperry Corporation and the Kerr-McGee Oil Co. He had also been director of the Bureau of the Budget between 1946 and 1950 and undersecretary of state, 1950-1952. ("James E. Webb," biographical file, NASA Historical Reference Collection.)

Alfred John ("Jack") Westland (1904-1982) (R-WA) was first elected to Congress in 1952 and was reelected to each succeeding Congress through the 88th (1963-1965).

Harry A. Wexler (1911-1962) earned a B.S. in mathematics from Harvard in 1932 and joined the Weather Bureau in 1934, earning an Sc.D. in meteorology from MIT in 1939. While with the Army Air Force's weather service during World War II, with Col. Floyd Wood as pilot he made the first penetration of an Atlantic hurricane. He became chief of the Weather Bureau's science services division in 1946 and its director of research in 1955. One of the first scientists to foresee the use of satellites for meteorological purposes, he became known as the father of the Tiros satellite. ("Harry A. Wexler," biographical file, NASA Historical Reference Collection.)

Anne W. Wheaton had the title of associate press secretary under Press Secretary James G. Hagerty. She began as a reporter in New York, 1912-1921, then became a public relations consultant for several national women's organizations from 1924 to 1939. She next served as director of women's publicity for the

Republican National Committee, 1939-1957. President Eisenhower states in his memoirs (*Waging Peace, 1956-1961* [New York: Doubleday, 1965], p. 320n) that he appointed her in early 1957 and that her experience with the Republican National Committee improved his communications with that office after she joined the White House.

Thomas D. White (1901-1965) was a career Air Force officer who served in a succession of increasingly responsible positions until his retirement in 1961. He was director of legislation for the secretary of the Air Force between 1948 and 1951; USAF deputy chief of staff for operations, 1951-1953; USAF vice chief of staff, 1953-1957; and chief of staff, 1957-1961. ("Gen. T.D. White," biographical file, NASA Historical Reference Collection.)

Alexander Wiley (1884-1967) (R-WI) was first elected to the Senate in 1938 and served until 1962. At this time, he was the ranking Republican member of the Senate Judiciary and Foreign Relations Committees and was a member of the Aeronautical and Space Sciences Committee.

John Bell Williams (1918-1983) (D-MS) was first elected to Congress in 1946 and served in the House until 1968 when he became governor of Mississippi for four years. As an Army Air Forces pilot in World War II, he lost an arm in an airplane crash.

Walter C. Williams (1919-) earned a bachelor's degree in aerospace engineering from LSU in 1939 and went to work for the NACA in 1940, serving as a project engineer to improve the handling, maneuverability, and flight characteristics of World War II fighters. Following the war, he went to what became Edwards Air Force Base to set up flight tests for the X-1, including the first human supersonic flight by Capt. Charles E. Yeager in October 1947. He became the founding director of the organization that became Dryden Flight Research Facility. In September 1959 he assumed associate directorship of the new NASA space task group at Langley, created to carry out Project Mercury. He later became director of operations for the project, then associate director of the NASA Manned Spacecraft Center in Houston, subsequently renamed the Johnson Space Center. In 1963 Williams moved to NASA Headquarters as deputy associate administrator of the office of manned space flight. From 1964 to 1975, he was a vice president for Aerospace Corporation. Then from 1975-1982 he served as chief engineer of NASA, retiring in the latter year. ("Walter C. Williams," biographical file, NASA Historical Reference Collection.)

William Willner was in charge of construction in the procurement and supply division of NASA Headquarters' office of business administration. By August 1960, he had moved to the office of research grants and contracts. (Headquarters telephone directories, Aug. 1959, p. 3; Aug. 1960, pp. 7, 9, NASA Historical Reference Collection.)

Dean E. Wooldridge (1913-) was a member of the technical staff of Bell Telephone Laboratories from 1936-1946. He was co-director of the research and development labs of Hughes Aircraft Company from 1946-1951, rising through the directorship to become vice president for research and development, 1952-1953. He served as president and director of Ramo-Wooldridge Corp. from 1953-1958 and of Thompson-Ramo-Wooldridge from 1958-1962, when be became director of the firm's Space Technology Labs.

John David Wright (1905-) had gone to work for Thompson Products, Inc. in 1933 and rose through the ranks of it and its successor organizations until he became chairman of the board of Thompson-Ramo-Wooldridge, Inc. (later, TRW, Inc.) in 1958.

DeMarquis D. Wyatt (1919-), a graduate in mechanical engineering from the University of Missouri, Rolla, joined the Lewis Laboratory in 1944, where he specialized in supersonic research in propulsion system installations. In 1958 he transferred to NASA Headquarters, where he held a series of positions as research engineer; assistant administrator for programming, program plans and analysis, planning, and for policy and university affairs. He retired in 1973. ("DeMarquis Wyatt," biographical file, NASA Historical Reference Collection.)

E(lmer) E. Yeomans (1902-1983) was a career naval officer who advanced through the ranks to rear admiral. At this time, he was superintendent of the U.S. Naval Postgraduate School at Monterey, California.

Herbert F. York (1923-) had been associated with scientific research in support of national defense since World War II. He was director of the Livermore Radiation Laboratory for the University of California before moving to the Department of Defense in March 1958 as chief scientist of the Advanced Research Projects Agency. He became the DOD's director of research and engineering in December 1958, during a DOD reorganization; this was the third-ranking civilian office after the secretary and deputy secretary of defense. He served as director of defense research and engineering until 1961. He then moved to the University of California, San Diego, as chancellor and professor of physics. He also served as a member of the President's Science Advisory Committee under both Eisenhower and Johnson and was later chief negotiator for the camprehensive test ban during the Carter administration. ("Dr. Herbert F. York," biographical file, NASA Historical Reference Collection; Herbert F. York, *Making Weapons, Talking Peace: A Physicist's Odyssey from Hiroshima to Geneva* [New York: Basic Books, 1987].)

John ("Jack") Donald Young (1919-) earned an M.S. from Syracuse in 1943 and served as an officer in the Marine Corps from 1942-1945. He worked for various government agencies in the next few years and then became a management consultant with McKinsey & Co. from 1954-1960. He served as NASA's director of management analysis from 1960-1961, then became successively deputy director for administration and deputy associate administrator at NASA Headquarters. He left NASA in 1966 for a series of management positions in the Bureau of the Budget and the Department of Health, Education, and Welfare. Thereafter, he became a professor of public management at American University. ("John D. Young," biographical file, NASA Historical Reference Collection.)

Stephen M. Young (1889-1984) (R-OH) was elected to Congress in 1932 and served from then until 1937, again 1941-1943 and 1949-1951. Elected to the Senate in 1958, he served there through 1971.

Eugene M. Zuckert (1911-) received an L.L.B. from Yale in 1937 and worked from then until 1940 as an attorney for the U.S. Securities and Exchange Commission. He taught and became an assistant dean at the Harvard Graduate School of Business Administration from 1940-1944 and then held a variety of positions in the government including membership on the AEC, 1952-1954, and secretary of the Air Force, 1961-1965.

Index

Vanguard Project, 1, 97, 256n
Van Horn, Estelle, 268
Van Horn, Kent, 268
Van Keuren, Katherine M., 182
Veterans Administration, 306
Victory, John F., 189
von Braun, Wernher, 9, 22-23, 40, 53,
    60, 64, 82, 87, 104, 108, 123, 129-
    130, 167, 172, 183, 206, 225-226,
    249, 261
    speeches, 151

Wade, Maj. (later Lt.) Gen. David, 84
Wakelin, James H., Jr., 45
Walker, Roy (see Walter, L. Rohe)
Wallace, Al, 115
Waller, Fletcher, 120
Wasielewski, Eugene W., 241
Wallops Island (see under NASA)
Walter, L. Rohe, 212n
War Production Board, 76
Washburn, Abbott M., 99
Waterman, Alan T., 7, 42, 44, 230, 243
Watson, Mark S., 10
Watson, Dick, 261
Watson, Thomas J., Jr., 90, 261
Watts, Fred, 55, 57, 275, 277
Weather Bureau (see also Reichelderfer),
    116, 202, 243
Weaver, Warren, 233
Webb, James E., 23, 108n, 251, 311-317
    passim
Welch, Edward C., 314
Werner, Bishop Hazen G., 158
Western Reserve University, 164
Westinghouse Electric Corporation, 208,
    230, 283, 287
Westland, Jack, 64
Wexler, Harry A., 116
Wheaton, Anne W., 47
Whitcover, Jules, 52
White, Gen. Thomas D., 19, 124, 171-
    172, 206, 284, 287, 300

interest in becoming NASA adminis-
    trator, 283
Whitman, Ann, 224
Whitman, Walter G., 196, 254
Whitney, Jack, 96
Whittier, Sumner, 306
Wiesner, Jerome, 273
    Wiesner report, 304, 307
Wiggins, James Russell, 233, 286, 290
Wilcox, Francis O., 161
Wiley, Sen. Alexander, 107, 201, 211,
    226
Williams, A. L., 261
Williams, Frank, 242
Williams, Gov. G. Mennen, 281
Williams, John Bell, 89
Williams, Ross Norman, 33n
Williams, Walter C., 176
Williamsburg (see it, project manage-
    ment, and executive development un-
    der NASA)
Willner, William, 59
Wilson, Meredith, 169
Wilson, V. Adm. R. E., 19n
Wilson, Robert E., 232, 297, 298
Wilson, Lt. Gen. Roscoe, 283
Wisconsin State College, 163
Wooldridge, Dean E., 109, 164
Wooldridge, Jack, 196
Worthington, Leslie B., 237
Wright, John David, 108, 109, 164
Wright Orville and Wilbur, 1
Wright Memorial Trophy, 291
Wriston, Henry, 157, 280, 286
Wyatt, DeMarquis D., 36, 144, 182, 206

X-15 program, 83, 99-100, 187, 188n,
    204

Yale University, 235, 242-243
    engineering school, 242
Yates, Lt. Gen. Donald N., 277
Yeomans, R. Adm. E[lmer] E.

Author of the Introduction

Roger D. Launius is chief historian of the National Aeronautics and Space Administration, Washington, D.C. A graduate of Graceland College, he received his Ph.D. from Louisiana State University in 1982. He is the author of numerous articles on the history of aeronautics and space, which appeared in *Prologue: The Journal of the National Archives, Journal of Third World Studies, Journal of the West, Airpower Journal, Aerospace Historian, Air Power History, Airlift: The Journal of the Airlift Operations School,* the *American Aviation Historical Society Journal, Airman,* the *Air Force Journal of Logistics,* and the *Utah Historical Quarterly.* He is author of *Joseph Smith III: Pragmatic Prophet* (University of Illinois Press, 1988); co-editor of *Different Visions: Dissenters in Mormon History* (University of Illinois Press, 1988); co-author of *MAC and the Legacy of the Berlin Airlift* (USAF, 1989) and *Anything, Anywhere, Anytime: An Illustrated History of the Military Airlift Command, 1941-1991* (USAF, 1991), as well as the author or editor of six other books. He is presently conducting research for a book-length study of aviation in the American West, 1903-1945, and completing a one-volume overview of the U.S. civilian space program for the Robert E. Krieger Publishing Co.

The Editor

J.D. Hunley is a historian in the NASA History Office. He earned B.A. (1963) and Ph.D. (1973) degrees from the University of Virginia and an M.A. from Harvard (1968), all in European history. At Virginia, he was elected to Phi Beta Kappa and received the C.S. Ashby Henry Prize in history. He attended Harvard on a Woodrow Wilson fellowship. After teaching for five years (1972-1977) at Allegheny College, he worked for more than a decade as a civilian historian in the Air Force History Program, where among other awards he won the program's monograph award in 1990. He began working in the NASA History Office in 1991. He has written *Boom and Bust: Society and Electoral Politics in the Düsseldorf Area, 1867-1878* (Garland, 1987) and *The Life and Thought of Friedrich Engels: A Reinterpretation* (Yale University Press, 1991). His articles have appeared in *Social Theory and Practice, Societas—A Review of Social History, Essays in History,* and *Comparative Bibliographic Essays in Military History.*

NASA HISTORY PUBLICATIONS

Reference Works, NASA SP-4000:

Grimwood, James M. *Project Mercury: A Chronology*. (NASA SP-4001, 1963).

Grimwood, James M., and Hacker, Barton C., with Vorzimmer, Peter J. *Project Gemini Technology and Operations: A Chronology*. (NASA SP-4002, 1969).

Link, Mae Mills. *Space Medicine in Project Mercury*. (NASA SP-4003, 1965).

Astronautics and Aeronautics: A Chronology of Science, Technology and Policy. (NASA SP-4004 to SP-4025, a series of annual volumes continuing from 1961 to 1985, with an earlier summary volume, *Aeronautics and Astronautics, 1915-1960*.)

Ertel, Ivan D., and Morse, Mary Louise. *The Apollo Spacecraft: A Chronology, Volume I, Through November 7, 1962*. (NASA SP-4009, 1969).

Morse, Mary Louise, and Bays, Jean Kernahan. *The Apollo Spacecraft: A Chronology, Volume II, November 8, 1962-September 30, 1964*. (NASA SP-4009, 1973).

Brooks, Courtney G., and Ertel, Ivan D. *The Apollo Spacecraft: A Chronology, Volume III, October 1, 1964-January 20, 1966*. (NASA SP-4009, 1973).

Van Nimmen, Jane, and Bruno, Leonard C., with Rosholt, Robert L. *NASA Historical Data Book, Vol. I: NASA Resources, 1958-1968*. (NASA SP-4012, 1976, rep. ed. 1988).

Newkirk, Roland W., and Ertel, Ivan D., with Brooks, Courtney G. *Skylab: A Chronology*. (NASA SP-4011, 1977).

Ertel, Ivan D., and Newkirk, Roland W., with Brooks, Courtney G. *The Apollo Spacecraft: A Chronology, Volume IV, January 21, 1966-July 13, 1974*. (NASA SP-4009, 1978).

Ezell, Linda Neuman. *NASA Historical Data Book, Vol II: Programs and Projects, 1958-1968*. (NASA SP-4012, 1988).

Ezell, Linda Neuman. *NASA Historical Data Book, Vol. III: Programs and Projects, 1969-1978*. (NASA SP-4012, 1988).

Management Histories, NASA SP-4100:

Rosholt, Robert L. *An Administrative History of NASA, 1958-1963.* (NASA SP-4101, 1966).

Levine, Arnold S. *Managing NASA in the Apollo Era.* (NASA SP-4102, 1982).

Roland, Alex. *Model Research: The National Advisory Committee for Aeronautics, 1915-1958.* (NASA SP-4103, 1985).

Fries, Sylvia D. *NASA Engineers and the Age of Apollo* (NASA SP-4104, 1992).

Project Histories, NASA SP-4200:

Human Space Flight Programs:

Swenson, Loyd S., Jr., Grimwood, James M., and Alexander, Charles C. *This New Ocean: A History of Project Mercury.* (NASA SP-4201, 1966).

Hacker, Barton C., and Grimwood, James M. *On Shoulders of Titans: A History of Project Gemini.* (NASA SP-4203, 1977).

Benson, Charles D. and Faherty, William Barnaby. *Moonport: A History of Apollo Launch Facilities and Operations.* (NASA SP-4204, 1978).

Ezell, Edward Clinton, and Ezell, Linda Neuman. *The Partnership: A History of the Apollo-Soyuz Test Project.* (NASA SP-4209, 1978).

Brooks, Courtney G., Grimwood, James M., and Swenson, Loyd S., Jr. *Chariots for Apollo: A History of Manned Lunar Spacecraft.* (NASA SP-4205, 1979).

Bilstein, Roger E. *Stages to Saturn: A Technological History of the Apollo/Saturn Launch Vehicles.* (NASA SP-4206, 1980).

Compton, W. David, and Benson, Charles D. *Living and Working in Space: A History of Skylab.* (NASA SP-4208, 1983).

Compton, W. David. *Where No Man Has Gone Before: A History of Apollo Lunar Exploration Missions.* (NASA SP-4214, 1989).

Satellite Space Flight Programs:

Green, Constance McL., and Lomask, Milton. *Vanguard: A History.* (NASA SP-4202, 1970; rep. ed. Smithsonian Institution Press, 1971).

Hall, R. Cargill. *Lunar Impact: A History of Project Ranger.* (NASA SP-4210, 1977).

Ezell, Edward Clinton, and Ezell, Linda Neuman. *On Mars: Exploration of the Red Planet, 1958-1978.* (NASA SP-4212, 1984).

Scientific Programs:

Newell, Homer E. *Beyond the Atmosphere: Early Years of Space Science.* (NASA SP-4211, 1980).

Pitts, John A. *The Human Factor: Biomedicine in the Manned Space Program to 1980.* (NASA SP-4213, 1985).

Naugle, John E. *First Among Equals: The Selection of NASA Space Science Experiments* (NASA SP-4215, 1991).

Center Histories, NASA SP-4300:

Hartman, Edwin, P. *Adventures in Research: A History of Ames Research Center, 1940-1965.* (NASA SP-4302, 1970).

Hallion, Richard P. *On the Frontier: Flight Research at Dryden, 1946-1981.* (NASA SP-4303, 1984).

Muenger, Elizabeth A. *Searching the Horizon: A History of Ames Research Center, 1940-1976.* (NASA SP-4304, 1985).

Rosenthal, Alfred. *Venture into Space: Early Years of Goddard Space Flight Center.* (NASA SP-4301, 1985).

Hansen, James R. *Engineer in Charge: A History of the Langley Aeronautical Laboratory, 1917-1958.* (NASA SP-4305, 1987).

Dawson, Virginia P. *Engines and Innovation: Lewis Laboratory and American Propulsion Technology.* (NASA SP-4306, 1991).

General Histories, NASA SP-4400:

Corliss, William R. *NASA Sounding Rockets, 1958-1968: A Historical Summary.* (NASA SP-4401, 1971).

Wells, Helen T., Whiteley, Susan H., and Karegeannes, Carrie. *Origins of NASA Names.* (NASA SP-4402, 1976).

Anderson, Frank W., Jr., *Orders of Magnitude: A History of NACA and NASA, 1915-1980.* (NASA SP-4403, 1981).

Sloop, John L. *Liquid Hydrogen as a Propulsion Fuel, 1945-1959.* (NASA SP-4404, 1978).

Roland, Alex. *A Spacefaring People: Perspectives on Early Spaceflight.* (NASA SP-4405, 1985).

Bilstein, Roger E. *Orders of Magnitude: A History of the NACA and NASA, 1915-1990.* (NASA SP-4406, 1989).

New Series in NASA History, Published by The Johns Hopkins University Press:

Cooper, Henry S. F., Jr. *Before Lift-Off: The Making of a Space Shuttle Crew.* (1987).

McCurdy, Howard E. *The Space Station Decision: Incremental Politics and Technological Choice.* (1990).

Hufbauer, Karl. *Exploring the Sun: Solar Science Since Galileo.* (1991).

McCurdy, Howard E. *Inside NASA: High Technology and Organizational Change in the U.S. Space Program.* (1993).